ACS SYMPOSIUM SERIES **280**

Polymer Stabilization and Degradation

Peter P. Klemchuk, EDITOR
CIBA-GEIGY Corporation

Based on a symposium sponsored by
the Division of Polymer Chemistry
at the 187th Meeting
of the American Chemical Society,
St. Louis, Missouri,
April 9–12, 1984

American Chemical Society, Washington, D.C. 1985

Library of Congress Cataloging in Publication Data
Polymer stabilization and degradation.
(ACS symposium series, ISSN 0097–6156; 280)

"Based on a symposium sponsored by the Division of
Polymer Chemistry at the 187th meeting of the
American Chemical Society, St. Louis, Missouri, April
9–12, 1984."

Bibliography: p.
Includes indexes.

1. Polymers and polymerization—Deterioration—
Congresses. 2. Stabilizing agents—Congresses.

I. Klemchuk, Peter P., 1928–   . II. American
Chemical Society. Division of Polymer Chemistry.
III. American Chemical Society. Meeting (187th: 1984:
St. Louis, Mo.) IV. Series.

QD380.P653   1985   547.7   85–9011
ISBN 0–8412–0916–2

ACS Symposium Series

M. Joan Comstock, *Series Editor*

Advisory Board

FOREWORD

The ACS SYMPOSIUM SERIES was founded in 1974 to provide a medium for publishing symposia quickly in book form. The format of the Series parallels that of the continuing ADVANCES IN CHEMISTRY SERIES except that, in order to save time, the papers are not typeset but are reproduced as they are submitted by the authors in camera-ready form. Papers are reviewed under the supervision of the Editors with the assistance of the Series Advisory Board and are selected to maintain the integrity of the symposia; however, verbatim reproductions of previously published papers are not accepted. Both reviews and reports of research are acceptable, because symposia may embrace both types of presentation.

CONTENTS

PREFACE

POLYMER DEGRADATION AND STABILIZATION have become increasingly of interest to an international audience, and contributions to the body of knowledge have come from scientists worldwide. The symposium upon which this book is based was billed as international to encourage the participation of scientists from many countries, especially because it was the first Polymer Degradation and Stabilization Meeting of the Polymer Division of the American Chemical Society since March 1977 in New Orleans, and also because a major subject of the symposium, hindered amine stabilizers, had resulted from extraordinary international cooperation.

The initial call for papers for this symposium asked for reports in the following areas, among others: new stabilizers, stabilizers of high permanence, new findings of degradation and stabilization mechanisms, stabilizing polymers for severe environments, the impact of new polymerization processes on polymer stability, the complexities contributed to stabilization by the solid state, and polymer thermooxidation at ambient conditions. The response was excellent. As a result, we were able to organize a comprehensive program for the symposium. We are grateful to the many contributors of papers to the symposium and especially to those individuals who submitted complete papers for inclusion in this volume.

A major objective of the symposium and of this symposium volume was to attempt a review of the most important developments in polymer stabilization of the last decade. The development of hindered amine stabilizers was a major accomplishment in that decade and was truly an international achievement. Early work on stable free radicals in the United States, Germany, England, and the Soviet Union, among others, led in 1962 to the first synthesis of 2,2,6,6-tetramethyl-4-oxopiperidine-N-oxyl, the N-oxyl of triacetonamine. That first synthesis by Neiman, Rozantsev, and Mamedova of the Soviet Union triggered interest and excitement among scientists in many countries—Japan, France, England, the United States, Switzerland, Italy, and Canada—and ultimately led to the development of the hindered amine stabilizers. We had invited many of the scientists who were involved in the early history of hindered amine stabilizers to come to the symposium and to help us relive the hindered amine story. All could not come, but many did participate and helped us to review the story of the development of hindered amines and to gain an appreciation of that development.

Three papers had been offered by scientists in the Soviet Union. There were to have been papers on "The Discovery of Hindered Amine Stabilizers" by Rozantsev, on "The Reactions of Aminyl Radicals and Mechanism of Amine Regeneration as Inhibitors of Oxidation" by E. T. Denisov, and on "Nitroxyl Radicals and Polymer Stabilization" by A. L. Buchachenko. None of the three authors was able to attend the symposium. However, Rozantsez and Denisov have submitted their papers for this symposium volume.

This volume should serve as a good source for information on the status of current developments on polymer degradation and stabilization. It contains chapters on hindered amine light stabilizers from the major centers that have contributed to the development of this important class of polymer stabilizers. It contains contributions on stabilizers of increased permanence from the two major centers engaged in studying polymer stabilizers of reduced volatility and increased permanence—the laboratories of Scott at the University of Aston in Birmingham, England, and the laboratories of Vogl, formerly at the University of Massachusetts and currently at the Polytechnic Institute of New York in Brooklyn. Chapters on phosphites by Spivack and mechanisms of stabilization by aminic antioxidants by Pospisil highlight those important stabilizer classes. Chapters on the photodegradation and stabilization of poly(phenylene oxide) and polycarbonate by Pickett and Pryde, respectively, on the thermal degradation of poly(vinyl chloride) by Hjertberg and the late Danforth, on the thermal oxidation and stabilization of polyethylene by Horng, on the thermal oxidation of UV-cured coatings by Heyward, and on the photooxidation of polyester-styrene graft polymers by Ranby all provide state-of-the-art information on these important substrates. The chapter by Carlsson on γ-irradiation of polypropylene provides information on a subject that has gained importance due to the advent of γ-ray sterilization of plastic articles. Chapters on a computer model to study the degradation and stabilization of polyethylene by Guillet, on chemiluminescence in studying polymer degradation and stabilization by Zlatkevich and Mendenhall, and on techniques for studying heterogeneous degradation on polymers by Clough contribute significant information on these important and timely subjects. The book closes with a paper on the timely subject, biodegradable polyethylene, by Bailey.

Many people have contributed to the preparation and publication of this volume. The efforts of the authors to prepare the manuscripts and share the findings of their work are acknowledged with appreciation. The editorial assistance of Bonnie B. Sandel of Olin Chemical Co., the typing and proofreading assistance of Nancy Lovallo, Dianna DiPasquale, and Derry Lounsbury, and the efforts of Suzanne Roethel and others at the American Chemical Society Books Department are all acknowledged with appreciation. The help of all these individuals and many others not named has been crucial to the success of this project.

The ACS Petroleum Research Fund and the Polymer Division, Inc., of the American Chemical Society provided financial assistance for the symposium, which was used for the most part to help with the travel expenses of symposium speakers from overseas. The Additives Department of the CIBA-GEIGY Corp. also contributed to the success of the symposium by providing secretarial, typing, mailing, and telephone support. The contributions of all these sources are gratefully acknowledged.

PETER P. KLEMCHUK
Additives Department
CIBA-GEIGY Corporation
Ardsley, NY 10502

December 7, 1984

Introduction to Hindered Amine Stabilizers

PETER P. KLEMCHUK

CIBA-GEIGY Corporation, Ardsley, NY 10502

The hindered amine story is an interesting success story involving participation by scientists throughout the world. The success of hindered amines has partly been due to the free exchange of information among scientists, so that the accomplishments of one group, studying stable free radicals, could be picked up by those in other groups, in other parts of the world, who were interested in stable free radicals as polymer stabilizers. The mechanisms of stabilization by hindered amines were somewhat mysterious in the beginning since these compounds didn't fit any of the known stabilizer types and known mechanisms. Since stable free radicals had been of interest for their antioxidant activity as trappers of chain-propagating peroxy radicals the triacetoneamine-N-oxyl and derivatives thus were, at first, of interest as radical-trapping antioxidants. However, it wasn't long before the light stabilizing activity of that class of materials was recognized and not much longer before practical versions of hindered amine light stabilizers were synthesized, tested, and introduced for market development. Their high order of effectiveness as light stabilizers encouraged many scientists to investigate the mechanisms by which they functioned. We now know much about those mechanisms and can marvel at the effectiveness of hindered amines which apparently are dependent on what are very small concentrations of N-oxyls for their activity.

There is much to learn and admire in the hindered amine story. Chemists can take pride in how effectively they have worked together across national boundaries to make hindered amine stabilizers an important product group for the stabilization of polymers. This introduction is a modest effort to review some of the early history of stable-free radicals including triacetoneamine-N-oxyl. This chapter was intended to serve primarily as an introduction to the hindered amine review which took place at the symposium and intentionally avoids covering material which other participants were expected to present. It is a "light-touch" overview.

0097–6156/85/0280–0001$06.00/0
© 1985 American Chemical Society

Early Free Radical Research

Radical processes occur in many important chemical reactions including:

polymerization	electrical discharge reactions
halogenation	photochemical reactions
combustion	high energy-initiated reactions
autooxidation	heterogeneous catalysis
	enzyme processes

The radical nature of these reactions was identified through mechanistic and kinetic analyses, but the lifetimes of the free radicals involved are usually too brief to permit them to have been studied directly. Thus, stable free radicals offer an opportunity to gain basic information about free radicals and free radical reactions, in particular:

1. elucidation of the nature of the unpaired electron; its behavior; properties of its molecular orbitals; reasons for its stability;
2. elucidation of the kinetics and mechanisms of radical reactions with stable radicals which can be observed and measured directly;
3. elucidation of structures of liquids and solids containing stable free radicals by the use of EPR and ESR techniques.

Stable free radicals are of particular importance to those who are engaged in polymer stabilization because they play a key role in the inhibition of autooxidation reactions.

The mechanisms of autooxidation reactions were elucidated through the landmark research carried out at the British Rubber Producers' Research Association, where the kinetics of autooxidation of olefins were studied in the 1940's and early 1950's. Some of the key researchers engaged in that work were L. Bateman, J. L. Bolland, G. Gee, A. L. Morris, P. Ten Have, among others. They contributed enormously to our understanding of autooxidation reactions of organic materials. Re-reading their papers produces appreciation of their important work and emphasizes the debt we, in polymer stabilization work, owe them. That work established the following:

1. mechanisms and kinetics of autooxidation reactions of organic materials;
2. free radical nature of autooxidation;
3. chain reaction nature of autooxidation;
4. hydroperoxides as the main oxidation products;
5. production of autooxidation initiation radicals from homolysis of hydroperoxides;
6. inhibition of autooxidation by hydroquinone; the stability of resultant semiquinone radicals terminated oxidation chains with no chain transfer.

The work at the British Rubber Producers' Association and subsequent work by Bickel and Kooyman, Thomas and Tolman, among others, led to the appreciation that a key requirement of oxidation inhibi-

tors was their ability to form stable radical intermediates. These
radical intermediates should be sufficiently stable so that they
don't participate in hydrogen abstraction from the substrate they
are stabilizing and therefore, not lead to chain transfer. In
Scheme I are shown the key steps in inhibition by the two main
classes of oxidation inhibitors, hindered phenols and secondary
aromatic amines.

Stable Free Radicals

The effectiveness of compounds yielding stable radical intermediates in the stabilization of organic materials led to interest
in stable free radicals in general. A review of the literature on
stable free radicals indicates that nitric oxide and nitrogen dioxide have been known to be stable free radicals for a very long
time. In 1845, Fremy first described his salt which is an inorganic
N-oxyl derivative. The first report of an organic N-oxyl was by
Wieland and Offenbacher in 1914 when they described the synthesis
and properties of diphenyl-N-oxyl. Diphenyl picrylhydrazyl was
recognized as a stable free radical in 1922 by Goldschmidt and Renn.
Structures of other early stable free radicals may be seen in
Figure 1. Those include several N-oxyl derivatives as well as galvinoxyl, a stable-free radical derived from a hindered phenol.
These are only a few of the stable free radicals, especially N-oxyl
radicals, which have been discovered. Dr. Murayama of the Sanyko
Company included many such structures in a paper published in the
early 1970's (1).

The first synthesis of 2,2,6,6-tetramethyl-4-oxypiperidine-N-
oxyl was described in a paper by M. B. Neiman, E. G. Rozantsev, and
Y. G. Mamedova (2). The synthesis was achieved by the oxidation of
triacetoneamine with hydrogen peroxide in the presence of sodium
tungstate. The key properties of the tetramethyloxypiperidine-N-
oxyl, which subsequently led to the whole family of hindered amine
stabilizers, were outstanding thermal and chemical stability.
Details of its properties are summarized in Figure 2. The compound
melted at 36°C. It was stable to melting, recrystallization, and
distillation. It underwent reactions of the carbonyl functionality
and had 6.1 x 10^{23} spins/mole. The N-oxyl was reduced with hydrogen on palladium catalyst to the corresponding hydroxylamine.
The N-oxyl remained intact after reactions with 1) hydroxylamine to
form the oxime, 2) semicarbazide to form the semicarbazone and 3)
2,4-dinitrophenylhydrazine to form the 2,4-dinitrophenylhydrazone.

The discovery of the 2,2,6,6-tetramethyl-4-oxylpiperidine-N-
oxyl was an important milestone in stabilization technology but the
triacetoneamine-N-oxyl was not useful by itself as a polymer stabilizer because it is orange and would impart objectionable color to
polymers and because it has a relatively low molecular weight and
too high a vapor pressure for practical uses. In order to commer-
cialize a practical stabilizer based on triacetoneamine-N-oxyl, the
following had to be achieved:

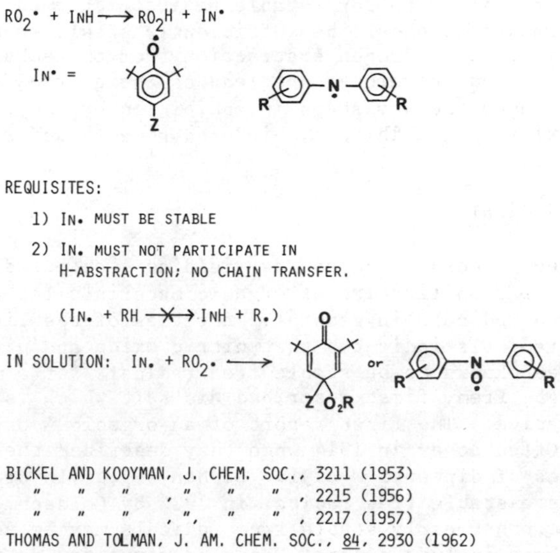

$RO_2{}^\bullet + INH \longrightarrow RO_2H + IN^\bullet$

$IN^\bullet =$

REQUISITES:

1) $IN^\bullet$ MUST BE STABLE

2) $IN^\bullet$ MUST NOT PARTICIPATE IN
   H-ABSTRACTION; NO CHAIN TRANSFER.

$(IN^\bullet + RH \overset{\times}{\longrightarrow} INH + R^\bullet)$

IN SOLUTION: $IN^\bullet + RO_2{}^\bullet \longrightarrow$ or

BICKEL AND KOOYMAN, J. CHEM. SOC., 3211 (1953)
" " " , " " " , 2215 (1956)
" " " , ", " " , 2217 (1957)
THOMAS AND TOLMAN, J. AM. CHEM. SOC., 84, 2930 (1962)

Scheme 1. Key requirement of oxidation inhibitors: stable radical intermediates.

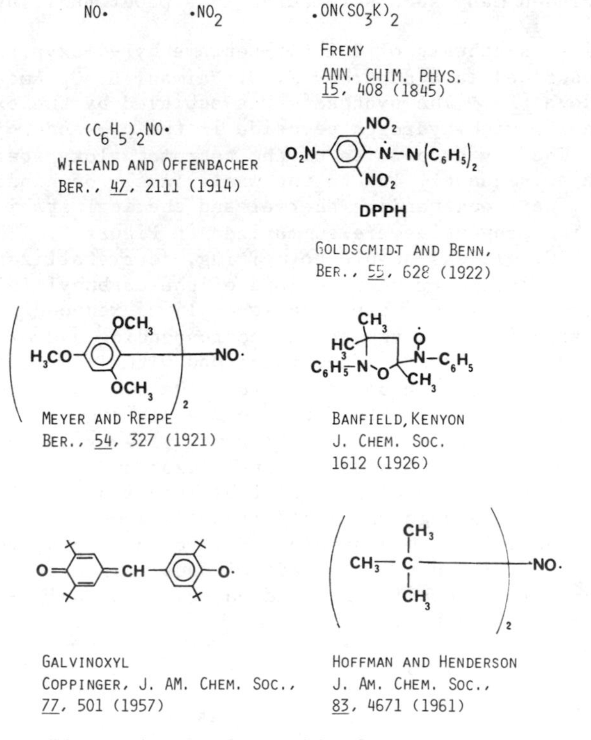

$NO^\bullet$ $^\bullet NO_2$ $^\bullet ON(SO_3K)_2$

FREMY
ANN. CHIM. PHYS.
15, 408 (1845)

$(C_6H_5)_2NO^\bullet$

WIELAND AND OFFENBACHER
BER., 47, 2111 (1914)

DPPH

GOLDSCMIDT AND BENN,
BER., 55, 628 (1922)

MEYER AND REPPE
BER., 54, 327 (1921)

BANFIELD, KENYON
J. CHEM. SOC.
1612 (1926)

GALVINOXYL
COPPINGER, J. AM. CHEM. SOC.,
77, 501 (1957)

HOFFMAN AND HENDERSON
J. AM. CHEM. SOC.,
83, 4671 (1961)

Figure 1. Early stable free redicals.

1. a suitable derivative with desired properties had to be found
   a. high stabilizing activity
   b. low color and low tendency to discolor in polymers
   c. stable to polymer processing conditions
   d. low volatility;

2. an economical manufacturing process had to be developed.

All these objectives have now been achieved and we are able to enjoy the availability of hindered amine stabilizers with outstanding light stabilizing and, in many instances, thermostabilizing properties. The discovery by Dr. Murayama and his co-workers at the Sankyo Company in Japan that the N-oxyl functionality was not critical to the effectiveness of the triacetoneamine-N-oxyl as a light stabilizer but that the N-H compounds of that class also possessed impressive light stabilizing activity, was a key finding which led to the subsequent development of polymer stabilizers which were colorless and had low discoloration tendencies.

Stable Free Radicals in Polymer Stabilization

The effectiveness of free radicals, such as N-oxyls, hydrazyls, and phenazyls, as stabilizers for polymers was described in a patent issued in 1952 (3). Various compounds of these classes were found to be effective as thermal and light stabilizers in nylon fabric, nylon film, polychloroprene, polyethylene, and poly(methyl methacrylate). Some of the structures described in that patent are shown in Figure 3.

Brownlie and Ingold (4) were among the first to show the effectiveness of N-oxyls and hydroxylamine analogs in stabilizing organic substances against autooxidation. The effectiveness of N-oxyls in trapping carbon-free radicals and peroxy-free radicals was clearly demonstrated in that work. The N-oxyls of that work, shown in Figure 4, were also effective in inhibiting the AIBN-initiated polymerization of styrene.

As shown in Figure 5, hindered hydroxylamines can function as polymer stabilizers by the donation of a hydrogen atom to a peroxy radical to yield an N-oxyl. The N-oxyl is in turn able to trap carbon-free radicals and in that way terminate the oxidation cycle at the first stage, which is the ideal point at which to terminate chain propagating free radicals – before they have an opportunity to react with oxygen to form peroxy radicals. Non-hindered hydroxylamines, with H-atoms on α-carbon atoms, usually give rise to nitrones as a result of hydrogen atom donation and do not trap carbon free radicals to yield alkoxyamino products. The thermal stability associated with hydroxylamines and the cost of converting a triacetoneamine derivative to a hydroxylamine derivative are disadvantages for the development of hydroxylamines as practical commercial polymer stabilizers.

The effectiveness of dibenzylhydroxylamine as a inhibitor of AIBN-initiated tetralin oxidation can be seen from the data in Figure 6. A comparison between dibenzylhydroxylamine and butylated hydroxytoluene (BHT) can be made from those data. The lower rate of oxygen consumption in the solution containing dibenzylhydroxyl-

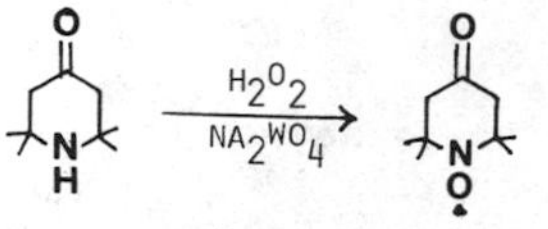

M.P. 36°C; STABLE TO
MELTING, RECRYSTALLI-
ZATION, DISTILLATION,
REACTIONS OF CARBONYL,
SPIN PER MOLE: 6.1×10^{23}

KEY PROPERTY - OUTSTANDING STABILITY.

REACTIONS

1. Pd/H_2 REDUCTION TO HYDROXYLAMINE

2. N-OXYL INTACT AFTER REACTIONS WITH:

    A. HYDROXYLAMINE ⟶ OXIME (ALSO BECKMANN REARRANGEMENT)
    B. SEMICARBAZIDE ⟶ SEMICARBAZONE
    C. 2,4-DINITROPHENYLHYDRAZINE ⟶ 2,4-DINITROPHENYL-
                                            HYDRAZONE

M. B. NEIMAN, E. G. ROZANTSEV, Y. G. MAMEDOVA, NATURE,
196, 472 (1962).

Figure 2.   First synthesis of 2,2,6,6-tetramethyl-4-oxo-
piperidine-N-oxyl.

COMPOUNDS EVALUATED:

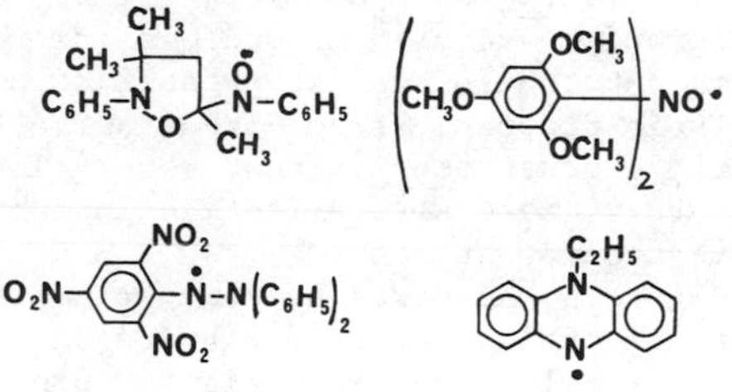

POLYMERS TESTED (THERMAL AND LIGHT STABILITY):

NYLON FABRIC, NYLON FILM, POLYCHLOROPRENE,
POLYETHYLENE, PMMA

D. MC QUEEN, U.S. 2,619,479; APPLIED 12/28/50;
ISSUED 11/25/52

Figure 3.   Stabilization of polymers with N-oxyls, hydrazyls,
and phenazyls.

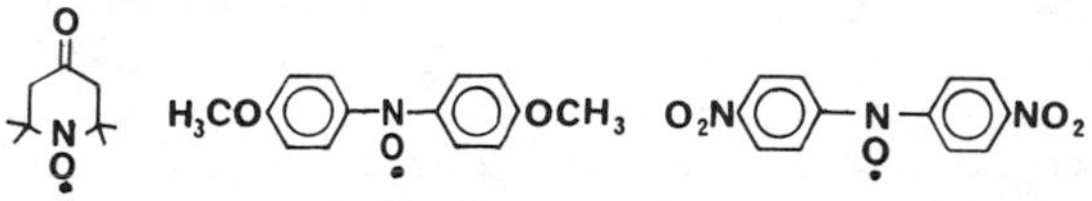

I, II, III, AND CORRESPONDING HYDROXYLAMINES INHIBITED. THE
AUTOOXIDATION OF STYRENE. AROMATIC N-OXYLS REACTED WITH
R. AND RO_2. ALIPHATIC N-OXYL REACTED ONLY WITH R..

$$R. + O_2 \xrightarrow{k_1} RO_2.$$
$$R. + IN. \xrightarrow{k_2} NOR$$

$$\frac{k_1}{k_2} \simeq 10 \text{ FOR } I$$

I, II, AND III INHIBITED AIBN-INITIATED STYRENE POLYMERIZATION

BROWNLIE AND INGOLD   CAN. J. OF CHEM., 45, 2427 (1967).

Figure 4.   Inhibition of autoxidation of styrene by N-oxyls
and hydroxylamines.

ADVANTAGES:

GENERATE N-OXYL ON HYDROGEN ABSTRACTION
$$>NOH + RO_2. \longrightarrow >NO. + RO_2H$$

$$>NO. + R. \longrightarrow >NOR$$
EFFECTIVE IN THERMAL OXIDATION INHIBITION
EFFECTIVE IN PHOTOOXIDATION INHIBITION

DISADVANTAGES:

THERMAL STABILITY

COSTLY TO MAKE N-HYDROXY-TRIACETONEAMINE

Figure 5.   Hydroxylamines as polymer stabilizers.

amine over that for BHT can be seen clearly from the data plotted
in Figure 6. Calculations of oxidative chain length – the number of
molecules of oxygen consumed per free radical generated – shows the
higher efficiency of dibenzylhydroxylamine over BHT as a radical
trapping inhibitor. Both dibenzylhydroxylamine and BHT were found
to trap about two radicals per inhibitor molecule to the end of the
induction period. Beyond the induction period both additives were
essentially without effect and the oxidation proceeded as if uninhibited.

And so, the work on mechanisms of autooxidation at the British
Rubber Producers' Association, the early work on the synthesis and
reaction of stable free radicals, the recognition of the rɔle of
stable free radicals in polymer stabilization, the discovery of stable triacetonamine-N-oxyl, and the search for practical candidates
for commercialization, have led to the development of hindered amine
stabilizers, a new class of polymer stabilizers. They are effective
in many polymers against photodegradation and also are effective
against thermooxidation in some polymers. The structures of the
current commercially available products for polymer stabilization
may be seen in Figure 7. These compounds are effective in meeting
the stabilizer requirements in many commercial polymers; however,
others are under development to satisfy requirements not being met
by them.

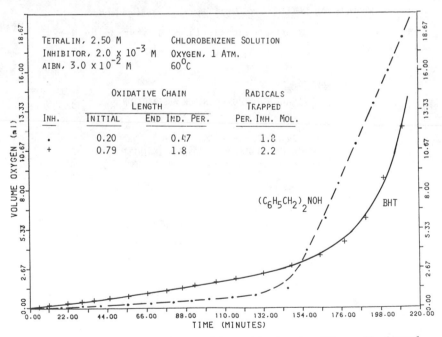

Figure 6. Inhibition of tetralin oxidation by dibenzylhydroxylamine, AIBN-inhibited.

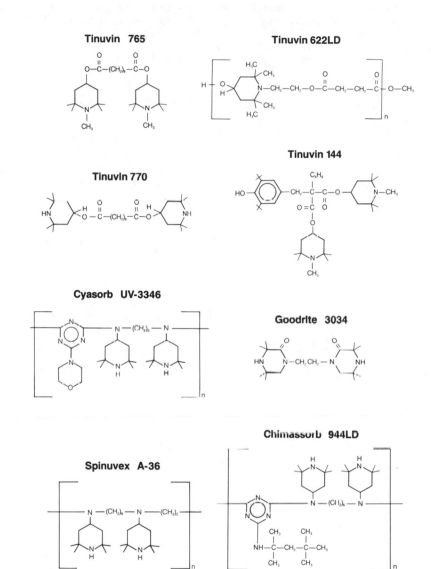

Figure 7.   Commercially available hindered amine stabilizers.

Literature Cited

(1) K. Murayama, Yuki Gosei Kagaku Kyoka Shi, 29, 366 (1971).
(2) M. B. Neiman, E. G. Rozantsev, Y. G. Mamedova, Nature, 196, 472 (1962).
(3) D. McQueen, U. S. Patent 2,619,479.
(4) I. T. Brownlie and K. U. Ingold, Can., J. of Chem., 45, 2427 (1967).

RECEIVED December 7, 1984

Discovery, Chemistry, and Application of Hindered Amines

E. G. ROZANTSEV[1], E. SH. KAGAN[2], V. D. SHOLLE[3], V. B. IVANOV[3], and V. A. SMIRNOV[2]

[1] Department of Biochemistry, MTIMMP, Moscow, USSR
[2] Polytechnic Institute of Novocherkassk, Novocherkassk, USSR
[3] Institute of Chemical Physics of the Academy of Science of the USSR, Moscow, USSR

This chapter is a comprehensive overview of the progress in the field of generation, chemistry, and application of nitroxyl radicals and their precursors, for example, hindered amines of the 2,2,6,6-tetramethyl-piperidine series. Because of the importance of nitroxyl radicals to polymer stabilization, this application is discussed at length, while the others are touched upon briefly.

Discovery

It is well known from chemical history that the discoveries of the first stable organic radicals, such as triphenylmethyl, diphenyl-picrylhydrazyl, tri-tert-butylphenoxyl, and nitroxides are very significant contributions to theoretical chemistry. The relative stabilities of these radicals were attributed by chemists to the participation of an unpaired electron in conjugated π-electron systems. Classical stable radicals can thus be thought of as a superposition of many resonance structures with different localizations of an unpaired electron. The first stable radical obtained by Pilotti and Schwerin in 1901 in the pure state can be described by a variety of tautomeric and resonance structures as shown in Scheme 1.
The development of the physical organic chemistry of stable radicals stimulated the investigation of relationships between their structures and reactivity. As a result, there developed the widely held concept that organic free radicals which do not possess a system of conjugated multiple bonds, such as

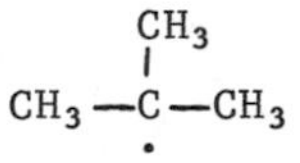

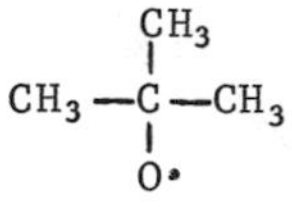

$$CH_3 -\overset{\underset{\displaystyle \cdot}{\overset{\displaystyle CH_3}{|}}}{C} -CH_3$$

(1)

(2)

$$CH_3 -\overset{\underset{\displaystyle O\cdot}{\overset{\displaystyle CH_3}{|}}}{\underset{|}{C}} -CH_3$$

(3)

cannot exist under ordinary conditions in a chemically pure state.
Due to their high reactivity, the concentrations of such radicals
diminish very rapidly, and their lifetimes at room temperature do
not often exceed fractions of a second.

Yet, paradoxically enough, in the late Fifties and early
Sixties, the first free radicals in a chemically pure state were
obtained and though they did not possess conjugated multiple bonds,
they were extremely stable under ordinary conditions. These bright-
ly colored crystalline radicals did not change their physical and
chemical characteristics for months, even when stored in air at room
temperature.

In these new radicals, the unpaired electron does not
participate in multiple bond systems and, therefore, it is essentially localized at the nitrogen-oxygen bond (1,2).

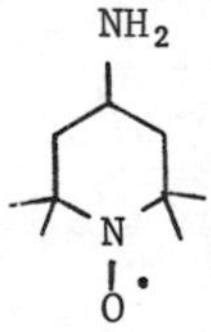

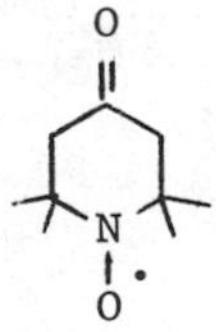

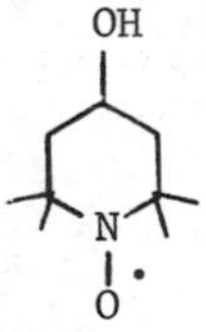

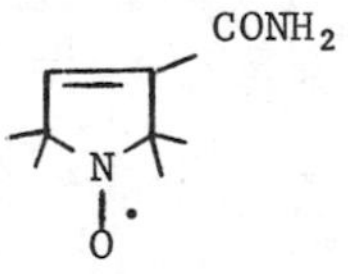

(4)   (5)   (6)   (7)

Thus, it was experimentally demonstrated that a whole class of
stable free radicals does exist, even though "theoretically" they
are unlikely to exist under ordinary conditions. From the Sixties
the interest in the new radicals by chemists, physicists and
biologists has been rapidly increasing.

After the first publication of his paper, one of the authors of
the present communication received more than a thousand inquiries
from many European countries and America, and in 1965 several
independent research groups and laboratories dealing with different
aspects of the new radical chemistry appeared.

Unlike the classical nitroxides such as porphyrexide and other
derivatives of quadrivalent nitrogen, the new radicals, called
nitroxyls (or iminoxyls) could undergo reactions without involvement
of the radical site. In 1961, E. G. Rozantsev and Yu. G. Mamedova
were the first to realize the "non-radical reactions of free
radicals" (3,4); e.g., the formation of 2,2,6,6-tetramethyl-4-
oxopiperidine-1-oxyl oxime without chemical involvement of the free

radical site:

$$\text{(8)}$$

This and some other non-radical reactions of radicals were theoretically predicted by Neiman back in the fifties. But, to obtain the starting stable radicals (5,6), it was necessary above all to obtain various sterically hindered amines. Consequently, the development of stable nitroxyl chemistry inspired great interest in hindered amines.

It is interesting to note that the first sterically hindered amine, called "triacetonamine," was described as far back as 1874 by two Russian chemists, Sokoloff and Lachinoff (7) and by a German chemist, Heintz (8), whose works were published almost simultaneously. The first syntheses of triacetonamine were based on the reaction of acetone with ammonia. Thus, triacetonamine became accessible even though the yield did not exceed 8%. At present, along with acetone, some other starting materials (9) such as diacetone alcohol, 2,2,4,4,6-pentamethyl-2,3,4,5-tetrahydropyrimidine and phorone are used for the synthesis of triacetonamine. It is not surprising that at the outset almost all hindered amines necessary for the synthesis of various hindered nitroxyl radicals were obtained from triacetonamine (10-12).

In 1962, utilization of the new nitroxyl radicals was suggested for inhibition of oxidative degradation of thermoplastic polymers (13). In 1965, McConnell published his early work which laid the foundation for application of the new radicals and their non-radical reactions in the spin-label method (14).

Synthesis

Actually, triacetonamine still remains the only starting compound for the synthesis of 2,2,6,6-tetramethylpiperidine derivatives. The main methods of its preparation are given in Scheme 2. A summary of methods of synthesis for triacetonamine and other hindered piperidines are summarized in (15).

A highly efficient method of producing triacetonamine from acetone and ammonia through Equation (4) was developed in the USSR (16). The cost of triacetonamine produced on the basis of this technology on a pilot plant scale is approximately $10 per kilogram which is significantly lower than that in the Aldrich catalog (1982) of chemicals and intermediates.

By now a large number of 2,2,6,6-tetramethylpiperidine derivatives has been prepared. These are mostly compounds having substituents in the four-position of a piperidine ring, and their

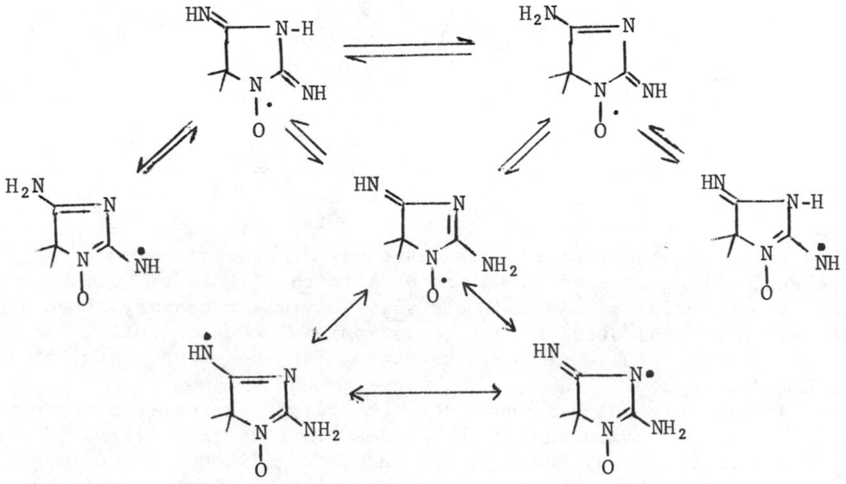

Scheme 1. Tautomeric and resonance structures of a stable
free radical.

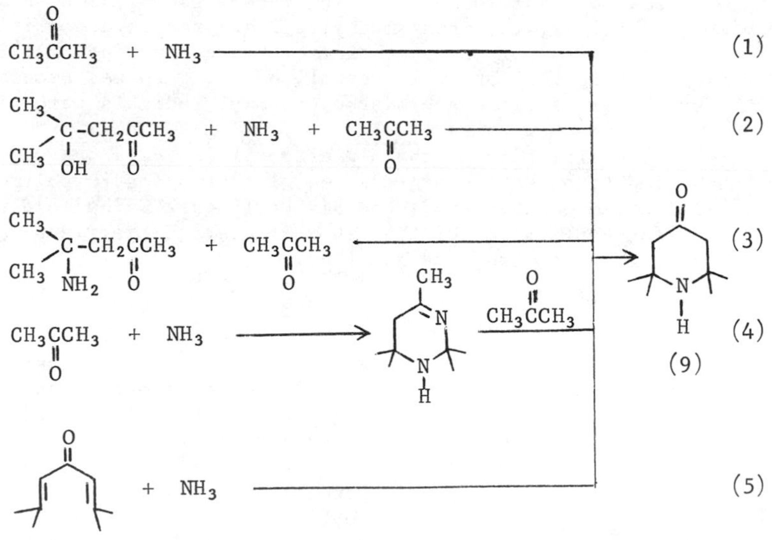

Scheme 2. Reactions for the synthesis of triacetonamine.

structures show their close structural relationship to triacetonamine. Although triacetonamine has some structural peculiarities, it undergoes many reactions typical of ordinary piperidines (17). Detailed information on triacetonamine chemistry and synthesis methods for 2,2,6,6-tetramethylpiperidine derivatives can be found elsewhere (18,19). The most important 2,2,6,6-tetramethylpiperidine derivatives (Scheme 3) were synthesized from triacetonamine (9) or ketone-radical (5) in one or more steps through typical transformations of the ketone carbonyl group.

The majority of the well known compounds in the 2,2,6,6-tetramethylpiperidine series was generated from (10, 11, 14). The most numerous ones are alcohol esters (10), amides and other amine derivatives (11).

Electrochemical methods for the synthesis of several important triacetonamine derivatives have been elaborated recently; e.g., the electrode reactions shown below:

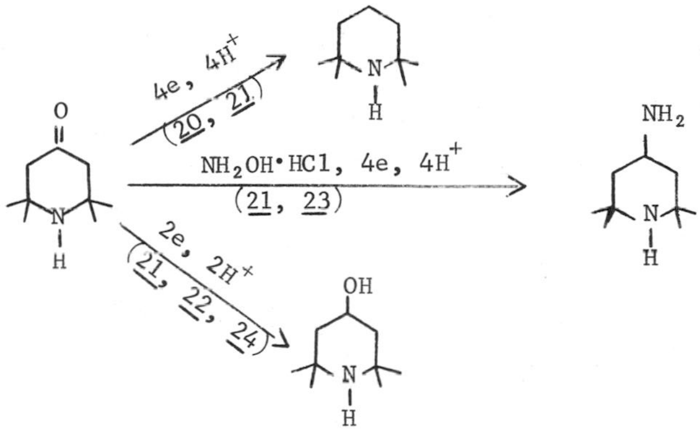

Electrochemical methods for obtaining the simplest 2,2,6,6-tetramethylpiperidine derivatives shown in the scheme were advantageously used by the authors. Electrochemical methods of 2,2,6,6-tetramethylpiperidine preparation have undeniable advantages over the method based on the reduction of triacetonamine by hydrazine at 150 - 200°C (25).

The reactions of the α-methylene group (positions 3 and 5), unlike the reactions of the carbonyl group, are unusual and occur under forcing conditions. Some of these reactions at position 3 are given in Scheme 4. More detailed information is given in (26,27).

Reactions of enamine (20) with Mannich bases, with esters of α,β-unsaturated acids and with acrylonitrile take place under more severe conditions as compared to the similar reactions of cyclohexane enamines.

Enamines (20) do not undergo the reactions of acylation and alkylation; but the Mannich reaction, that of bromination and the Favorskii rearrangement are unexpectedly easy. No other methods for introducing substituents into position 3 of triacetonamine, but those indicated in Scheme 4, are known.

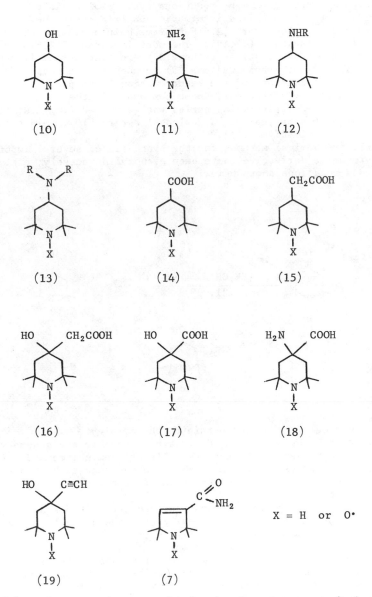

Scheme 3. Most important hindered amine structures derived
from triacetonamine.

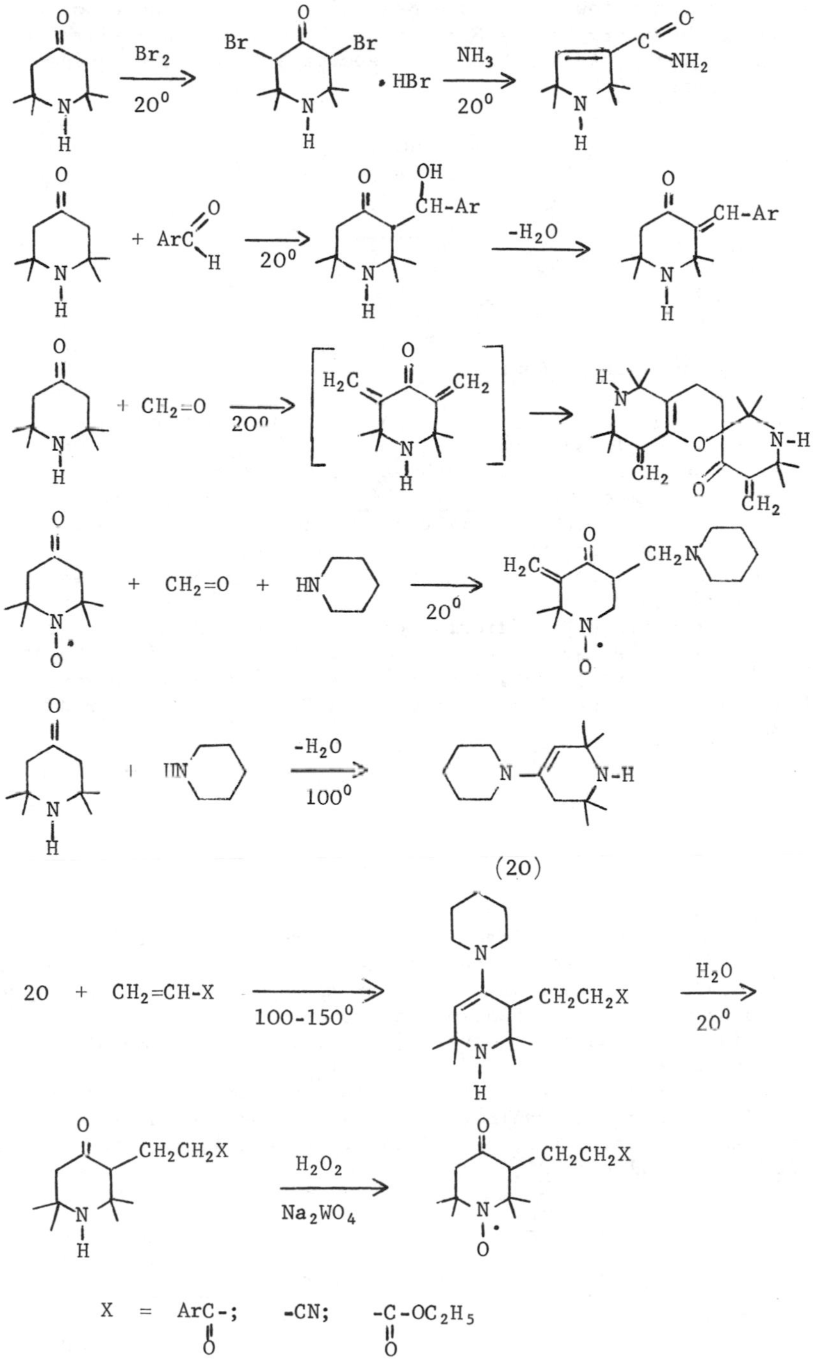

(20)

X = ArC-; -CN; -C-OC$_2$H$_5$
         ‖   ‖
         O   O

Scheme 4. Reactions of hindered amines at the 3- and 5-methylene groups.

3-Formyl-2,2,6,6-tetramethylpiperidine-1-oxyl obtained by the reaction given below could have been of great importance for the synthesis chemistry of 2,2,6,6-tetramethylpiperidine derivatives. But, after its synthesis had been reported, there is no evidence of its utilization (28).

Chemistry of Nitroxyl Radicals

There is no need to give a thorough description of the chemistry of hindered nitroxyl radicals as it has been fully discussed in a number of summaries including the papers written by the authors of the present communication (18,19,29-37). This paper is aimed at giving some general characteristics of the chemical properties of nitroxyl radicals and the methods for their synthesis.

Nitroxyl radicals have extraordinarily high stability. Nevertheless, under some conditions they behave as free radicals.

Hindered nitroxyl radicals recombine with active alkyl radicals to give hydroxylamine ethers. The rate of this reaction amounts to 10^6-10^8 L/mol. sec.; i.e., nitroxyls are efficient acceptors of alkyl radicals; they also react vigorously with hydroxyl radicals (37-38).

The most characteristic reaction of hindered nitroxyls is their reduction which results either in the corresponding hydroxylamine or amine. Hydroxylamines, as a rule, are readily oxidized to the corresponding radicals. Oxidation can be accomplished with atmospheric oxygen in the presence of catalytic amounts of heavy metal salts; e.g., cupric salts. PbO_2 and $K_3[Fe(CN)_6]$ can also be used as oxidizing agents. This easily occuring transformation-radical $\longrightarrow$ hydroxylamine $\longrightarrow$ radical-is the basis for several important syntheses, in particular, electrochemical syntheses of the nitroxyl radicals (4) and (6) (39,40):

(6)

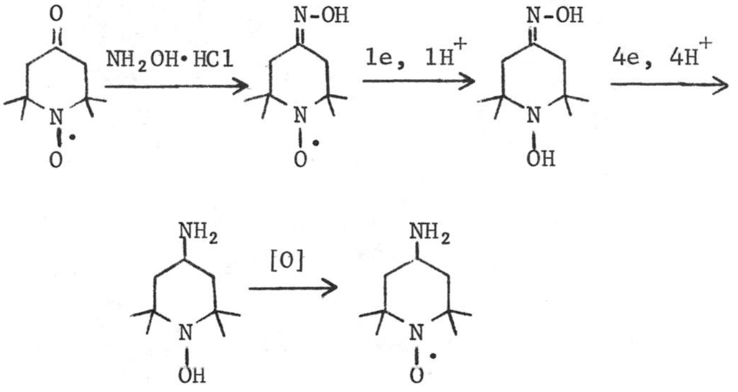

(4)

Photoexcitation increases nitroxyl reactivity appreciably. The photochemical reaction of the radical (6) with toluene yields quantitatively the corresponding hydroxylamine and its benzyl ether (41). This reaction simulates to a certain extent the process of polymer stabilization by nitroxyl radicals.

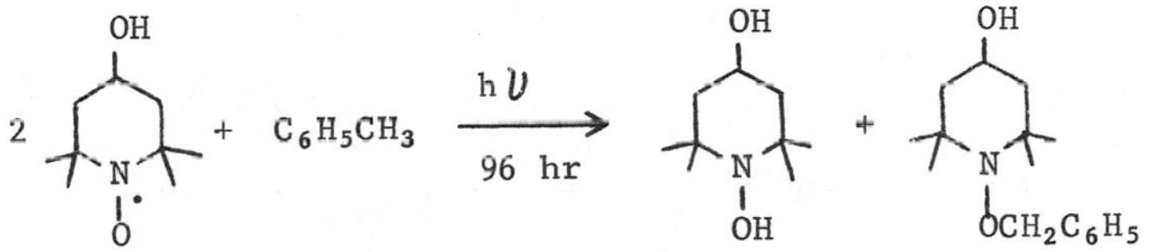

(5)

Though reduction of the nitroxyl group occurs quite readily, there are agents which achieve reduction of other groups without involving paramagnetic centers. Examples of such reactions are given below:

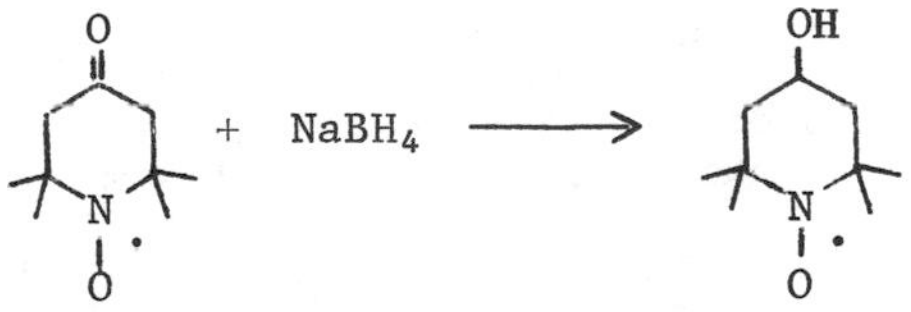

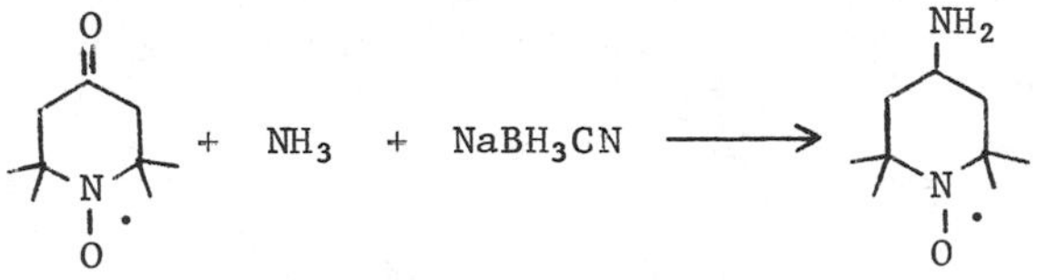

In acid medium, nitroxyl radicals disproportionate to produce hydroxylamines and oxoammonium salts.

The mechanism of this reaction is considered in (42). At pH 1, the equilibrium of this reaction is shifted to the right; at pH 3, to the left. This means that nitroxyl radicals are weak bases, much weaker than the corresponding amines. This property facilitates the separation of amines from radicals by extracting an amine with dilute hydrogen chloride (pH 2-3). This method of radical purification is used for preparatory purposes. Oxoammonium salts are strong oxidizing agents; in fact, the oxidizing properties of nitroxyl radicals in acid medium are associated with oxoammonium salts. For example, in acid medium, radicals oxidize alcohols to the corresponding aldehydes.

Strong oxidants such as chlorine and bromine oxidize nitroxyl radicals to corresponding oxoammonium salts (30):

$$2 \ >N\text{-}O \ + \ Cl_2 \longrightarrow \ >N\overset{+}{=}O \ \ \bar{C}l$$

Developments in Nitroxyl Radical Synthesis

The unusual characteristics of the nitroxyl group determine the methods of nitroxyl radical synthesis. Here are the major ways of their synthesis.

1. Oxidation of the corresponding hindered piperidines. This synthesis technique can be used only when the starting piperidine contains an H-atom on nitrogen. This condition restricts the application of the method to the most simple nitroxyl radicals of the piperidine series. The hydrogen peroxide-sodium tungstate system is most often used for piperidine oxidation. The mechanism and kinetics of this reaction are described in (43).
    The techniques of hindered piperidine synthesis are constantly improving. Take, for example the method of 2,2,6,6-

tetramethyl-4-piperidinol oxidation which makes it possible to obtain the corresponding radical with 98% yield in two hours (44).

Triacetonamine and 2,2,6,6-tetramethyl-4-piperidinol are oxidized by the hydrogen peroxide-sodium carbonate system very selectively, giving practically a quantitative yield (45). For amine oxidation, the hydrogen peroxide-acetonitrile system is often effective enough (46,47), while for hindered piperidine oxidation, peracids can be also used.

2. <u>Reactions which do not involve the radical site</u>. This is the most significant class of hindered nitroxyl reactions. These reactions represent a chemical basis for the spin-label method. The simplest hindered nitroxyls, such as those indicated in Scheme 3, are commonly used as the starting compounds. Some of these radicals are commercially available in several countries. Many selective reagents are known at present which do not inter- act with a nitroxyl group. So, to carry out a new synthesis, you have only to choose a proper reagent for the given reaction using recommendations reported in the scientific literature and carry out the reaction.

3. <u>Hydroxylamine utilization</u>. In those cases where it is impossible to preserve the radical center through all the stages of synthesis, it can be converted to the corresponding hydroxyl- amine. The latter as mentioned earlier, unlike nitroxyl radicals, is stable in acidic medium and is reduced at a higher potential. Several examples of hydroxylamine utilization for radical synthesis from the authors' experience are given on p. 18 and below (48).

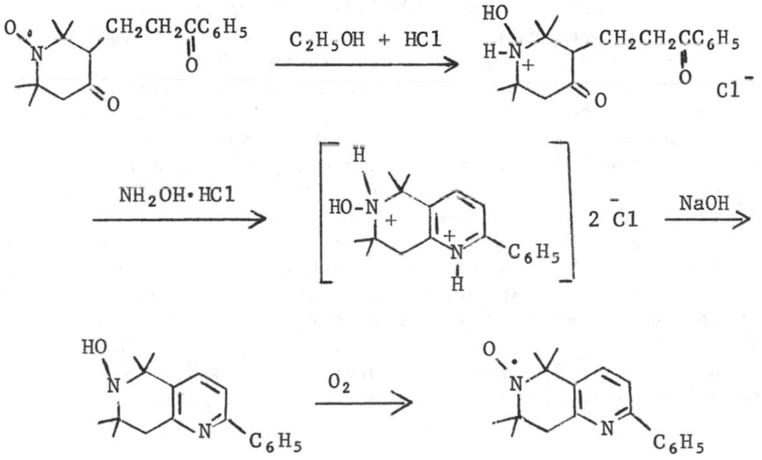

Application

The investigations undertaken with the purpose of studying the characteristics of radicals resulted in wide practical application of the latter. Nitroxyl radicals are used in biochemistry, organic chemistry, high molecular weight compound chemistry, analytical chemistry and medicine, and the range of their application is steadily increasing.

Hindered nitroxyls have found their widest application in the spin-label method widely used in biophysical research. As has already been mentioned, this method was devised in the course of nitroxyl chemistry development when the reactions of the radicals of this class, not involving a nitroxyl group, were discovered. Several monographs and many review papers with a total number of more than one thousand pages are devoted to the spin-label method (35-39,49,50). These communications cover all the aspects of hindered nitroxyl application in this method. Therefore, the spin-label method is not discussed here at length.

Another important field of the application of nitroxyl radicals and their precursors, hindered amines, is the protection of polymers and other organic compounds against the factors causing their thermo- and photo-destruction. In the present work, an attempt has been made to describe briefly the main achievements in this field.

In conclusion, we shall consider some other fields of nitroxyl radical application.

Polymer Stabilization

"Hindered amine stabilizers have been the single most important development of the last decade in polymer stabilization" - Peter P. Klemchuk.

The possibility of stabilizing polymers with hindered nitroxyl radicals is based on their reaction with polymer radicals which are engaged in propagating polymer oxidation. It was established that hindered amines, the precursors of nitroxyl radicals, are also effective photostabilizers and, as such, they are more effective than most of the common photostabilizers for polymers.

Complete information on structure and action mechanisms of stabilizers based on hindered amines can be found in (38,52-56). Here are presented only modern trends in the chemistry of photo- and thermostabilizers based on hindered amines, the mechanisms of their action and the methods of selecting stabilizing compositions, including mixtures of hindered amines with other classes of stabilizers.

In Scheme 5 are given the structures of some 2,2,6,6-tetramethylpiperidine derivatives which are effective polymer stabilizers. The analysis of the structures of these compounds allows the formulation of the design principles of polymer stabilizers based on hindered amines in general and hindered piperidines in particular (38,51).

Data on hindered amine structures as polymer stabilizers lead to the following conclusions about the development of research in this field:

1. Methods are being elaborated for synthesizing multifunctional stabilizers comprising in one molecule the fragments of either hindered amine or a nitroxyl radical and UV absorber, organic phosphite, hindered phenol, thioether, metal derivative, especially those of nickel, etc.

    The introduction of a UV absorber fragment usually results in increased effectiveness of photoprotective action because a different synergistic mechanism operates between UV absorber and antioxidant, and the UV absorber fragment can quench the excited state of the nitroxyl responsible for photochemical transformation. Introduction of an additional antioxidant group is connected mainly with the necessity to increase the stability of the material in the process of its utilization. Among such stabilizers is Tinuvin 144, a Ciba-Geigy stabilizer (21). At 110°-120°C, it performs comparably to thermostabilizers such as Irganox 1010 (22).

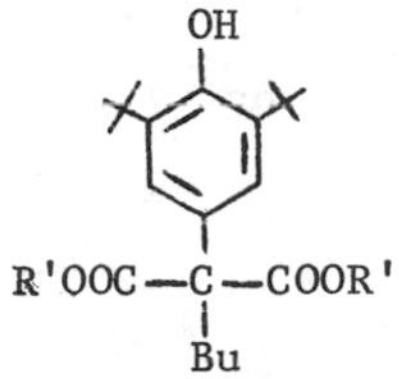

(21)

R', See Scheme 5

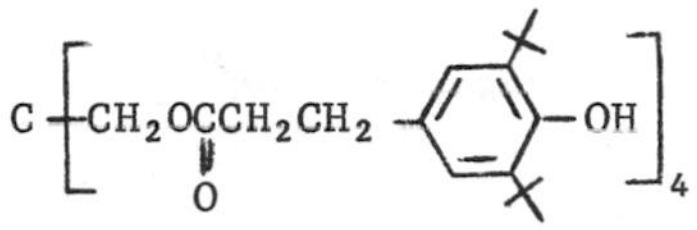

(22)

2. Work is carried out to obtain compounds comprising several hindered amine or nitroxyl fragments. The mechanism by which the effectiveness of photostabilizers with two or more hindered amine fragments is increased is not established. It was empirically observed that nitroxyl effectiveness during the inhibition of polypropylene thermooxidation increases with an increase in the number of paramagnetic centers in a molecule. The effectiveness of thermostabilizers of the bi-radical type increases

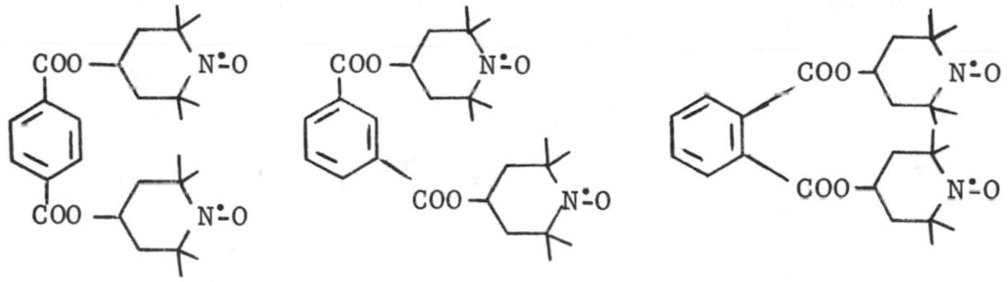

when the distance between paramagnetic centers decreases. This phenomenon may be due to a cage reaction of the alkyl radical, which is formed during the interaction of a polymer with nitroxyl with another nitroxyl group. This reaction hinders the side initiation of oxidation by the nitroxyl radical. There isn't a satisfactory explanation to the increase in photostabilizer effectivity with the increase in the number of NH groups because mono- rather than polyradicals are formed during photooxidation of hindered amines.

3. Stabilizers of high compatibility with polymers are being synthesized. To improve their compatibility with polymers, some fragments are introduced into the stabilizer. For example, alkyl substituents provide high effectiveness to photostabili-

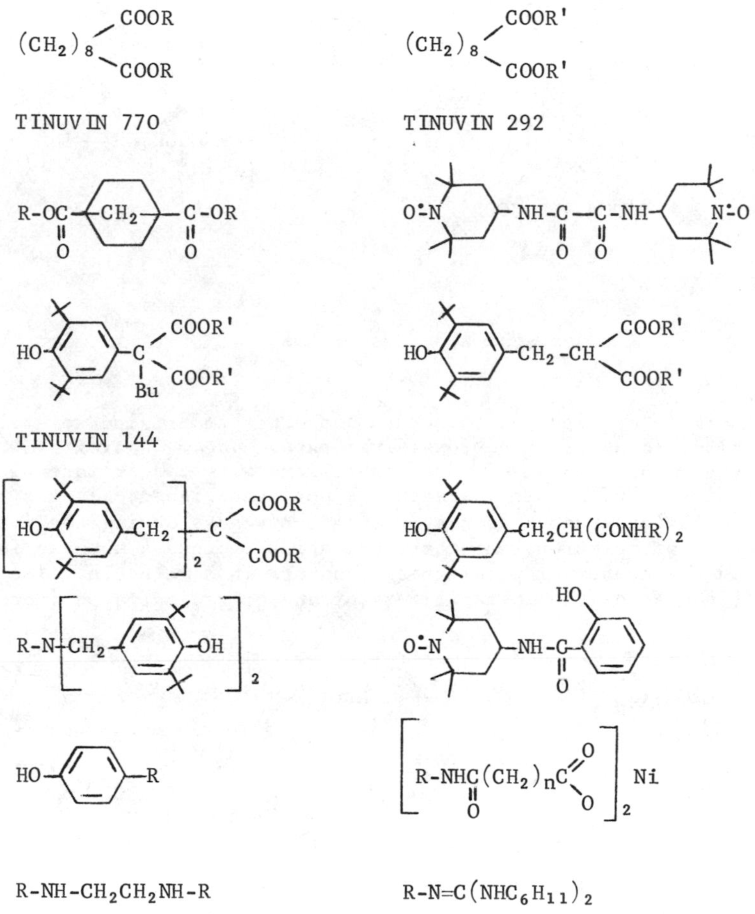

$(CH_2)_8$ COOR / COOR

TINUVIN 770

$(CH_2)_8$ COOR' / COOR'

TINUVIN 292

R-OC—(CH$_2$)—C-OR
  ‖   ‖
  O   O

O⁼N——NH-C—C-NH——N⁼O
         ‖ ‖
         O O

HO——C COOR' / COOR'
    |
    Bu

TINUVIN 144

HO——CH$_2$-CH COOR' / COOR'

[HO——CH$_2$]$_2$ C COOR / COOR

HO——CH$_2$CH(CONHR)$_2$

R-N[CH$_2$——OH]$_2$

O⁼N——NH-C——HO
         ‖
         O

HO——R

[R-NHC(CH$_2$)$_n$C O / O]$_2$ Ni
     ‖
     O

R-NH-CH$_2$CH$_2$NH-R

R-N=C(NHC$_6$H$_{11}$)$_2$

Scheme 5. Structures of hindered amine stabilizers for polymers.
Continued on next page.

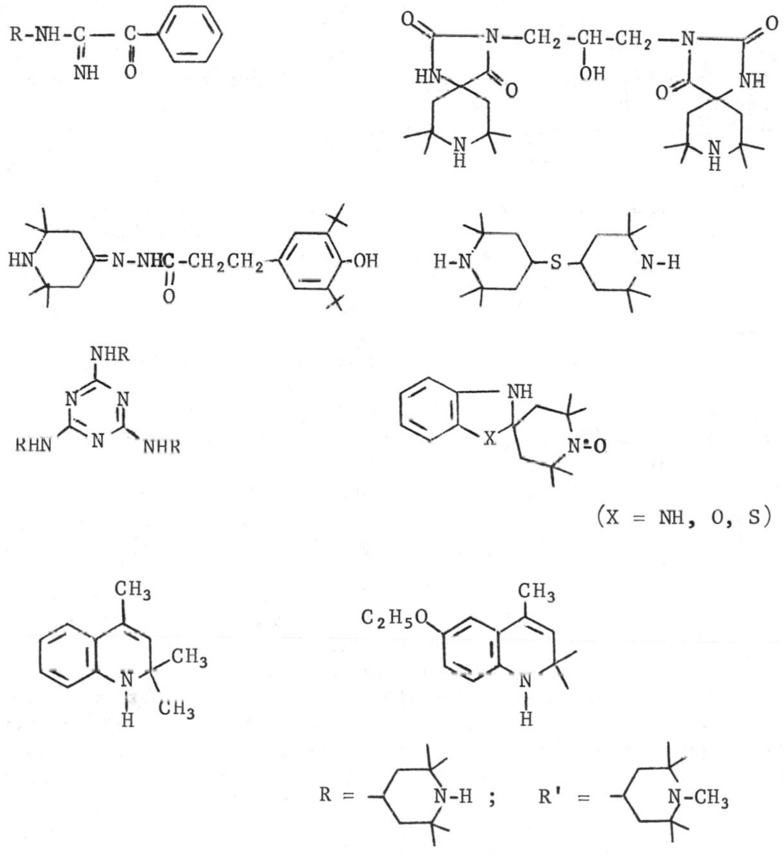

Scheme 5. Continued.

zers for polyolefins. Some characteristics of stabilizers'
structures appear to contribute to their accumulation in those
sites where photooxidation occurs. This seems to be particularly applied to polypropylene which is nonuniformly oxidized.
4. Methods of hindered amine stabilizer synthesis are being devel-
    oped with the purpose of reducing cost and using available raw
    materials.
The action mechanism of polymer photostabilizers based on hindered
amines attracted great attention after numerous reports on their
high effectiveness had appeared. The main conclusions from the
scientific literature on the mechanisms of hindered amine action are
offered below. More detailed information can be found in numerous
reviews, the ones compiled by the authors of the present article
including (38,51-56).

The following methods are used at present to protect polymers
against photooxidation (59):
a. The use in polymer composition of substances which at first
    absorb UV radiation and then release the absorbed energy in a
    form which is not harmful to the polymer. These compounds,
    called UV absorbers, protect the polymer against UV radiation
    by direct absorption.
b. Incorporation into polymer of compounds which can deactivate
    photo excited states (either singlet or triplet).
c. Incorporation into polymers of substances capable of catalyzing
    the non-radical decomposition of hydroperoxides.
d. Make use of compounds (antioxidants) which are capable of
    trapping oxidation-propagating free radicals thereby breaking
    the oxidation chains.
Hindered amines, nitroxyl radicals, and their transformation products cannot act by the mechanisms of UV light screening or quenching of photo excited states. Their action is due mainly to their
involvement with the oxidation propagating chemical reactions taking place during polymer photooxidation; so, they act as radical
trapping antioxidants. But a number of important features essentially distinguish them from common aromatic antioxidants (phenols
and amines) which are poor polymer photostabilizers.

Several reactions may be involved in providing protection to
polymers against photodecomposition destruction. The contribution
of each of the processes depends on the nature of the polymer and
the oxidation conditions. The time during which a hindered amine
is transformed into nitroxyl in polypropylene is short compared to
the induction period. Therefore, piperidines and corresponding
nitroxyl radicals are almost equally effective for polymer stabilization. In rubber, which is more rapidly oxidized than polypropylene, nitroxyl radicals are fairly more effective than hindered
amines. Subsequent transformations of nitroxyl radicals, formed
through Reactions 6-8 are shown in Scheme 6 (53).

The antioxidant action of nitroxyl radicals is due to their
ability to react with alkyl radicals as in Reaction 9. Because of
their reactivity with alkyl radicals, nitroxyls are better photooxidation inhibitors than the inhibitors of other classes. The
rate constant for the reaction of nitroxyls with hydroxyl radical
(Equation 10) is 10^9 L/mol. sec. This reaction can be of particular
importance for the inhibition of polymer photooxidation.

Reactions 11-13 initiate polymer photooxidation. However,

they do not have an appreciable influence on hindered amine efficiency because Reactions 11 and 12 occur only at elevated temperatures, and Reaction 13 occurs only under the action of light in a
narrow wavelength band, corresponding to nitroxyl radical absorption, and this is the region where nitroxyls absorb light more
poorly than phenols and aromatic amines. Moreover, hydroxylamines
formed from Reactions 11-13 are efficient antioxidants while the
products of the photooxidation of aromatic antioxidants (phenols
and amines) often function as sensitizers for initiating polymer
photooxidation.

Reaction 13 was studied with model compounds. Radical (6)
undergoes the following photolytic reaction:

$$2 \; HO{-}\!\!\!\!<\!\!\!{-}N{\cdot}O \; + \; C_6H_5CH_3 \; \xrightarrow{h\nu}$$

$$HO{-}\!\!\!\!<\!\!\!{-}N{-}OH \; + \; HO{-}\!\!\!\!<\!\!\!{-}N{-}OCH_2C_6H_5$$

50%   50%

Under similar conditions, the five-membered nitroxyl (7) decomposes
to eliminate NO. Poor stability of five-membered nitroxyls is
probably one of the reasons why examples of that type are not
offered commercially as polymer stabilizers.

Hydroxylamines and their ethers, formed during the aforesaid
Reactions 9-13 also inhibit photooxidation (Scheme 7). Hydroxylamines, unlike nitroxyl radicals, react with peroxide radicals
(Reaction 11). The rate constant of this reaction is close to the
corresponding constants estimated for most active phenols.

Hydroxylamine ethers are poorer inhibitors than commonly used
phenols and aromatic amines (rate constant for their reaction with
Me_2CNCOO radical in chlorobenzene at 65° is 1-20 L/mol. sec.)

Under thermal conditions, hydroxylamine ethers can reversibly
decompose (Reaction 15). The radicals formed disproportionate to
eliminate olefins and yield hydroxylamine (Reaction 16). In the
presence of sufficiently effective acceptors of alkyl radicals
(e.g., oxygen), the reaction rate of peroxy radical formation is
much higher than that of hydroxylamine formation. Thus, in the
process of polymer photooxidation, nitroxyl radicals regenerate and
can break multiple oxidative chains.

Analysis of reaction rate constants in model systems shows
that at room temperature, the main reaction leading to regeneration
of nitroxyl radicals is their interaction with peroxide radicals,
(Reaction 11) and at elevated temperatures (more than 80°) the main
reaction is that of hydroxylamine ether decomposition (Reaction 15)
(53).

It has been proposed, mainly in the patent literature, to use
hindered amines in mixtures with other classes of stabilizers. A
short survey of literature data along these lines is given in (38).
The mixtures containing hindered amines and UV absorbers 2-(hydroxy-

$$>\text{N-O} + \text{P}\cdot \longrightarrow >\text{N-OP} \qquad (6)$$

$$>\text{N-O} + \cdot\text{OH} \rightleftharpoons >\text{N} \underset{\text{O}}{\overset{\text{OH}}{\diagdown}} \qquad (7)$$

$$>\text{N-O} + \text{PH} \rightleftharpoons >\text{N-OH} + \text{P}\cdot \qquad (8)$$

$$>\text{N-O} + \text{POOH} \rightleftharpoons >\text{N-OH} + \text{PO}_2^\cdot \qquad (9)$$

$$>\text{N-O} + \text{PH} \overset{h\nu}{\rightleftharpoons} >\text{N-OH} + \text{P}\cdot \qquad (10)$$

Scheme 6. Transformations of nitroxyl radicals.

$$>\text{N-OH} + \text{PO}_2\cdot \longrightarrow >\text{N-O} + \text{POOH} \qquad (11)$$

$$>\text{N-OH} + \text{POOH} \longrightarrow >\text{N-O} + \text{PO}\cdot + \text{H}_2\text{O} \qquad (12)$$

$$>\text{N-OP} + \text{P}'\text{O}_2 \longrightarrow >\text{N-O} + \text{P}'\text{OOH} + >\text{C=C}< \qquad (13)$$

$$>\text{N-OP} + \text{P}'\text{O}_2\cdot \longrightarrow >\text{N-O} + \text{P}'\text{OOP} \qquad (14)$$

$$>\text{N-OP} \rightleftharpoons >\text{N-O} + \text{P}\cdot \qquad (15)$$

$$>\text{N-O} + \text{P}\cdot \longrightarrow >\text{N-OH} + >\text{C=C}< \qquad (16)$$

Scheme 7. Reactions of hydroxylamines and hydroxylamine ethers in polymer stabilization.

benzophenones, salicylates, 2-(hydroxy-2-phenyl)benzotriazoles) have been shown to be synergistic in many instances. A UV absorber reduces considerably the rate of polymer photooxidation by absorbing incident light. Photooxidation and antioxidant consumption usually take place mainly in the surface layer of polymer samples. Due to concentration gradients, the antioxidant diffuses to the polymer surface, protecting it against oxidation until the stabilizer is entirely consumed. Accordingly, a synergistic effect is observed only in thick polymer samples at relatively weak light intensity.

Hindered phenols, sulfides and phosphites act as antagonists and decrease or, at best, do not change the effectiveness of hindered amines. Several examples of the use of mixtures of hindered amines with other classes of stabilizers are given in (60). These data do not contradict the fact that mixtures comprising other classes of stabilizers exhibit high efficiency. According to B. Ranby and J. Rabek, when these stabilizers are present in a constant ratio, an autosynergistic mechanism is observed.

Synergism with hindered amines has been less investigated. To all appearances, the first example of the synergistic mechanism of this type in the case of inhibited polypropylene oxidation was pointed out by one of the authors of the present paper as far back as 1964 (61).

Other Fields of Hindered Amine and Nitroxyl Radical Application

Other fields of nitroxyl radical application are not as far advanced as the spin-label method and polymer stabilization technique. Therefore, they will be discussed only briefly or just mentioned in passing.

Adhesion. A promising trend in nitroxyl radical application is associated with the modification of polymer surfaces to promote adhesion. Bonds which are formed as a result of the interaction of radicals with a surface and an adhesive are known to be of great importance for the forces determining the adhesion of the system. The direct way to increase the number of radicals on a substrate surface is to treat it with stable radical solutions. Nitroxyl radicals of the 2,2,6,6-tetramethylpiperidine series have been found to considerably improve durability of adhesive bonds (62,63). For instance, when sticking together high pressure polyethylene, fluoroplastic F-4 and polyvinylchloride, pretreatment of their surfaces with nitroxyl radical solutions improved the durability of adhesive bonds by 160-170%. This result is considerably better than that obtained by treatment of the aforesaid polymer surfaces with the most active adhesion modifiers known (64).

Nitroxyl radicals can also be used to improve the mechanical properties of filled polymers, resins in particular, by increasing the adhesive quality of the fillers by treating their surfaces with solutions of nitroxyl radicals. The efficiency of nitroxyl radicals as modifiers is higher than that of well-known modifiers; e.g., 4-toluene isocyanate.

Pretreatment of polymer surfaces with nitroxyl radicals of the 2,2,6,6-tetramethylpiperidine series in the process of chemically forming coatings leads to a considerable improvement in fastening nickel to the substrate and a two to three-fold increase in the

process rate. The latter is probably connected with the reducing
properties of nitroxyl radicals (65,66).

Nitroxyl radicals can be used not only to increase durability
of adhesive bonds but also to decrease it. This effect can be
achieved by thermal treatment of radical-modified surfaces at
100-160°C. This is probably due to the fact that at elevated
temperatures nitroxyl radicals react with substrate active groups of
surfaces thereby sharply decreasing surface adhesion.

Pharmacology. One of the features that distinguishes cancer cells
from normal ones is that the adhesive quality of the former is more
than an order of magnitude poorer than that of normal cells. Some
authors have assumed that modification of cancer cells' outer
membranes with the purpose of improving their adhesive quality leads
to inhibition of malignant tissue growth. Perhaps this can explain
a higher effectiveness of a number of paramagnetic analogs of anticarcinogenic drugs (67).

Medico-biological properties of nitroxyl radicals have been
intensively studied. Their application as anticarcinogenic drugs is
based on the investigations headed by N. M. Emanuel at the Institute
of Chemical Physics, Academy of Sciences, USSR, which revealed the
role of free radicals in the process of tumor growth and the ability
of oxidation inhibitors to suppress this process.

The simplest radicals of the 2,2,6,6-tetramethylpiperidine
series have been found to possess antitumor activity. They caused
inhibition of tumor development (1[a] strain leucosis) but proved to
be of little merit compared to the drugs used in practice (35).

Substances containing moities other than nitroxyl radical,
operating by another mechanism, exhibited high anticarcinogenic
activity. In the scientific literature, there are comparative data
on triethyleneiminyl thiophosphine (Thio TEPh) and Thio TEPh modified with a nitroxyl radical fragment. The latter proved to be more
effective, and its toxicity less than 1/10 that of Thio TEPh (35).

Similar data have been obtained for the spin-labelled antibiotic,
Rubomizine (35).

A number of communications indicate that nitroxyl radicals in
the absence of oxygen increased coliform bacterium sensitivity to
irradiation. This property of nitroxyl radicals has been
subsequently intensively studied. It was established that the
nitroxyl action mechanism appears to be due to irreversible reaction
with irradiation-induced DNA radicals to form radical-DNA reaction
products, which result in disruption of DNA synthesis. Moreover,
nitroxyl radicals are electron acceptor compounds, capable of
trapping an electron from the molecule of the target which brings

about its oxidation and damage. Though the action mechanism of nitroxyl radicals is determined by the availability of a paramagnetic center, nitroxyl radicals of even the same series can exhibit different activities. Biradicals, for instance, bis-(2,2,6,6-tetramethyl-1-oxyl-4-piperidinyl) succinate, proved to sensitize tissue to irradiation more effectively than monoradicals of the 2,2,6,6-tetramethylpiperidine series.

In general, nitroxyl radicals exhibit their sensitizing properties only in the absence of oxygen. This phenomenon can be used for malignant tumor therapy. Tumor cells are heterogeneous in many parameters. There are colonies among them, poorly supplied with oxygen. During irradiation, these cells are not damaged because oxygen sensitizes the irradiation action. Nitroxyls increase the sensitivity of tumor cells to irradiation thereby causing their destruction.

The practical application of nitroxyl radicals to sensitize tissues to irradiation is connected with many difficulties. Both in a cell culture and in a living organism, nitroxyl radicals are rapidly reduced to the corresponding hydroxylamines and at the same time the sensitivity of the tumor to irradiation disappears. A full summary of data on how nitroxyl radicals influence the response of normal and tumor cells to irradiation is presented in monograph (68).

Nitroxyl radicals are of particular significance for the synthesis of paramagnetic analogs of biologically active materials making it possible to follow them by EPR. By this means, a scientific trend in pharmacology was developed. Literature data on the synthesis, properties and application of paramagnetic biologically active compounds are given in monograph (35).
Analysis. Among other fields of nitroxyl radical application, mention should be made of their in the analysis of organic and inorganic substances and their utilization as reagents in organic chemistry. Analysis of inorganic cations is based on the application of nitroxyl radicals containing chelate-forming moieties such as 1,3-diketone fragments. Metal chelates are separated by extraction and determined by EPR. Methods of analysis for mercury, nickel, cobalt and other metals have been devised. Their detection range is $10^7 - 10^6$ m (69).

The method of spin-trap has been widely used for detection and identification of active radicals formed during polymerization photolysis, radiolysis and other procedures. This method is based on the reactions yielding nitroxyl radicals; e.g.,

$$R-N-O + R'\cdot \longrightarrow R-N-R'$$
$$\underset{O}{\overset{\displaystyle |\cdot}{}}$$

Complete information on application of the spin-trap method is
given in summaries (70-72).
Geophysics. Utilization of dynamic proton polarization phenomenon
in nitroxyl radical solutions facilitated the improvement of spin-
precision magnetometers by synchronizing the processes of nuclear
polarization and precision measurement. Deuterated radical
solutions serve as sensors in such instruments. Instruments of this
type have already been produced in France. They find application in
geophysical research, geological explorations and in searches for
metallic objects under water.
Oil Recovery. Nitroxyl radicals have found significant application
in the oil production industry to control the rate of water
encroachment of an oil pool. The method consists of the following:
nitroxyl radicals are introduced into the water-injection well
together with water, and water encroachment of the oil pool is
followed by sampling and determining the nitroxyl radical content by
EPR. Introduction of this method will permit increases in oil
recovery (73).
Diamond Quality Assessment. The physical-chemical characteristics
of nitroxyl radicals make it possible to evaluate crystal quality of
diamonds, for example, with high sensitivity and reliability. It is
possible to detect cracks on the crystal surface, to introduce
quantitative criteria for estimation of the quality of diamonds, and
to carry out defect differentiation (74).
Antiwear. It has been shown that the addition of nitroxyl radicals
of the 2,2,6,6-tetramethylpiperidine series to lubricating compositions consisting of mineral oil and alkyl adipates (20%) considerably improves their lubricating properties (75). When utilizing the
suggested compositions, the maximum load can be increased by 1.2-1.6
times for the same and sometimes even smaller friction factor. At
the same time, the weight wearout rate is reduced by 1.2-2.0 times
for bronze samples and by 2-8 times for steel samples.
    In conclusion, it is worth noting that nitroxyl radical
applications have been limited heretofore by unjustifiably high
prices. The prices of the simplest radicals in the Aldrich chemical
catalog of 1982 were in the range of $1,000-$15,000 a kilogram. At
present, they are being steadily reduced. In the authors' opinion,
the prices can be reduced even for small volumes of production at
least by 1/50. This will lead to a wider investigation of their
properties and, naturally, will extend the range of their
application.

Literature Cited

1. Rozantsev, E. G.; Lebedev, O. L.; Kararnovsky. S. N. "The
    Discovery of the Phenomenon of Stable Localized Free-Radical
    Center Formation," Otkrytie SSSR N 246 August 6, 1981.
    Otkrytiya SSSR, 18, Moscow.
2. Rozantsev, E. G. "Free Iminoxyl Radicals," Khimiya, Moscow,
    1970.
3. Rozantsev, E. G. Izv. Akad. Nauk SSSR, Ser. Khim. 1963,
    1669.
4. Neiman, M. B.; Rozantsev, E. G.; Mamedova, Yu. G. Nature
    1963, 200, 256.

5. USSR Patent 166 032, 1962.
6. Rozantsev, E. G. Izv. Akad. Nauk SSSR, Ser. Khim. 1964, 2218.
7. Sokoloff, N.; Latschinoff, P. Ber. 1874, 7, 1384.
8. Heintz, W. Ann. 1874, 175, 133.
9. Rozantsev, E. G. Ivanov, V. P. Khim.-Pharm. Zh. 1971, 47.
10. Rozantsev, E. G. Izv. Akad. Nauk SSSR, Ser. Khim. 1964, 2187.
11. Rozantsev, E. G.; Krinitzkaya, L. A. Tetrahedron 1965, 21, 491.
12. Rozantsev, E. G.; Neiman, M. F. Tetrahedron 1964, 20, 131.
13. USSR Patent 166 133, 1964.
14. Ohnishi, S.; McConnell, H. M. J. Am. Chem. Soc. 1965, 87, 2293.
15. Sosnovsky, G.; Konieczny, M. Z. Naturforsch. 1977, 32B, 328.
16. Kedik, S. A.; Rozantsev. E. G.; Usvyatsov, A. A. Dokl. Akad. Nauk SSSR 1981, 257, N 6, 1382.
17. Prostakov, N. S.; Gayvaronskaya, L. A. Usp. Khim. 1978, 47, 859.
18. Rozantsev. E. G.; Kagan. E. Sh.; Sholle, V. D. in "Paramagnetic Models of Drugs and Biochemicals," Ed. Zdanov, R. I., C.R.C. Press Boca Raton, Florida 1985.
19. Rozantsev, E. G.; Dagonneau, M.; Kagan, E. Sh.; Sholle, V. D.; Michailov, V. I. Synthesis 1984.
20. Kagan, E. Sh.; Avrutskaya, I. A.; Kondrashov, S. V.; Novikov, V. T.; Fioshin, M. Ya.; Smirnov, V. A. Electrochemical Synthesis of 2,2,6,6-tetramethylpiperidine, Khimiya geterociclicheskih soedinenii 1984, 3, 358-359.
21. Tomilov, A. P.; Smirnov, V. A.; Kagan, E. Sh. "Electrochemical Synthesis of Organic Compounds," Rostov University Ed. 1981, 59-61.
22. USSR Patent 908.017, 1982.
23. Surov, I. I.; Avrutskaya, I. A.; Fioshin, M. Ya. Electrochimia 1983, 19, 1561-1565.
24. Tsarkova, T. G.; Avrutskaya, I. A.; Fioshin, M. Ya. Electrochimia 1984, 20, N3, 404-407.
25. Rubtsov, M. V.; Baichikov, A. G. "Synthetic Pharmaceutical Chemicals, Meditsina," Moscow, 1971, 194.
26. Rozantsev, E. G.; Dagonneau, M.; Kagan, E. Sh.; Mikhailov, V. I. J. Chem. Res. (S) 1979, 260; J. Chem. Res. 1979, 2901.
27. Mikhailov, V. D. "Enamines in the Synthesis of Nitroxyl Radicals of the 2,2,6,6-tetramethylpiperidine Series," Cand. Dis. RGU, Rostov-on-the-Don, 1975.
28. Briere, R.; Espil, J. C.; Ramasseul, R.; Rassat, A.; Rey, P. Tetrahedron Let. 1979, 941.
29. Rozantsev, E. G. "Free Nitroxyl Radicals," Plenum Press, New York, London, 1970.
30. Rosantsev, E. G.; Sholle, V. D. Organitscheskaya Khimia Svobodnych Radicalov, Khimia, Moscow 1979.
31. Rozantsev, E. G.; Sholle, V. D. Synthesis 1971, 190-202.
32. Rozantsev, E. G.; Sholle, V. D. Synthesis 1971, 401-414.
33. "Spin-Labelling, Theory and Application," Ed. by L. Bezlinez, Academic Press, New York, London, 1976.
34. Keana, I. F. W. Chem. Rev. 1978, 78, 37-63.
35. Zhdanov, R. I. "Paramagnitnye Modeli Biologitscheski Aktivnych Soedinenii," Nauka, Moscow, 1981.

36. Forrester, A. K.; Hay, G. M.; Thomson, R. H. "Organic Chem-
     istry of Stable Free Radical," Academic Press, London, 1967,
     405.
37. Buchachenko, A. L.; Vasserman, A. M. "Stabilnye Radicaly,"
     Khimia, Moscow, 1973.
38. Dagonneau, M.; Ivanov, V. B.; Rozantsev, E. G.; Sholle, V. D.;
     Kagan, E. Sh. "Sterically Hindered Amines," J. M. S.-Rev.
     Macromol. Chem. Phys. 1982-1983, c. 22(2), 169-202.
39. Kagan, E. Sh.; Kondrashov, S. V.; Selivanov, V. N. in "All-
     Union Conference on Electrochemistry" June 21-25, 1982,
     Moscow, 1982, 208.
40. Bogdanova, N. N.; Surov, I. I.; Avrutskaya, I. A.; Fioshin,
     M. Ya. Electrochimia 1983, 19, 1286.
41. Keana, I. F. W.; Dinerstein, R. I.; Baitis, F. J. Org. Chem.
     1970, 36, 209-211.
42. Golubev, V. A.; Zhdanov, R. I.; Gida, V. M.; Rozantsev, E. G.
     Izv. Akad. Nauk SSSR Ser. Khim. 1971, 853-855.
43. Sen, V. D.; Golubev, V. A.; Efremova, N. M. Izv. Akad. Nauk
     SSSR, Ser. Khim. 1982, 61-72.
44. Sosnovsky, G.; Konieczny, M. Z. Naturforshc. 1976, Teil B.,
     31, 1376-1378.
45. Levina, T. M.; Rozantsev, E. G. A. S. Chegolya. Dokl. Akad.
     Nauk SSSR 1981, 261, 109-110.
46. USSR Patent 391 137, 1973.
47. Raucman, E. I.; Rosen, G. M.; Abon-Donia, M. B. Synth. Comm.
     1975, 5, 409-413.
48. Kagan, E. Sh.; Mikhailov, V. I.; Sholle, V. D.; Rozynov,
     B. V.; Rozantsev, E. G. Izv. Akad. Nauk SSSR, Ser. Khim.
     1977.
49. Likhtenshtein, G. I. "Spin-Labelling Methods in Molecular
     Biology," Nauka, Moscow, 1977.
50. Kuznetsov, L. N. "The Spin-Probe Method," Nauka, Moscow,
     1976.
51. Toda, T.; Kurumada, T. Sankyo Kenkyosho Nempo "Research and
     Development of Hindered Amine Light Stabilizers," Ann. Rep.
     Sankyo Res. Lab. 1983, 35, c. 1-37.
52. Shlyapintokh, V. Ya. "Photochemical Transformations and Poly-
     mer Stabilization," Khimia, Moscow, 1979.
53. Shlyapintokh, V. Ya.; Ivanov, V. B. "Developments in Polymer
     Stab.," Ed. G. Scott, Appl. Sci. Publ. London, 1982, 5, 41.
54. Allen, N. S. "Developments in Polym. Photochemistry," Ed.
     N. S. Allen. Appl. Sci. Publ., London, 1981, 239.
55. Carlsson, D. J.; et. al. "Developments in Polymer Stabiliza-
     tion," Ed. G. Scott, Appl. Sci. Publ., London, 1979, 1, 219.
56. Gugumus, F., "Developments in Polymer Stabilization,"
     Ed. G. Scott, Appl. Sci. Publ., London, 1979, 1, 261.
57. Shlyapintokh, V. Ya.; Bystritskaya, E. B.; Shapiro, A. B.;
     Smirnov, V. A.; Rozantsev, E. G.; Izv. Akad. Nauk SSSR, Ser.
     Khim. 1973, 1915.
58. Shlyapintokh, V. Ya.; Ivanov, V. B.; Hvostach, D. M.; Shapiro,
     A. B.; Rozantsev, E. G. Dokl. Akad. Nauk SSSR, 1975, 225,
     1132.

59. Randy, B.; Rabek, J. F. "Photodegradation, Photooxidation and Photostabilization of Polymers," Wiley-Interscience, New York, 1975.
60. Allen, N. S.; McKeller, I. F.; Wilson, D. Chem. Ind. 1978, 887-889.
61. Rozantsev, E. G.; Krinitskaya, L. A.; Troitskaya, L. S. Khim. Prom. 1964, 20, 180.
62. Vakula, V. A.; Pritykin, L. H. "Foundation of Physical Chemistry of Polymer Adhesion," Khimia, Moscow, 1984.
63. USSR Patent 600 162, 1978.
64. Pritykin, L. M.; Genel, L. S.; Gerenrot, V. G.; Shapiro, A. B.; Bakula, V. L. Plastics 1981, 36-39.
65. USSR Patent 732 405, 1980.
66. Danjushina, G. A.; Kagan, E. Sh.; Smirnov, V. A.; Pritykin, L. M. Plasticheskie Massy 1982, 83.
67. Pritykin, L. M. Biofisika 1976, 21, 1059.
68. Pelevina, I. I.; Afanasjev, G. G.; Gotlib, V. Ya. "Klectochnye Faktory Reaktsii Opukholei na Obluchenie i Khimikoterapevticheskie Vozdeistvija," Nauka, Moscow, 1978, 304.
69. All-Union Conference on Nitroxyl Radicals, Chernogolovka, May 12-14, 1982, Tezisy Dokladov.
70. Pedersen, I. A.; Torssel, K. Acta. Chem. Scand. 1971, 25, 3151-62.
71. Yansen, E. G. Acc. Chem. Res. 1971, 4, 31-40.
72. Lagercrantz, C. J. Phys. Chem. 1971, 75, 3466.
73. Bukin, I. I.; Rozantsev, E. G. et. al. Neftyanoe Khozyaistvo 1978, 46.
74. Bukin, I. I.; Kagarmanov, N. F. et al. Sverkhtverdye Materialy 1983, 27, 23.
75. USSR Patent 802 358, 1981.

RECEIVED January 31, 1985

Progress in the Light Stabilization of Polymers

TOSHIMASA TODA, TOMOYUKI KURUMADA, and KEISUKE MURAYAMA

Chemical Research Laboratories, Research Institute, Sankyo Co., Ltd., 2–58, Hiromachi 1-chome, Shinagawa-ku, Tokyo 140, Japan

Since their initial production by our laboratories about ten years ago, hindered amine light stabilizers (HALS) have become established as excellent light stabilizers of polymers. This review deals with their synthesis, evaluation and development.

Hindered Amine Light Stabilizers (HALS) are a new class of highly efficient stabilizers protecting polyolefins and other polymers against light-induced deterioration. They were initially developed into commercial products in our laboratories. In this review we describe the details of how they were synthesized, evaluated and developed.

In order to find a new potential stabilizer, our initial efforts were directed toward synthesizing new stable nitroxyl radicals and evaluating their stabilizing activity. Although these stable radicals were good light stabilizers, they could not be used commercially since they contributed color to the polymers. Finally, intensive studies led to the discovery that hindered amine compounds derived from 2,2,6,6-tetramethyl-4-oxopiperidine may be converted into the corresponding stable nitroxyl radical and are excellent light stabilizers for polymers. Derivatives of 2,2,6,6-tetramethylpiperidine were synthesized and tested, and as a result, esters of 4-hydroxy-2,2,6,6-tetramethylpiperidine were selected as practical light stabilizers, particularly for polyolefins.

Degradation and Stabilization

Oxidation reactions are generally believed to involve a free radical chain reaction as proposed by Bolland and Gee (1). The main steps of this reaction are:

$$RH \xrightarrow{\text{heat or light}} R\cdot \tag{1}$$

$$R\cdot + O_2 \longrightarrow ROO\cdot \tag{2}$$

$$ROO\cdot + RH \longrightarrow ROOH + R\cdot \tag{3}$$

$$ROOH \xrightarrow{\text{heat or light}} RO\cdot + \cdot OH \tag{4}$$

$$2\ ROOH \longrightarrow RO\cdot + ROO\cdot + H_2O \tag{5}$$

$$RO\cdot + RH \longrightarrow ROH + R\cdot \tag{6}$$

$$HO\cdot + RH \longrightarrow HOH + R\cdot \tag{7}$$

$$2\ ROO \longrightarrow \text{Product ketones, alcohols, etc.} \tag{8}$$

Radicals formed during initiation react with oxygen leading to chain reactions. Generally, Reaction 2 is faster than Reaction 3. The decomposition of hydroperoxides by heat or UV light (Equation 4) causes formation of alkoxy and hydroxy radicals leading to chain branching. UV absorbers can absorb this harmful energy, slowing initiation reactions, while antioxidants (InH) act to interfere with propagation according to the following reaction:

$$ROO\cdot + InH \longrightarrow ROOH + In\cdot \tag{9}$$

Finally, peroxide decomposers can decompose hydroperoxides in a non-radical process to interfere with Reaction 4.

Typical examples of the stabilizers generally used to prevent the above chain reaction are: (a) UV absorbers -- 2(2-hydroxy-3-tert-butyl-5-methylphenyl)-5-chlorobenzotriazole and 2-hydroxy-4-octoxybenzophenone; (b) Antioxidants -- 3,5-di-tert-butyl-4-hydroxytoluene and octadecyl 3-(3,5-di-tert-butyl-4-hydroxyphenyl)propanoate; (c) Peroxide decomposers -- dilauryl thiodipropionate. In addition quenchers such as the organic nickel complex, Ni(II) bis-(diisopropyl dithiocarbamate) are used for the deactivation of the excited states of the chromophoric groups responsible for light initiation.

Since polyolefins readily undergo oxidative deterioration on exposure to UV light, conventional stabilizers were found to be unsatisfactory for long-term outdoor applications. Therefore, more effective stabilizers for polyolefins were desired.

Stable Nitroxyl Radicals

Since UV absorbers cannot completely prevent the initiation reactions above, we felt that better antioxidants were necessary to obtain improved long-term photo-oxidative stability of polyolefins. We reasoned that stable radicals which scavenge radicals formed in the oxidative process might compensate for the imperfections of the absorbers. On the basis of the assumption that the diaryl nitroxyl radicals, rather than their parent diarylamines, are the active antifatigue species in rubber protection, we first attempted to obtain new stable nitroxyl radicals.

The first radicals we looked at, the alkylaminotropone nitroxy radicals (1), were unstable (2). However, since Neiman et al. (3) had reported that 2,2,6,6-tetramethyl-4-oxopiperidine-1-oxyl (2) was extremely stable, we turned to nitroxyl radicals derived from fully hindered amines. In this section we will describe the synthesis of new stable nitroxyl radicals and their evaluation in polypropylene.

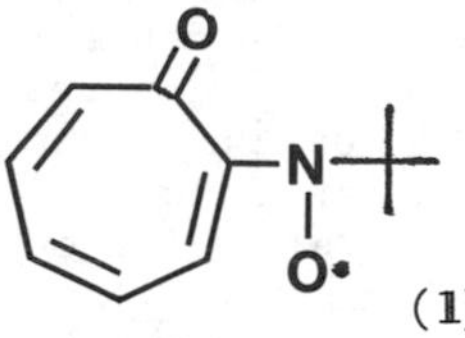

(1)

2,2,6,6-Tetramethyl-4-substituted piperidine-1-oxyls. Neiman's stable nitroxyl radical (2) was prepared in our laboratory and tested for its stabilizing activity on polymers. Its light stabilizing effect was superior to that of conventional UV absorbers; however, it was unstable at elevated temperatures. Heating crystals of (2) at 105-110°C for 3 hours under nitrogen gave 1-hydroxy-2,2,6,6-tetramethyl-4-oxopiperidine (3) and a trace of 2,6-dimethyl-2,5-heptadien-4-one (4). On the other hand, when crystals of (2) were allowed to stand at room temperature for six months, 1-hydroxy-2,2,6,6-tetramethyl-3-(2,2,6,6-tetramethyl-4-oxopiperidinoxy)-4-oxopiperidine (5) was obtained as a precipitate from the liquefied substance. We believe the decomposition of (2) proceeds as shown in Chart 1 (4).

The heat stability of additives is an important factor since polymer processing is done at high temperatures (150–280°C). Therefore we considered ways to obtain more thermally stable nitroxyl radicals. If the hydrogen alpha to the keto group of (2) is involved in its decomposition as shown in Chart 1, analogous compounds having no keto group should be more stable than (2). Indeed, after heating under the conditions described above, both 2,2,6,6-tetramethylpiperidine-1-oxyl (6) and 4-hydroxy-2,2,6,6-tetramethylpiperidine-1-oxyl (7) were almost completely recovered.

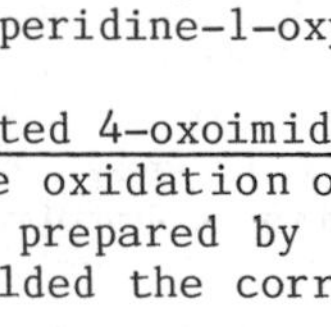

(6)

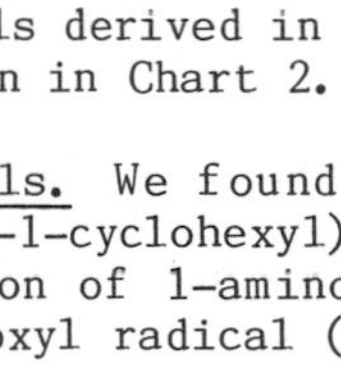

(7)

Additional examples of stable radicals derived in our laboratory from 4-ketopiperidine-1-oxyl (2) are shown in Chart 2.

Substituted 4-oxoimidazolidine-1-oxyls. We found that the hydrogen peroxide oxidation of 2,5-bis(spiro-1-cyclohexyl)-4-oxoimidazolidine (17) (9) prepared by self-condensation of 1-amino-1-cyanocyclohexane (16) yielded the corresponding nitroxyl radical (27) (10).

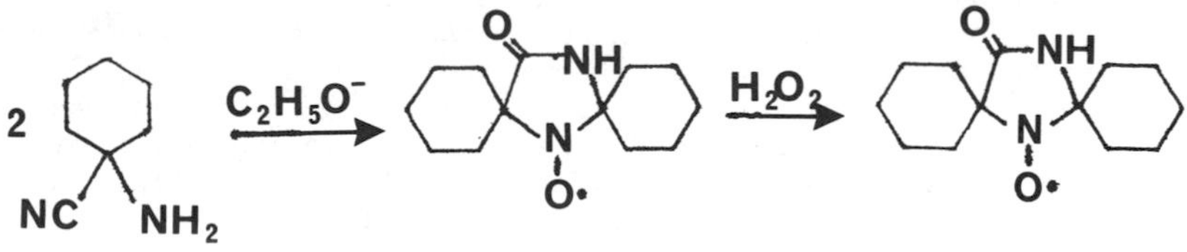

(16)   (17)   (27)

When the new stable radical (**27**) showed excellent light stabilizing activity for polypropylene, syntheses of analogous compounds
were planned. From a self-condensation reaction only two different
substituents may be introduced on the imidazolidine ring. Mechanistic studies of this reaction using the ^{15}N-labeled compound (**16a**)
showed that the nitrogen of the nitrile group appears only in the 3-
position of the imidazolidine ring (11). The self-condensation of
(**16a**) yielded the corresponding 4-oxoimidazolidine(3-^{15}N) (**17a**), as
confirmed by its mass spectrum (molecular ion at m/e 223.171 corresponding to $C_{13}H_{22}O^{15}NN$) and the presence in esr of the corresponding
nitroxyl radical (**27a**) triplet. The reaction of (**16a**) with cyclo-
hexanone also yielded (**17a**). The reaction is considered to proceed
as shown in Chart 3 (11).

Therefore, it is possible to obtain compounds with four different substituents at the 2- and 5-positions of the 4-oxoimidazoli-
dine ring by reacting the appropriate carbonyl compound with an alpha
aminonitrile formed in a Strecker reaction from a selected ketone.
The oxidation of these imidazolidines by hydrogen peroxide in the
presence of catalytic amounts of sodium tungstate in acetic acid gave
new stable nitroxyl radicals (12) shown in Chart 4.

<u>Light Stabilizing Activity of Nitroxyl Radicals.</u> The polymer stabi-
lizing activity of the stable radicals was evaluated as follows: (i)
The stabilizer (0.25 weight %) was incorporated into polypropylene
powder by wet blending with ethanol. The resulting mixture was
preheated to 215°C for 2 minutes, and then compression-molded, i.e.,
at 215°C for 0.5 minutes to yield 0.5 mm sheets. As a control,
unstabilized polypropylene sheet was prepared in a similar manner.
(ii) For testing of light stability each test specimen was exposed to
light in a Fade Meter (carbon arc lamp) at a black panel temperature
of 63±3°C and examined periodically by bending test to determine the
time to embrittlement. (iii) For testing of thermal stability each
test specimen was placed in a forced air circulation oven at 150°C
and the embrittlement time was measured by bending tests. Results
are summarized in Table I.

As shown in Table I, these stable radicals showed strikingly
higher light stabilization activity in polypropylene than that of the
UV absorber tested. We felt that their activity was related to their
radical scavenging ability. This hypothesis is supported by the
observation that the coupled products (**32**) and (**33**) were obtained by
the reaction of the nitroxyl radicals (**2**) and (**27**), respectively,
with a C-radical derived from AIBN (10). The radical scavenging
ability of the stable nitroxyl radicals is now well known to play a
major role in the mechanism of light stabilization by hindered amine
compounds (13).

Unfortunately, the stable nitroxyl radicals caused yellowing in
polymers. The yellowing may be due to the weak absorption of visible
light by the nitroxyl radicals themselves or to formation of colored
reaction products in the polymers. We observed that yellowing
occured when nitroxyl radicals were used in combination with a phenolic antioxidant such as 3,5-di-tert-butyl-4-hydroxytoluene (**34**).
This may be due to the reaction of (**2**) and (**34**) which gives a highly
colored product (**35**) as shown in Chart 5.

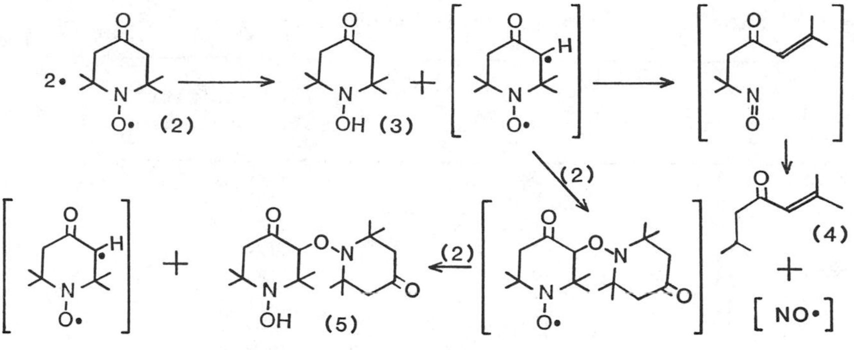

Chart 1. Decomposition of 2.

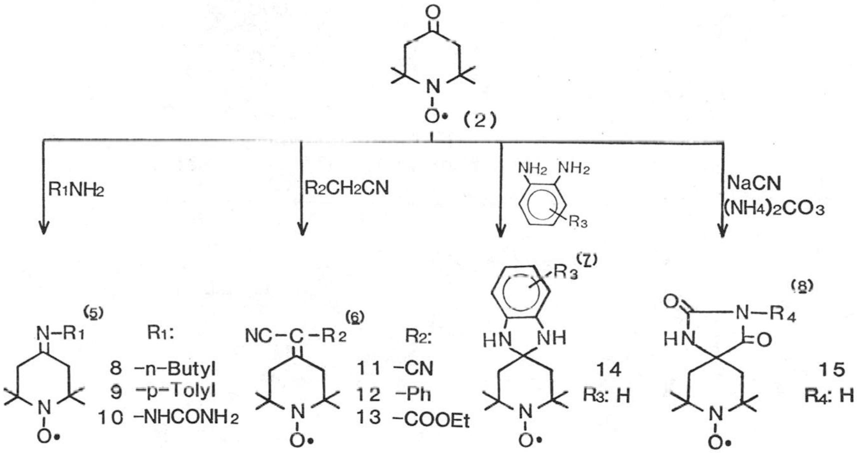

Chart 2. Examples of stable radicals derived from 2.

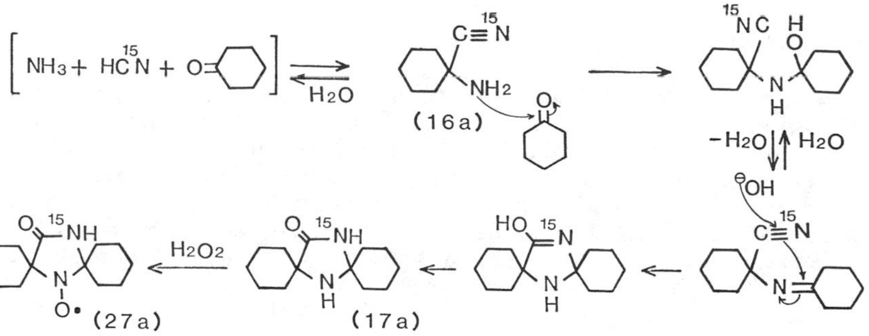

Chart 3. Reaction of 16a with cyclohexane.

Table I. Stabilizing Activity of Nitroxyl Radicals

Compound Number	m.p. (b.p.) °C	Light Stability[a] embrittlement time (hours)
4	36	400
8	(113/1mmHg)	300
9	113	140
10	195	300
11	95	240
12	132	260
13	(125/0.1mmHg)	380
14	160	260
15	248	400
27	228	320
28	226	600
29	115	320
31	175	260
UV absorber		80
None		40

a) Thermal embrittlement times of all test sheets were less than one day.

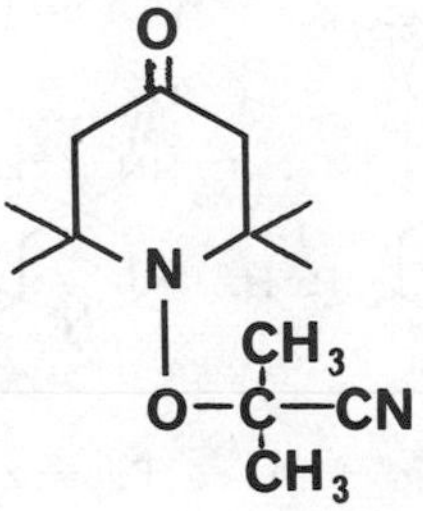

(32)

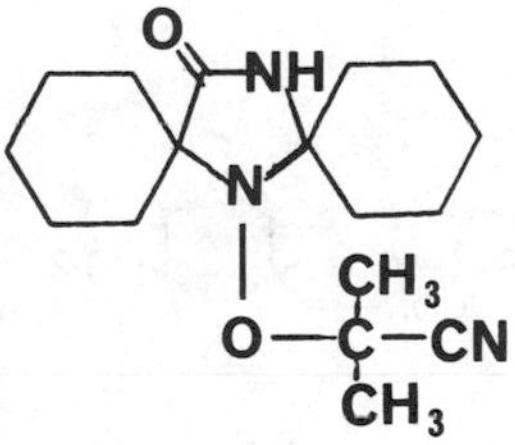

(33)

Hindered Amine Compounds

Elimination of the yellowing introduced by the stable nitroxyl radicals was essential for commercial development of these excellent stabilizers. Since phenolic antioxidants are necessary for the thermal stabilization of polymers during processing, we turned our attention to ways in which unfavorable interactions between the hindered phenols and the stabilizing nitroxyl radicals could be avoided. In this section we describe the discovery of the light–stabilizing activity of hindered amine compounds, an improved synthetic method for these compounds, the synthesis of a number of derivatives, and the evaluation of their stabilizing activity.

Nitroxyl Radical Precursors. Since the dehydrogenating ability of the nitroxyl radicals toward phenols caused a colored product in polymers, we hypothesized that stable radicals that showed poor dehydrogenating activity might be non-yellowing light-stabilizers. It has been reported (14) that dialkylamino radicals are relatively poor dehydrogenating agents. For example, dimethylamino radicals give dimethylamine and N-methylformaldimine by disproportionation at high temperatures. We observed the esr spectra of radicals generated by photolysis of solutions of N-chloramines (36) and (37) in the esr cavity (15). In the absence of oxygen, the N-chloramines gave the spectra of the corresponding amino radicals (38) and (39). In the presence of oxygen, the spectra of the corresponding nitroxyl radicals (28) and (6) were obtained. Formation of the amino radicals was further confirmed by: (a) photolysis of a mixture of (37) and dibenzylmercury in benzene under argon to give 1-benzyl-2,2,6,6-tetramethylpiperidine (40), and (b) the formation of the same product from amine (41) under the same reaction conditions (Chart 6).

These results suggest that hindered amine compounds could be converted to stable nitroxyl radicals through the corresponding amino radicals in polymers; that is, they are the precursors of the stable radicals. In fact a key compound, 2,2,6,6-tetramethyl-4-oxopiperidine (42), showed high light-stabilizing activity in polypropylene, comparable to that of the nitroxyl radicals, lending support to this interpretation.

Synthesis of 2,2,6,6-tetramethyl-4-oxopiperidine. Two syntheses of 2,2,6,6-tetramethyl-4-oxopiperidine (triacetonamine, (42)) had been reported: (a) reaction of acetone with ammonia in the presence of calcium chloride (16) and (b) the reaction of 2,6-dimethyl-2,5-heptadien-4-one (phorone, 4) with ammonia (17). Method (a) required a long reaction time (9-15 days), and the yield was only 20-30% based on acetone. By method (b), triacetonamine was obtained in good yield (70% based on phorone), but the synthesis of phorone from acetone required 2-3 weeks and proceded in only 30% yield (17).

We reinvestigated the reaction of acetone and ammonia (19). Ammonia was introduced into a mixture of acetone and calcium chloride. The resulting oil was carefully distilled and three compounds, (43), (42), and (44), were obtained in 53, 24, and 0.6% yield, respectively (Chart 7). The formation of (44), 1,9-diaza-2,2,8,8,10,10-hexamethyl-4-oxospiro[5.5]undecane (or pentaacetondiamine), could be explained by reaction of 4-amino-4-methyl-2-pentanone (diacetonamine (45)) with (42). Formation of (45) had been reported to proceed almost quantitatively upon hydrolysis in dilute hydrochloric or oxalic acid of (43), (2,2,4,4,6-pentamethyl-2,3,4,5-tetrahydropyrimidine or acetonine); and acetonine was obtained in good yield from acetone and excess ammonia (20). We showed that the reaction of (43) with acetone or with (42) in the presence of calcium chloride gave the expected products, (42) and (44). Heating (43) in water in the presence of zinc chloride gave (42) and (44) in 35 and 6% yield, respectively. Therefore, triacetonamine (42) could be obtained in good yield via acetonine (43) which was obtained almost quantitatively from reaction of acetone with excess ammonia.

Later, we further improved this synthesis of triacetonamine (21). When conversion of (43) was carried out in acetone in the presence of water and an ammonium halide catalyst, (42) was obtained in unexpectedly high yield (more than 100% based on (43)). We propose that ammonia, from the ammonium halide or (43), is utilized for extra triacetonamine.

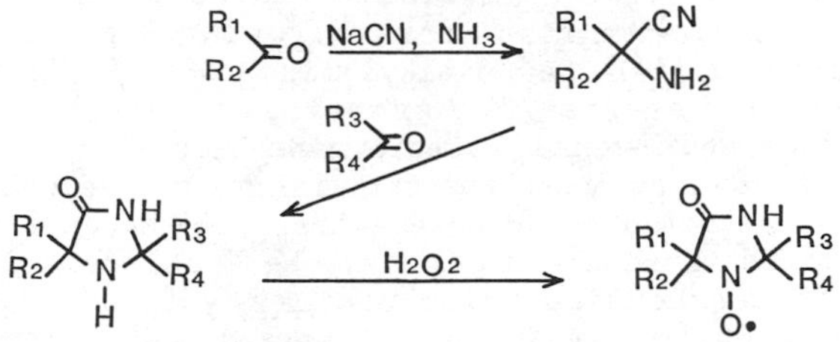

R₁	**R₂**		**R₃**	**R₄**			

	R₁	R₂	R₃	R₄
17	Cyclohexyl		Cyclohexyl	
18	Cyclohexyl		H,	n–Propyl
19	Me–,	Me–	Me–,	Me–
20	Me–,	Me–	Me–,	i-Butyl,
21	Me–,	Me–	Me–,	Et–
22	Me–,	Me–	Cyclohexyl	
23	Cyclohexyl		H,	n–Undecyl
24	Cyclohexyl		H,	Ph–
25	Me–,	Ph–	Cyclohexyl	
26	2–Me–Cyclohexyl		2–Me–Cyclohexyl	

	R₁	R	R₃	R₄
27	Cyclohexyl		Cyclohexyl	
28	Me–,	Me–	Me–,	Me–
29	Me–,	Et	Me–,	Et–
30	Me–,	Me–	Cyclohexyl	
31	2–Me–Cyclohexyl		2–Me–Cyclohexyl	

Chart 4. New stable nitroxyl radicals from the oxidation of imidazolidines by hydrogen peroxide.

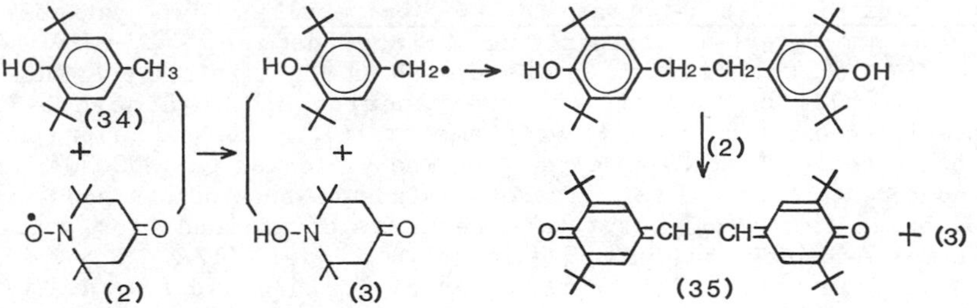

Chart 5. Reactions of 2 and 34.

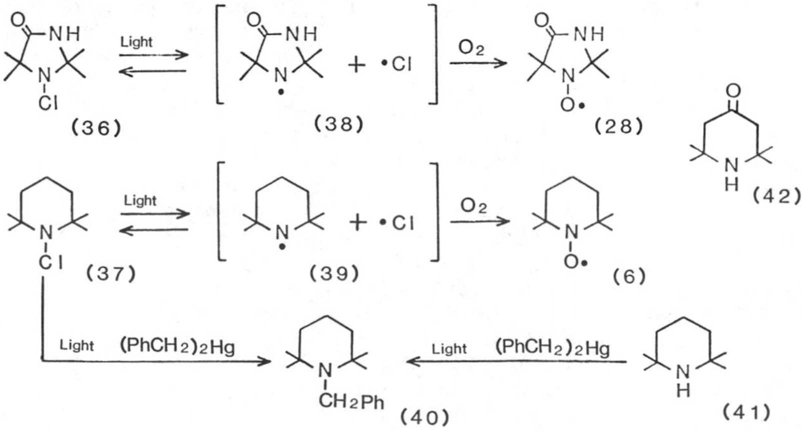

Chart 6. Radicals generated by photolysis of solutions of N–chloramines 36 and 37.

Synthesis and Stabilizing Activity of Hindered Amines. As mentioned previously, the hindered piperidine compounds showed excellent light-stabilizing activity in polypropylene. In order to find more efficient compounds, various derivatives of 2,2,6,6-tetramethyl-4-oxopiperidine (**42**) were synthesized and tested. In this section we describe some typical examples from the great number of derivatives prepared in our laboratory.

In analogy to the Bucherer reaction, we carried out the reaction of 4-amino-4-cyano-2,2,6,6-tetramethylpiperidine (**46**) with phenyl isocyanate. Heating the resulting ureido compound in benzene gave the imino-hydantoin (**47**), while heating in aqueous hydrochloric acid gave 3-phenylhydantoin (**48**). Similarly, reaction of 4-cyano-4-hydroxy-2,2,6,6-tetramethylpiperidine (**49**) with p-tolyl isocyanate gave the imino-oxazolidone (**50**) and the oxazolidinedione (**51**). Compounds having various substituents at the 3-position were prepared (Chart 8). The light-stabilizing activity of these compounds and the 4-oxoimidazolidines described earlier are listed in Table II (22,23).

Table II. Light-stabilizing Activity of Hindered Amines

Compound Number	m.p. (b.p.) °C	Embrittlement Time[a] (hours)
42	(105/18mmHg)	320
47	177	220
52	196	300
53	157	400
54	115	800
48	143	860
55	187	660
56	166	660
57	96	800
50	152	60
58	95	80
51	156	60
59	174	200
60	107	40
17	220	180
19	170	120
23	(170/0.0005mmHg)	380
25	131	600
UV absorber		80
None		40

a) Tested by the method described for Table I.

Syntheses of hydantoin compounds from 1,3,5-triaza-7,7,9,9-tetramethylspiro[4.5]decane-2,4-dione (61) were also planned, since derivatives of this type showed high activity compared to oxazolidones and 4-oxoimidazolidines. Further, three different types of

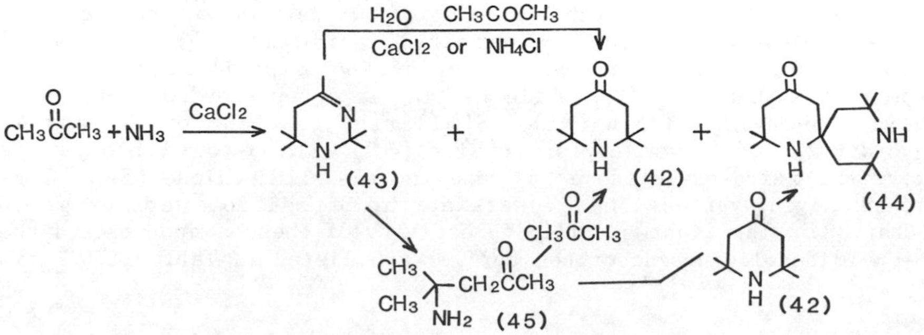

Chart 7. Compounds obtained from the reaction of acetone and ammonia.

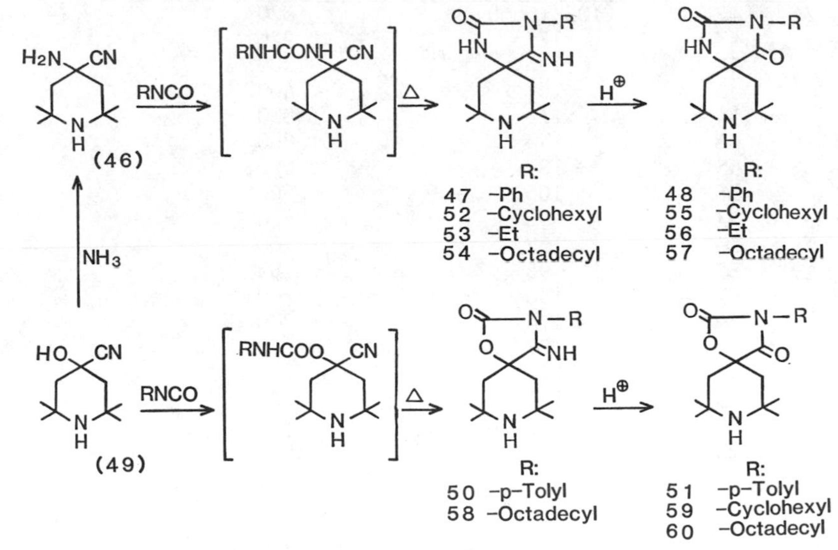

Chart 8. Compounds having various substituents at the 3-position.

piperidine acetals were synthesized (Chart 9), and their light-stabilizing activities are shown in Table III (24).

Table III. Light-stabilizing Activity of Piperidine-spiroacetals

Compound Number	m.p. (b.p.) °C	Embrittlement Time[a] (hours)
63	125	160
64	53	1540
65	139	800
66	(128/2mmHg)	460
67	(194/3mmHg)	600
68	137	220
69	203	300
70	212	1340
UV absorber		80
None		40

a) Tested by the method described for Table I.

Catalytic hydrogenation of (42) led to 4-hydroxy-2,2,6,6-tetramethylpiperidine (71) which was derivatized by known methods as shown in Chart 10.

Since it was expected that bifunctional compounds would exhibit more stabilizing activity than monofunctional ones, polycarboxylic acid piperidinol esters were synthesized. Their stabilizing activities are summarized in Table IV.

The light-stabilization activity of several esters was tested by an improved test method (25). This method was carried out as follows: (i) Preparation of the test specimen. A mixture of 100 parts of polypropylene powder, 0.2 parts of the antioxidant, octadecyl 3-(3,5-di-tert-butyl-4-hydroxyphenyl)propanoate, and 0.25 parts of the stabilizer was needed for 10 minutes at 200°C in a Brabender plasti-Corder to give a homogeneous material. This material was then pressed to a thickness of 2-3 mm in a laboratory press. A portion of this sheet was pressed for 6 minutes at 260°C in a hydraulic press and then immediately placed in cold water, yielding a 0.5 mm sheet. Following the same procedure, a 0.1 mm film was obtained from the 0.5 mm sheet. This film was cut into test pieces 50x120 mm. Polypropylene films without any stabilizers were prepared as above and used as controls. (ii) Testing of light-stability. The test specimens were aged as described for Table I above. The exposed films were subjected to tension tests at regular intervals, and the times recorded when the test pieces contracted to 50% of their original extension (Half Life Time, HLT).

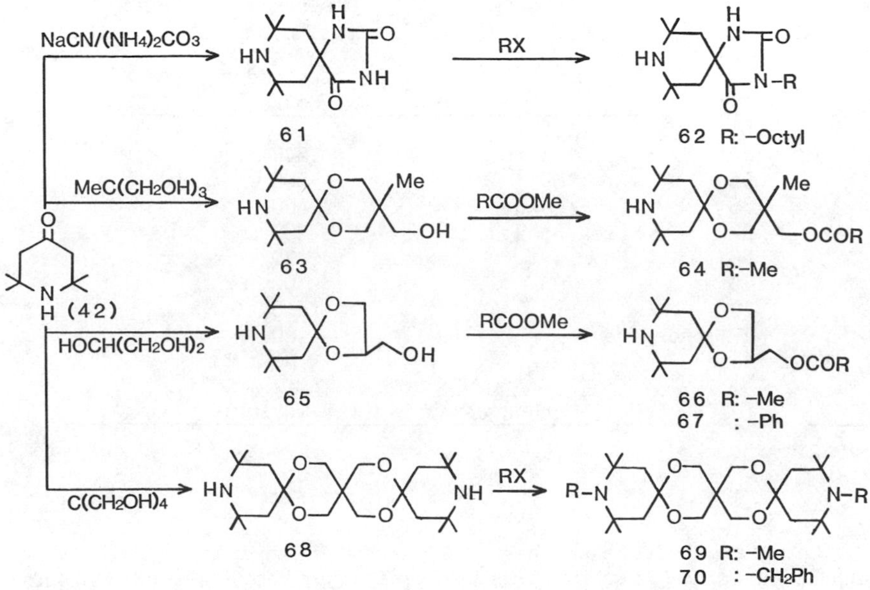

Chart 9. Synthesis of three different types of piperidine acetals and hydantoin compounds.

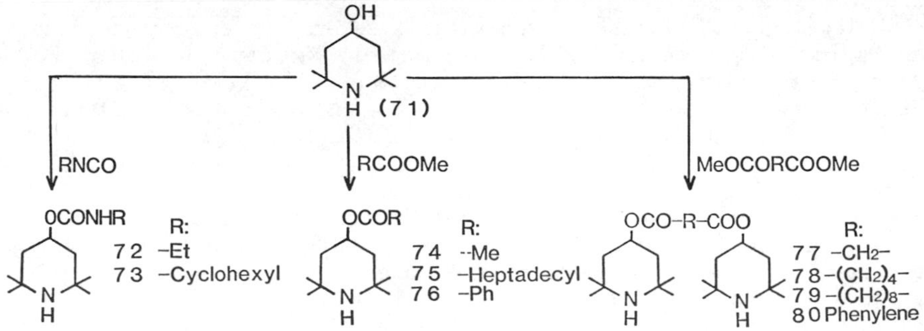

Chart 10. Derivatization of 71.

Table IV. Light-stabilizing Activity of 4-Piperidinols

Compound Number	m.p. (b.p.) °C	Embrittlement Time[a] (hours)	HLT[b] (hours)
72	128	800	
73	70	420	
74	125	700	
75	42	700	1260
76	98	500	470
77	108	920	600
78	91	500	1030
79	86	720	1300
80	208	500	
Control		40	180
UV absorber		80	360

a) Tested by the method described for Table I.
b) Tested by the improved method described above.

In addition, we attempted to prepare piperidine compounds bonded with a hindered phenol moiety, since phenolic antioxidants are essential for processing and long-term stabilization. The proposed compounds were expected to have light-stabilizing activity and at the same time thermal stabilizing activity in the polymers. Bifunctional compounds were also synthesized. The corresponding diols were obtained by the reaction of the amines with ethylene oxide (Chart 11). The stabilizing activities of the above compounds are summarized in Table V.

Table V. Stabilizing Activity of the Piperidine-phenols

Compound Number	m.p. (b.p.) °C	Light-stability[a] HLT(hours)	Thermal[b] embrittlement (hours)
81	78	1170	5
83	138	810	25
84	178	680	11
85	56	670	26
86	53	810	26
87	179	790	16
Control[c]		180	5
UV absorber[c]		360	5

a) Tested by the improved method described for Table IV.
b) 0.5 mm sheet, 150°C forced air oven.
c) Specimens prepared with antioxidant.

Selection of Practical Stabilizers

As described above, it has long been apparent that hindered amine
compounds have intrinsically high light-stabilizing activity in polymers. However, the effectiveness of any particular structural group,
such as the hindered amine function, may depend on its permanence in
the polymer. There are many factors which affect the permanence of
the stabilizer, among them its compatibility with the polymer. Physical loss of the stabilizer may be influenced by physical properties
such as melting point and molecular weight. Therefore, the stabilizer's effectiveness is a composite of not only intrinsic chemical
activity but also physical properties resulting from its chemical
structure. In addition, there other important factors for industrial
applications: (a) no initial discoloration and no color changes in
the stabilized polymer on exposure to heat and light; (b) the absence
of harmful interactions with the polymer and other additives; (c) low
toxicity; (d) easy handling; (e) low cost for high performance. The
physical state of the stabilizer at ordinary temperatures is another
important criterion. For almost all polymer applications solid stabilizers are desired; however, liquid stabilizers are preferred in
paints and coatings. Finally, the availability of a synthetic method
allowing low production costs is a very important consideration.
    The first compound chosen by the above criteria was 4-benzoyl-
oxy-2,2,6,6-tetramethylpiperidine (**76**). The second compound, bis-
(2,2,6,6-tetramethyl-4-piperidinyl) sebacate (**79**) was selected and
developed jointly by Sankyo Co., Ltd. and Ciba-Geigy Ltd. In addition we selected a third compound, 1-[2-(3,5-di-tert-butyl-4-hydroxy-
phenylpropionyloxy)ethyl]-4-(3,5-di-tert-butyl-4-hydroxyphenylpropio-
nyloxy)-2,2,6,6-tetramethylpiperidine (**83**), and developed it as a new
type of stabilizer having both thermal and light-stabilization activity. These compounds are widely used under the trade names SANOL
LS-744 (**76**), -770 (**79**), and -2626 (**83**) (Chart 12). Compound (**79**) is
also sold by Ciba-Geigy Ltd. as Tinuvin 770 under a licensing agreement from Sankyo.

Light-Stabilizing Activity of Commercial HALS.
The light-stabilizing
activity of the three commercial hindered amine light stabilizers
mentioned above is shown in Figure 1a for polypropylene plates. As
shown in the figure, HALS-III was slightly less effective than HALS-I
or HALS-II. In thin sections of polyolefins, HALS-I was a less
effective light-stabilizer than HALS-II or HALS-III. This may be
related to the problem of physical loss of the stabilizer through
migration and volatilization from the polymer. In Table V above, the
excellent thermal-stabilizing activity of HALS-III is shown. Figure
1b amplifies this point, showing HALS-III to be comparable to Irganox
1010, a highly effective antioxidant.

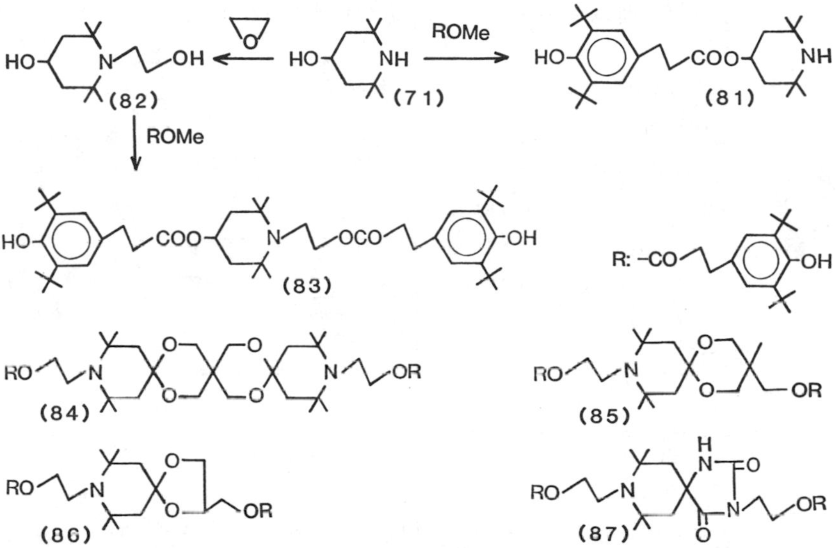

Chart 11. Diols obtained by reaction of the amines with ethylene oxide.

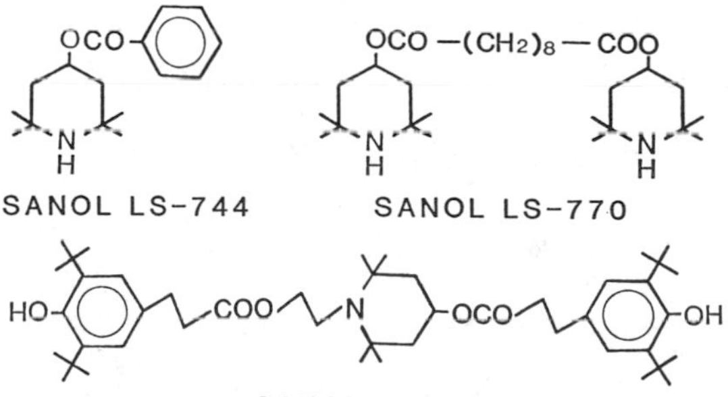

SANOL LS-744   SANOL LS-770

SANOL LS-2626

Chart 12. Structures of Sanol LS-744, Sanol LS-770, and Sanol LS-2626.

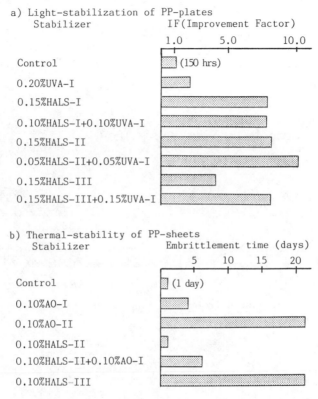

Figure 1.   Stabilizing activity of HALS.   Substrate: PP, MFR; 4.0.
Stabilizers: HALS-I; SANOL LS-744, HALS-II;   SANOL LS-770,
HALS-III;   SANOL LS-2626, UVA-I; Tinuvin 326, AO-I;   Irganox 1076,
AO-II; Irganox 1010.   (a)   Base stabilization: 0.10% AO-II +
0.10% Ca-stearate (HALS-III, without AO-II).   Test specimens:
compression-molded plates (2.0 mm thick).   Weathering test:
sunshine carbon arc lamp weather meter (WEL-SUN-HC, Suga Test
Instruments Co., Ltd.).   Black panel temperature: 63 ± 3 °C,
without water spray.   Failure criterion: time to 50% loss of
original elongation (half life time = H.L.T.).   IF = H.L.T. with
stabilizer/H.L.T. without stabilizer.   (b)   Test specimens:
compression-molded sheets (0.5 mm thick).   Aging test: forced
air circulation oven at 150 °C.

Acknowledgments

The work described in this review is the result of helpful cooperation on the part of many people at Sankyo Co., Ltd. The authors
would like to acknowledge the cooperation of their colleagues in the
research and development of hindered amine light stabilizers. The
authors also wish to express their thanks to Ciba-Geigy Ltd. for
their instructive joint cooperation in development of hindered amine
light stabilizers.

Literature Cited

1. Bolland, J. L.; Gee, G. Trans. Faraday Soc., 1946, 42, 236, 244.
2. Toda, T.; Mori, E.; Murayama, K. Bull. Chem. Soc. Jpn., 1972,
    45, 1852.
3. Neiman, M. B.; Rozantsev, E. G.; Mamedova, Yu. G. Nature
    (London), 196, 472.
4. (a) Murayama, K; Yoshioka, T. Bull. Chem. Soc. Jpn., 1969, 42,
    230.
    (b) Yoshioka, T.; Higashida, S.; Morimura, S.; Morayama, K.
    Bull. Chem. Soc. Jpn., 1972, 44, 2207.
5. Murayama, K.; Toda, T.; Akagi, S.; Kurumada, T.; Watanabe, I.;
    Kitaoka, A. U.S. Patent 3 431 232, 1969.
6. Kitaoka, A.; Murayama, K.; Morimura, S.; Akagi, S.; Kurumada,
    T.; Watanabe, I. U.S. Patent 3 436 369, 1969.
7. Murayama, K.; Morimura, S.; Akagi, S.; Kurumada, T.; Watanabe,
    I. U.S. Patent 3 431 233, 1969.
8. Murayama, K.; Morimura, S.; Tanaka, T. U.S. Patent 3 536 722,
    1970.
9. Noland, W. E.; Sundberg, R. J.; Michaelson, M. L. J. Org. Chem.,
    1963, 28, 3576.
10. Murayama, K.; Morimura, S.; Yoshioka, T. Bull. Chem. Soc. Jpn.,
    1969, 42, 1640.
11. Toda, T.; Morimura, S.; Murayama, K. Bull. Chem. Soc. Jpn.,
    1972, 45, 557.
12. Toda, T.; Morimura, S.; Horiuchi, H.; Murayama, K. Bull. Chem.
    Soc. Jpn., 1971, 44, 3445.
13. (a) Chakraborty, K. B.; Scott, G. Chem. Ind. (London), 1978,
    237.
    (b) Bolsman, T.A.B.M.; Blok, A. P.; Frijns, J.H.G. Recl. Trav.
    Chim. Pays-Bas, 1978, 97, 313.
    (c) Gratten, D. W.; Carlsson, D. J.; Wiles, D. M. Polym. Degrad.
    Stab., 1979, 1, 69.
14. Mackay, D.; Waters, W. A. J. Am. Chem. Soc., 1965, 92, 5235.
15. Toda, T.; Mori, E.: Horiuchi, H.; Murayama, K. Bull. Chem. Soc.
    Jpn., 1972, 45, 1802.
16. Francis, F. J. Chem. Soc., 1927, 2897.
17. (a) Guareschi, J. Chem. Ber., 1895, 28, 160R.
    (b) Mackay, D.; Waters, W. A. J. Chem. Soc. C, 1966, 814.
18. Claisen, L. Ann., 1876, 180, 4.
19. Murayama, K.; Morimura, S.; Amakasu, O.; Toda, T.; Yamao, E.
    Nippon Kagaku Zasshi, 1969, 90, 296.

20.   (a) Matter, E. Helv. Chim. Acta, 1947, 30, 1114.
      (b) Bradbury, R.; Hancox, N. C.; Hatt, H. H. J. Chem. Soc.,
      1947, 1394.
21.   Murayama, K.; Morimura, S.; Yoshioka, T.; Kurumada, T. U.S.
      Patent 3 959 298, 1976.
22.   Murayama, K.; Morimura, S.; Yoshioka, T.; Kurumada, T. US.
      Patent 3 639 406, 1972.
23.   Kitaoka, A.; Murayama, K.; Morimura, S.; Toda, T.; Akagi, S.;
      Kurumada, T.; Watanabe, I. U.S. Patent 3 448 074, 1969.
24.   Murayama, K.; Morimura, S.; Yoshioka, T.; Toda, T.; Mori, E.;
      Horiuchi, H.; Higashida, S.; Matsui, K.; Kurumada, T.; Ohta, N.;
      Ohsawa, H. U.S. Patent 3 899 464, 1975.
25.   (a) Murayama, K.; Morimura, S.; Yoshioka, T.; Matsui, K.; Kuru-
      mada, T.; Watanabe, I.; Ohta, N. U.S. Patent 3 640 928, 1972.
      (b) Murayama, K.; Morimura, S.; Yoshioka, T.; Horiuchi, H.;
      Higashida, S. U.S. Patent 3 840 494, 1974.
26.   (a) Soma, N.; Morimura, S.; Yoshioka, T.; Kurumada, T. U.S.
      Patent 4 237 294, 1980.
      (b) Rody, J. U.S. Patent 4 340 533, 1980.
27.   Gugumas, F. "Developments in Polymer Stabilization-1"; Applied
      Science Publishers Ltd.: London, 1979, pp. 261-308.

RECEIVED December 7, 1984

A Decade of Hindered Amine Light Stabilizers

HELMUT K. MÜLLER

Plastics and Additives Division, CIBA-GEIGY, Ltd., Basel, Switzerland

Shortly after the first synthesis of triacetoneamine-
N-oxyl, tetramethyl-hindered piperidines were found to
be effective light stabilizers for polymers. That
finding unleashed an extensive worldwide investigation
of those compounds which to date has resulted in the
filing of more than 600 patent applications. Hindered
amine light stabilizers have proven effective for the
weathering stabilization of a large number of major
polymers, especially polypropylene, polyethylene,
polystyrene, impact polystyrene, ABS, SAN, polyurethane as well as thermoplastic and thermosetting coatings. New applications for hindered amine stabilizers
are found regularly. They are clearly the most significant development of polymer stabilization of the
decade.

During the last 15 years, hindered amine light stabilizers, now
commonly referred to as HALS (1), have been one of the most
actively investigated classes of polymer additives. This statement
is easily substantiated by the fact that since the late Sixties,
more than 600 patent applications involving HALS have been filed by
more than 80 chemical companies or research institutes. The list
of companies involved reads like a "Who's Who" of major chemical
producers. As can be seen in Figure 1, after an induction period
of about five years, patents have been filed at a steady rate of
approximately 50 applications per year. By far, the biggest number
of patented HALS is based on derivatives of triacetoneamine, a
condensation product of acetone and ammonia (Figure 2). Some
examples of structural variants not based on triacetoneamines are
shown in Figure 3. As far as we know today, all but one of the
commercially available hindered amine stabilizers are derived from
triacetoneamine. The number of scientific and technical papers,
many of them dealing with the mechanisms of action of HALS, also
runs into the hundreds.

What started all this effort was the observation by a
Dr. K. Murayama (2) that triacetoneamine-N-oxyl, an unusually

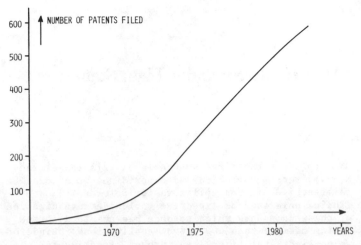

Figure 1. HALS-related patent applications from 1969 to 1983.

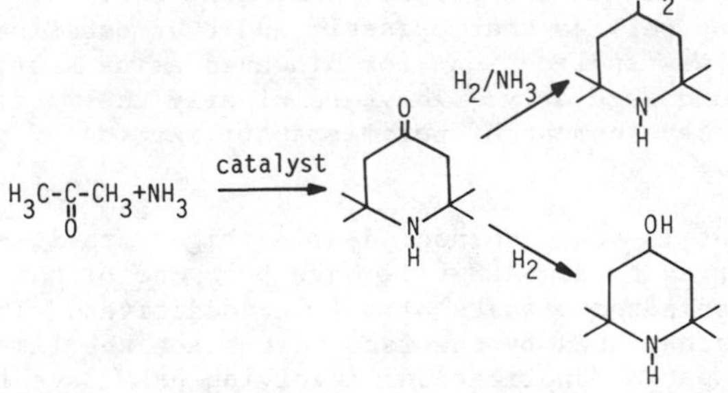

Figure 2. Synthesis scheme for important triacetonamine derivatives.

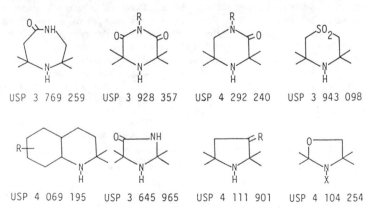

Figure 3. Structural variants of hindered amine light stabilizers.

stable free radical, synthesized for the first time by M. B. Neiman and coworkers (3), was found to be a highly efficient light stabilizer for a variety of polymers. However, triacetoneamine-N-oxyl did not lend itself to practical use on account of its physical and chemical properties. The compound easily undergoes thermally-induced disproportionation (4). It is volatile and colored and therefore imparts color to the substrate to be stabilized. Another drawback to the triacetoneamine-N-oxyl is its possible chemical reactions with phenolic antioxidants. Through elimination of the keto group in the 4-position, obviously responsible for the thermal instability (5), and by switching from the N-oxyl compounds to the parent amines as "stable radical precursors," which indeed proved to be highly effective light stabilizers (6), the drawbacks of triacetoneamine-N-oxyl have been overcome (Figure 4).

After the publication of the first HALS patents, synthesis and applicational work were started to check the merits of this new and unusual light stabilizer structure. The findings of the Japanese workers were confirmed in full. Especially in polypropylene, the performance of HALS was almost unbelievably high in comparison to the state-of-the-art; i.e., UV absorbers, Ni complexes and benzoates. Early findings of HALS effectiveness led to a cooperative effort between the Sankyo company of Japan and CIBA-GEIGY of Switzerland being initiated with the objective of commercializing this important class of light stabilizers. Despite the combined efforts, testing, selection and process development of suitable HALS candidates took time, because of the unusually long testing times required and the rapidly growing number of compounds available. Finally in 1973, three developmental HALS compounds were sampled on a worldwide basis as light stabilizers with emphasis on polyolefins and styrenic polymers (Figure 5). One of the three, bis(2,2,6,6-tetramethylpiperidinyl-4) sebacate (HALS I), survived the sampling stage and was commercialized by the second half of 1974 (7).

Early Results

The first papers on the performance of hindered amine light stabilizers were published by F. Gugumus (8) and K. Leu (9) in 1974. One of the reasons why nearly half a decade lapsed between the basic invention and the first commercialized HALS product is indicated by the data in Figure 6 as published by F. Gugumus. The delay in commercializing HALS was in part due to the worldwide recession in 1975 caused by the first oil crisis and also because more time was needed to gain a broad acceptance in the marketplace due to the fact that HALS-based systems impart three to four times higher stability to a number of plastic materials than had been achievable with light stabilizers previously available (9). Originally, our outdoor exposure studies were run in central Italy; later we switched to exposing samples in Florida in order to decrease the time of exposure. However, even in Florida, depending on the stabilizer concentrations, samples containing HALS may take years to fail. For example, a white pigmented polypropylene stretched tape with 0.6% bis(2,2,6,6-tetramethylpiperidinyl-4) sebacate has not failed yet after more than six years' outdoor weathering in Florida

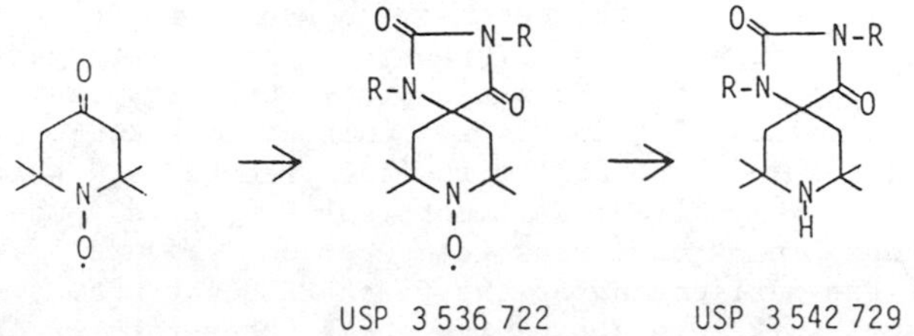

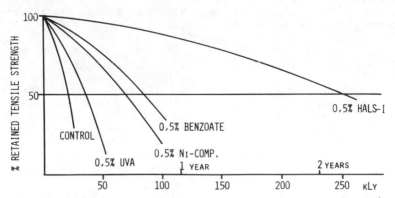

USP 3 536 722   USP 3 542 729

Figure 4. The discovery of HALS (transition from N—oxyls).

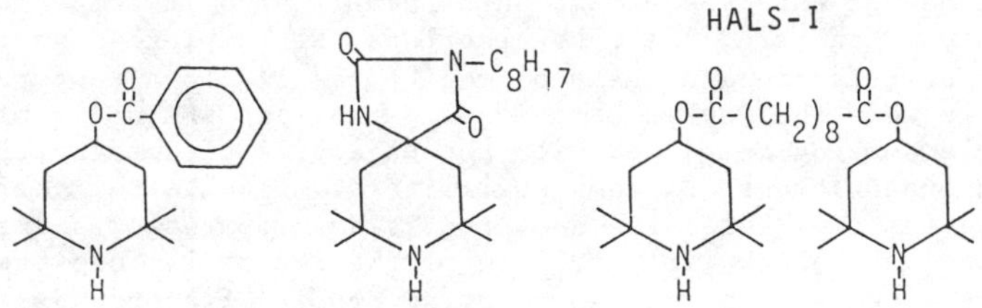

HALS-I

Figure 5. Developmental HALS sampled in 1973.

Figure 6. Outdoor weathering of PP stretched tape (Italy).

(Figure 7). The outstanding effectiveness of bis(2,2,6,6-tetra-
methylpiperidinyl-4) sebacate (HALS I) in stabilizing 2 mm polypropylene plaques on outdoor weathering in Arizona can clearly be
seen from the data shown in Figure 8.

For many years, polypropylene fiber has been the subject of
in-depth research aimed at improving its light stability. The
inherent light sensitivity of polypropylene and the high surface-
to-volume ratio of fibers are the prime reasons for the difficulties in stabilizing polypropylene fiber. The considerable improvement achieved with bis(2,2,6,6-tetramethylpiperidinyl-4) sebacate
was, however, drastically reduced when polypropylene fine fibers
(up to 15 dpf) or articles made thereof, were treated at elevated
temperatures in textile finishing operations involving latex backing and tentering. It was soon recognized that the performance
loss after thermotreatment was due to the physical depletion of
the stabilizer as a consequence of its small size, relatively
rapid rate of migration, and its relatively low molecular weight.
This experience initiated the development of higher molecular
weight compounds (10) and eventually of polymeric light stabilizers (11).

One answer to the performance loss on thermotreatment or
tentering of polypropylene fiber was a stabilizer with a molecular
combination of a sterically hindered phenol and a sterically
hindered amine. The structure of bis(1,2,2,6,6-pentamethyl-4-
piperidinyl) 1-butyl-1-(3,5-di-tert-butyl-4-hydroxybenzyl)malonate,
which exemplifies this combination, is shown in Figure 9. The idea
behind such a product was to combine the necessary increase in
molecular weight with an additional stabilizing function. The
phenolic moiety imparts thermal stability; the aminic part provides
the light stability as shown in Table 1. The light exposure data
in Table 1 were generated with a Xenotest 1200 light exposure
device.

Polymeric HALS. The final answer to the polypropylene fiber
stability problem was the development of polymeric HALS. The
structures of two such polymeric compounds, polymeric HALS-A and
polymeric HALS-B are shown in Figure 10. Polymeric HALS have to be
considered the stabilizers of choice for stabilizing polypropylene
fiber. They offer excellent performance as light and heat stabilizers at moderate use concentrations (12); some representative data
are presented in Table 2. Moreover, they solve the polypropylene
fine fiber problem related to tentering in the use of latex backing. Polypropylene fine fibers stabilized with the afore-mentioned
compounds are virtually unaffected by thermotreatment as shown in
Table 3. The acrylic latex treatment, the most severe exposure
condition, should simulate the experience of polypropylene fabrics
in upholstery and carpeting applications where latex application
and thermotreatment to eliminate fabric stresses are real experi-
ences. The data in Table 3 show clearly that the excellent light
stability conferred by the low molecular weight HALS-I,
bis(2,2,6,6-tetramethylpiperidinyl-4) sebacate disappears almost
entirely after latex treatment and tentering. It should be
mentioned that polymeric HALS-A, the polyester type, is broadly
approved for polyolefin articles in contact with food.

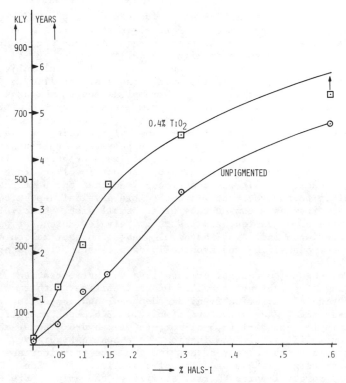

Figure 7. Light stability of polypropylene tapes (Florida);
KLY to 50% retention of tensile strength.

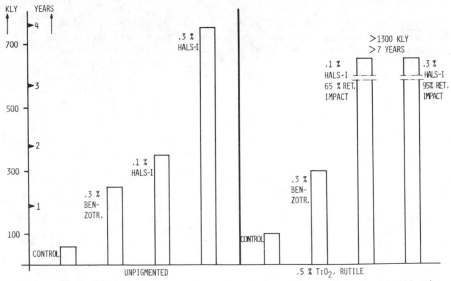

Figure 8. Light stability of 2 mm polypropylene plaques (Arizona);
KLY to 50% retention of impact strength.

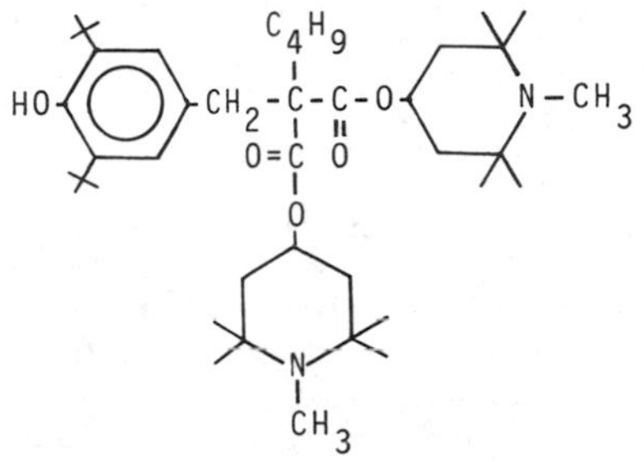

Figure 9. Molecular combination of a sterically hindered phenol and HALS (HALS-PH).

Table 1. Performance of a Molecular Hindered Phenol-HALS Combination in PP Fiber 120/12 DEN (Exposure: Xenotest 1200)

	OVEN AGING AT 110°C DAYS TO FAILURE	LIGHT STABILITY XO 1200 HOURS TO FAILURE
0.2% COMMERCIAL AO	12	500
0.1% HALS-PH	15	1200
0.2% HALS-PH	24	2000
0.4% HALS-PH	26	3000

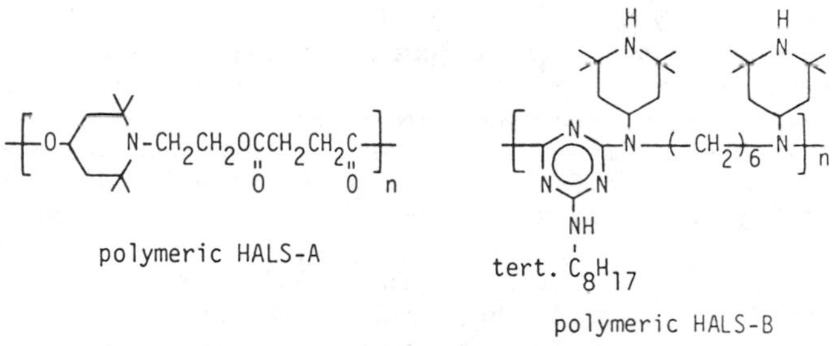

polymeric HALS-A

polymeric HALS-B

Figure 10. Structures of polymeric HALS.

The outstanding antioxidant activity of polymeric HALS in 200 μm
blown LDPE film (Table 4) confirms the thermal stabilizing activity
as previously demonstrated in polypropylene fiber. In LDPE as in
polypropylene fiber, HALS-B was found superior to HALS-A. The high
degree of interest in oligomeric and polymeric HALS is evidenced by
the numerous patent applications filed on compounds of that type in
recent years.

Novel Applications. The effectiveness of HALS is not restricted to
thermoplastics like polypropylene or polyethylene. Rapidly growing
areas of application are certain industrial paint systems as e.g.
two-coat metallic paints for automotive coatings. Figure 11 contains typical examples of premature failure of one or two-coat
metallic paints (13). Heavy cracking and delamination are easily
recognizable in the photo and demonstrate the need for improvement.
Such improvement is possible by stabilizing the crosslinked clear
topcoat with a synergistic system of a low volatility benzotriazole
UV-absorber and a suitable HALS compound. It is clearly evident
from the data in Figure 12 that only the combination of a HALS and
a UV absorber retained outstanding stability of the automotive
paint after more than two years' Black Box exposure in Florida.
     Distinctly synergistic effects of HALS/UV-absorber systems
have also been observed in styrenic polymers as shown in Figures 13
and 14 (14). The synergistic effect of the HALS/UV-absorber system
in SAN is very pronounced in that low values of the Yellowness
Index (YI) are maintained for long periods (Figure 13). The same
is true for the two-phase polymer ABS; Figure 14 shows that
retention of impact strength is markedly superior with the HALS/UV-
absorber combination. Analogous effects have been found in unpigmented polypropylene and polyethylene.
     The foregoing has been an attempt to show in an extremely
shortened version, how HALS were developed and how they have performed in the past. What will occur in the future is difficult to
predict. However, certainly it is expected that new HALS compounds
will be put on the market and that those that are already commercialized will enjoy high growth rates on account of the possibility
of saving energy and natural resources in using them. On the one
hand, there are certainly still unexplored or barely explored applicational areas for hindered amines; as an example of this is the
finding that a stabilizer of a hindered amine class has been found
that improves the light stability of the yellow image dye and the
dark stability of the cyan image dye in color photographic film
(15).
     On the other hand, fine tuning of additives or additive sys-
tems for specific polymers and applications still offers additional
possibilities. In this context it was found recently by F. Gugumus
that the polymeric HALS-B is the first choice if one wants to stabilize HDPE tapes or moldings most effectively, as demonstrated by
the data presented in Table 5 and Figure 15. Another example is
the stabilization of all kinds of greenhouse covers made from LDPE
or EVA films which presents an improvement application for light
stabilizers in Europe. Certain grades of these films may contain
inorganic materials like China clay, aluminum silicate or carbon

Table 2.   Performance of UV Stabilizers in PP Multifilaments 130/37 DEN

UV-STABILIZER	XENOTEST 1200 TO 50% RET. TENACITY	OVENAGING AT 110°C TO BRITTLENESS
CONTROL	650 HOURS	16 DAYS
0.15% POLYMERIC HALS-A	3200	70
0.3% POLYMERIC HALS-A	4600	86
0.15% POLYMERIC HALS-B	3600	134
0.3% POLYMERIC HALS-B	4500	158
0.5% NI-COMPOUND	1600	28
1.0% NI-COMPOUND	1950	43

Table 3.   Effect of Thermal Treatment on Light Stability of PP Multifilaments 130/37 DEN

STABILIZER	XENOTEST 1200 HRS TO 50% RETAINED TENACITY		
	NOT TREATED	20' AT 120° C	ACRYLIC LATEX 20' AT 120° C
CONTROL	490	550	460
0.3% LOW MOL. WT. HALS-1	4600	2300	650
0.6% LOW MOL. WT. HALS 1	7400	5200	750
0.3% POLYMERIC HALS-A	4250	4200	3700
0.6% POLYMERIC HALS-A	5500	5500	5600

Table 4.   Polymeric HALS as Antioxidants in LDPE Film: Days at 100 °C to 50% Retention of Elongation

ADDITIVES	DAYS TO FAILURE
BASE STABILIZATION	190
0.15 % POLYMERIC HALS-A	790
0.3 % POLYMERIC HALS-A	1050
0.075% POLYMERIC HALS-B	730
0.15 % POLYMERIC HALS-B	> 920 (100%)

Figure 11.   Typical failure of a two-coat metallic paint.

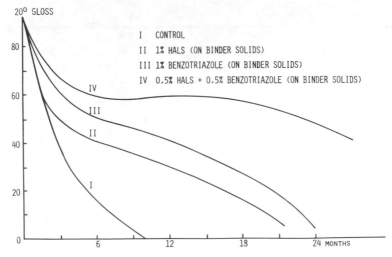

Figure 12.   Light stability of a two-coat high-solid paint
(silver metallic) with a benzotriazole and a HALS type
stabilizer. Exposure: Florida, 5° south, black box, unheated.

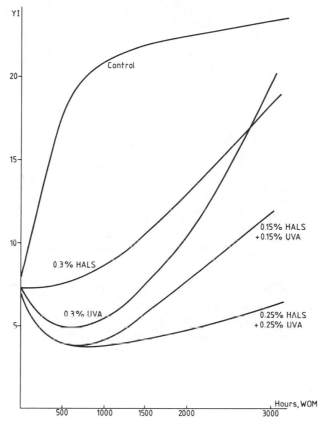

Figure 13.   Inhibition of light-induced discoloration of SAN.
Exposure: xenon arc weather-ometer.

Table 5.   Florida Exposure of HDPE Tapes: KLY to 50% Retention of
Tensile Strength

ADDITIVE	NATURAL	0.4 % TiO_2
CONTROL	100	95
.05% HALS-1	130	180
.05% POLYMERIC HALS-A	150	200
.05% POLYMERIC HALS-B	210	245
.1% HALS-1	210	260
.1% POLYMERIC HALS-A	180	225
.1% POLYMERIC HALS-B	275	345

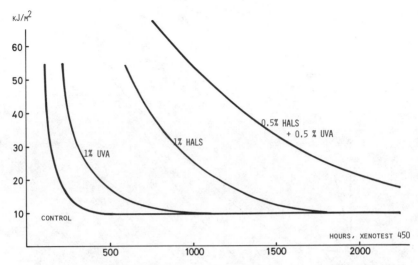

Figure 14.   Impact strength of ABS after accelerated light exposure.   Exposure: xenotest 450.

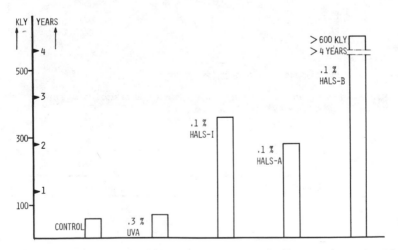

Figure 15.   Light stability of HDPE plaques (Ziegler).   Florida exposure, KLY to 50% retention of tensile impact strength.

black as infrared barriers in order to keep the temperature inside
the greenhouse during the cooler nights as high as possible.   The
influence of such materials on the light stability of blown LDPE
film exposed in Florida is shown in Table 6.   The data in Table 6
show very clearly the severely detrimental effect of China clay and
the somewhat lesser detrimental effect of aluminum silicate on the
film's light stability.   In Table 7 are shown the influences of
3% China clay on differently stabilized LDPE films.   In the un-
filled film, HALS-B is best, followed by HALS-A and a nickel
compound.   In the presence of the infrared barrier material, HALS-B
still performs best, however at a reduced level, whereas HALS—A and
the nickel compound are about equal to each other and less effective than HALS-B.

Hopefully, the data presented in this paper have been
convincing regarding the outstanding effectiveness of HALS in stabilizing polymers and convincing that HALS and HALS-based systems
are still good for surprises and new applications.

Table 6.   Influence of Fillers on Light Stability of LDPE Blown Film:
Florida Exposure; KLY to 50% Retention of Elongation

FILLER	CONTROL	.15% HALS-A .15% BENZOPH. UVA
NONE	50 KLY	255 KLY
3% CHINA CLAY	25	90
3% AL-SILICATE	31	135
3% CHALK	46	> 250

Table 7.   Influence of China Clay on Light Stability of LDPE Blown
Film: Florida Exposure; KLY to 50% Retention of Elongation

ADDITIVES	WITHOUT	WITH 3 % CHINA CLAY
.15% Ni-COMPOUND .15% BENZOPH.-UVA	210 KLY	85 KLY
.15% HALS-A .15% BENZOPH.-UVA	255	90
.15% HALS-B .15% BENZOPH.-UVA	> 400	220

Literature Cited

1. Mod. Plast. Intern. 13, October, 51 (1983).
2. K. Murayama, Pharmacia (Japan) 10, 573 (1974).
3. M. B. Neiman, E. G. Rozantsev and Y. G. Mamadova, Nature, 196, 472 (1962).
4. K. Murayama and T. Yoshioka, Bull. Chem. Soc. Japan, 42, 2307 (1969).
5. K. Murayama et al., USP 3,536,722 (1970) Sankyo Co. Ltd.
6. K. Murayama et al., USP 5,542,729 (1970) Sankyo Co. Ltd.
7. Preliminary Product Information: TINUVIN 770, Ciba-Geigy Ltd., 1975.
8. F. Gugumus, Fourth European Plastics and Rubber Conference, 1974, Paris.
9. K. Leu, The Plastics Institute of Australia, Residential Technical Seminar, 1974, Teorigal, Australia.
10. Preliminary Product Information: TINUVIN 144, Ciba-Geigy Ltd., 1978.
11. Preliminary Product Information: TINUVIN 622, Ciba-Geigy Ltd., 1978; CHIMASSORB 944, Ciba-Geigy Ltd., 1979.
12. F. Gugumus, Third International Conference, Polypropylene Fibers and Textiles, 1973, York, Great Britain.
13. W. Rembold and G. Berner, Fifth International Conference on Advances in the Stabilization and Controlled Degradation of Polymers, 1983, Zurich, Switzerland.
14. B. Gilg, Fifth European Plastics and Rubber Conference, 1978, Paris.
15. Technical Information on Image Dye Stabilizer G 15-994, Ciba-Geigy Ltd., 1982.

RECEIVED February 11, 1985

Mechanistic Studies of Sterically Hindered Amines
In the Photooxidation of Liquid Polypropylene Model Substances

B. N. FELDER

Plastics and Additives Division, CIBA-GEIGY, Ltd., Basel, Switzerland

It is a fact that while research on UV stabilizers has advanced to products which significantly improve the weatherability of polymeric materials, the exact processes by which the various classes of stabilizers fulfill their functions are not entirely certain. This especially holds for the very important class of Hindered Amine Light Stabilizers (HALS, tetramethylpiperidine derivatives, TMP).

2,2,6,6-tetramethylpiperidine derivatives

Symbol: $>$N—X

X : H, R, OR, O˙

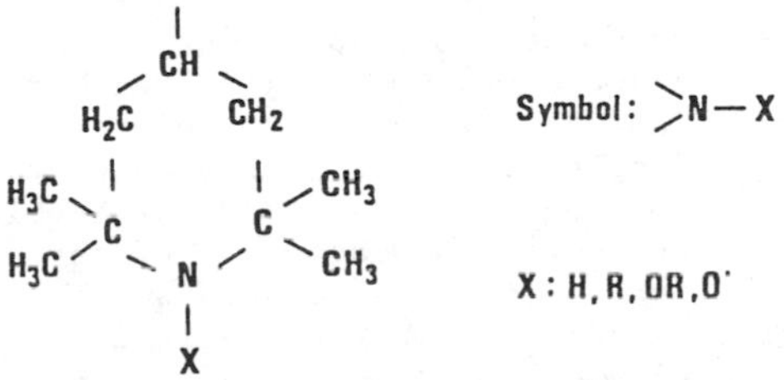

These compounds are efficient in many polymers and applications and show light stabilizing properties superior to those of conventional additives [see (1) for reference].

Considering the high effectiveness of these additives, a strong interest exists in understanding their mode of action. Indeed, over the last 15 years numerous investigations on this problem have been already carried out. Among the results having emerged from these research activities, the most important facts and ideas may be summarized as follows: tetramethylpiperidines are non-absorbing in the UV wavelength region of terrestrial solar radiation. Secondary and tertiary amines of this type also do not deactivate excited chromophores which are, as impurities, always present in polymers (e.g., carbonyl-groups, hydroperoxides, peroxides) (2). See (1) for additional reference.

Furthermore, most secondary tetramethylpiperidines do not, at least not at concentrations usually applied in practice, quench

0097–6156/85/0280–0069$06.00/0

singlet oxygen--a chemically aggressive species formed in many
photochemical reactions in the presence of oxygen--which, in the
view of some authors, might contribute to the photooxidation of
the substrate (3, 4).

     In addition, it has been shown that tetramethylpiperidines do
not, at ambient temperature, decompose common hydroperoxides.
This has been carefully examined in our laboratory, using various
aliphatic hydroperoxides as model compounds (1). On the other
hand, amine-induced decomposition of polypropylene hydroperoxides
has however been observed (14), and it was concluded by the authors
that this decomposition process appears to be sufficiently rapid
that it can play a role in the destruction of -OOH during dark
storage. The same authors believe, however, that the dominating
photoprotective role of hindered amines is their ability to destroy
macroradicals catalytically in the irradiated polymer. It has
indeed frequently been reported that TMP derivatives are reactive
to peroxy radicals, which are generally accepted as the essential
carrier of light induced oxidation of polyolefins (2, 5, 7). Considering the usually cited set of reactions for the photooxidation
of polyolefins shown in Scheme I, it is obvious that scavenging of
macroperoxy radicals would reduce the kinetic chain length of the
oxidation process which, if non-inhibited, causes a chain of newly
formed consecutive hydroperoxides by propagating peroxy radicals.

     A further point of discussion, however, is whether the often
cited reactions (6, 7, 8) of TMP derivatives (and their conversion
products) with peroxy radicals may sufficiently compete with the
propagation steps (5) in Scheme I. Indeed, TMP-derivatives including NOR are known to be rather weak scavengers of peroxy radicals
(7) in the liquid phase. Based on considerations which take into
account that rapid randomization of macroradicals is largely
restricted in the solid polymer, Carlson and Wiles, however, concluded that fast radical scavenging would in fact not be needed for
efficient inhibition of long chain polymer photooxidation processes
(2, 8, 9).

     Nevertheless, the question arises whether other processes,
specific to the structural unit of individual polymers, operate in
addition to the overall scavenging action of hindered amines mentioned above. Of special interest with respect to this question
is certainly the case of polypropylene, a polyolefin in which TMP
derivatives show outstanding stabilizing properties.

     When considering possible ways to explore the mode of action
of stabilizers, the controversal question arises whether investigations should either favor experiments using macromolecular samples
or well defined, low molecular model compounds of the polymer in
question.

     Experiments on polymers directly, on the one hand, throw light
on overall phenomenological aspects under conditions relevant to
practice. It is, however, extremely difficult to draw unambiguous
mechanistic conclusions from this kind of investigation.

     On the other hand, investigations on low molecular models of
the polymer seem, at first glance, to be much more advantageous in
order to obtain mechanistic information. They carry, however, a
great risk in that special observations and findings relevant to
the model might be overestimated. Due to the restricted mobility

of the polymer chain, it is, furthermore, rather difficult to
transfer kinetic results gained from measurements in solution
efficiently to the solid and partially crystalline powder.

Fully aware of this conflicting situation, we decided, nevertheless, to perform mechanistic studies on low molecular models of
polypropylene, in the hope of achieving on this basis some new
information and ideas with respect to the basic processes taking
place in this polymer. Isooctane, as a model for the structural
unit of polypropylene, seemed to us to be an appropriate substance
for this kind of investigation.

Model for polypropylene

Isooctane

$$H_3C \diagdown \atop H_3C \diagup \overset{\displaystyle |}{\underset{\displaystyle |}{C}} - CH_2 - \overset{\displaystyle CH_3}{\underset{\displaystyle CH_3}{\overset{|}{\underset{|}{C}}}} - CH_3$$

with H on the central carbon

PP - unit

$$-\overset{\displaystyle CH_3}{\underset{\displaystyle H}{\overset{|}{\underset{|}{C}}} - CH_2 -$$

Results of Model Investigations on Isooctane

Photooxidation of Isooctane. The overall rate of photooxidation
of isooctane was investigated using carefully selected conditions
and well-defined rates of radical initiation. Radicals were, at
constant rate, initiated in isooctane by photolysis or tertiary
butyl peroxide (tertiary BuOOBu) under oxygen flushing. The radical initiation rates applied, I_o, were varied over a relatively
wide range between the limits

$$10^{-2} < I_o < 10^{-5} \text{ M/h*}$$

The results obtained for the non-inhibited and N-H inhibited
photooxidation are shown in Figure 1. The amine used was the
tetramethylpiperidine-derivative:

(I)

with OCOPh substituent on tetramethylpiperidine, N-H

An experiment carried out with an   N-octyl derivative of tetramethylpiperidine showed quite analagous results.

According to the results shown in Figure 1, the kinetic chain
length of the photooxidation of isooctane is very low. Even for
the lowest rate of radical initiation applied, I_o 10^{-5} M/h, the
kinetic chain length of the non-inhibited photooxidation did not
exceed a value of 1. Radical termination, therefore, seems to
dominate over a peroxy radical chain reaction according to
equation (5) in Scheme I.

*M = Moles isooctyl radicals/l

$$\left.\begin{array}{c} ROOH \\ ROOR \end{array}\right\} \xrightarrow{h\nu \atop I} RO^{\cdot} + OH^{\cdot} \qquad \textbf{(1)}$$

$$\left.\begin{array}{c} RO^{\cdot} \\ OH^{\cdot} \end{array}\right\} + R\!-\!H \xrightarrow{k_o} R^{\cdot} + \left.\begin{array}{c} ROH \\ H_2O \end{array}\right\} \qquad \textbf{(2)}$$

$$R^{\cdot} + O_2 \xrightarrow{k_{o_2}} ROO^{\cdot} \qquad \textbf{(3)}$$

$$2\,ROO^{\cdot} \xrightarrow{2k_t} ROOR + O_2 \qquad \textbf{(4)}$$

$$ROO^{\cdot} + R\!-\!H \xrightarrow{k_p} ROOH + R^{\cdot} \qquad \textbf{(5)}$$
$$\downarrow O_2$$
$$ROO^{\cdot}$$

Inhibition:

$$ROO^{\cdot}\ \ {>}NH \xrightarrow{k_{NH}} ROH + {>}NO^{\cdot} \qquad \textbf{(6)}$$

$$R^{\cdot}\ \ {>}NO^{\cdot} \xrightarrow{k_{NO^{\cdot}}} {>}N\!-\!OR \qquad \textbf{(7)}$$

$$ROO^{\cdot}\ \ {>}NOR \xrightarrow{k_{NOR}} {>}NO^{\cdot} + ROOR \qquad \textbf{(8)}$$

Scheme I. Conventional scheme for polyolefin photooxidation.

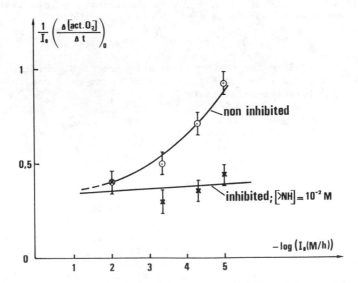

Figure 1. Photoixidation of isooctane initiated by tertiary BuO· radicals. Relative initial rate of iodometrically titratable "active oxygen" $1/I_o(\Delta[\text{act. } O_2]/\Delta t)$ against $-\log I_o$.

At this time, it should be pointed out that the iodometric method applied in our experiments was not sensitive to tertiary peroxides. Therefore, our measurements are not expected to include peroxides formed by self-termination of tertiary peroxy radicals.

Characterization of Photooxidation Products Formed in Isooctane. Among the iodometrically titratable species formed in the photooxidation of isooctane, a substance, developed in appreciable amounts, was observed which was easily distinguishable from ordinary alkyl-hydroperoxides. This substance is, at room temperature, readily destroyed by addition of olefins to the reaction medium. This is a reaction which is typical and specific for peracids (10) and can be employed for quantitative assessment of peracids in the presence of alkylhydroperoxides. In contrast to alkylhydroperoxides, peracids react with olefins forming the corresponding epoxide, according to the equation

$$R\text{-}CO\text{-}OOH + {>}C = C{<} \longrightarrow R\text{-}CO\text{-}OH + {>}C - C{<}$$

Products reacting in this way in photooxidized isooctane have therefore been assigned to peracids. Their rate of formation has been determined in our photooxidation experiments by use of 7-tetradecene, which turned out to be a convenient olefin for peracid-analysis. The results obtained for different rates of radical initiation are shown in Figure 2.

Peracid formation is proportional to irradiation time t. The slope of the lines in Figure 2 which is the rate of peracid formation relative to I_o, is definitely increasing with decreasing rate of radical initiation I_o:

Unit for $\dfrac{1}{I_o}$ $[R\text{-}C(=O)OOH]_t$: $\dfrac{M \text{ (peracids)}}{M \text{ (isooctylradicals)}}$ x hour

After removal of peracids by tetradecene addition, substantial amounts of iodometrically titratable oxidation products remained in photooxidized isooctane. Their relative rate of formation did not depend markedly on the rate of radical initiation and was found to be in the order of 40 to 45% of I_o. This is shown in Figure 3. Although the chemical composition of these products has not been identified, it is, however, assumed that these products were mainly hydroperoxides and probably dialklyperoxides other than those formed by self-recombination of tertiary peroxy radicals. This latter species is, as already mentioned, not expected to be titrated by our iodometric method.

Distribution of Radicals Formed in Isooctane by Tertiary BuO• Attack. The following investigation has clearly shown that the different C-H sites of isooctane (primary C-H, secondary C-H, tertiary C-H) are roughly statistically attacked by tertiary BuO radicals: Tertiary BuOOBu has, under N_2 flushing, photolytically been decomposed in isooctane in the presence of a relatively high excess of an ${>}NO•$ radical which, in the absence of O_2, quantitatively traps the alkyl radicals produced by H-abstraction.

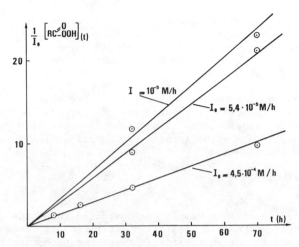

Figure 2.   Peracid formation in isooctane photooxidation against
time of irradiation, for different rates of radical initiation I_o.

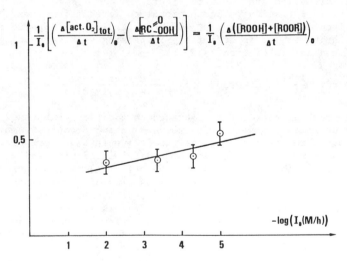

Figure 3.   Relative initial rate of formation of photooxidation
products other than peracids in isooctane against $-\log I_o$.

Gas chromatographic analysis of the reaction mixture yielded three product peaks associated with isomeric alkyl substituted hydroxyl-amines of the general formula

$$\text{>NO ---isooctyl}$$

The peak-area ratio of the three isomers was found to be:

$$A_1 : A_2 : A_3 = 0.77 : 0.16 : 0.07$$

In isooctane the ratio of primary, secondary and tertiary C-H sites is:

primary C-H : secondary C-H : tertiary C-H = 0.83 : 0.11 : 0.06

The isomer corresponding to the main peak, A_1, has been isolated from the reaction mixture and accumulated by preparative thin layer chromatography. ^{13}C-NMR analysis, carried out in our Analytical Department, has, without a doubt, proved that the product associated with this main peak $>NO-R_1$ was a mixture of two isomers corresponding to the <u>two primary radicals</u> derived from isooctane; i.e.,

The conclusion therefore is that in isooctane up to 80% of the alkyl radicals produced by tertiary BuO· attack were primary radicals, which, in the presence of O_2 would readily convert to primary peroxy radicals $RCH_2OO·$.

<u>Proposed Model of Isooctane Photooxidation.</u>   Based on our experimental results, the photooxidation of isooctane is not likely to proceed by a classical chain reaction of propagating peroxy radicals according to equation (5) in Scheme I.   The results rather indicate that the photooxidation proceeds by peracid radicals, which, by H-abstraction from the substrate, will lead to peracids. The question is, from where do these peracid radicals originate. A logical explanation seems to be a process in which primary peroxy radicals would be involved.   The best kinetic fit with our experimental results was obtained on the basis of the following model. In the first place, this model takes into account that the species mainly formed by tertiary BuO· radical attack of isooctane are primary peroxy radicals.   These radicals are assumed to decay

predominantly by recombination and not by H-abstraction from the substrate.

Now, this recombination of primary peroxy radicals is, according to our model, considered to be the immediate source for peracid radicals. This idea is-see Scheme II-that primary peroxy radicals recombine first to a tetroxide cage. Under steady state conditions, this tetroxide is supposed to decompose at the rate of its formation. One part of this decomposition, α , leads to peroxides. A second part, β , is, possibly by intermediate formation of aldehyde, further oxidized to peracid radicals and HOO· .

With regard to further reactions of the radicals $RC\underset{OO·}{\overset{O}{\lessgtr}}$ and HOO· , different possibilities have been considered. It turned out, that the following reactions seem to dominate:
- Formation of peracids by H-abstraction of isooctane
  by peracid radicals.   according
- Self recombination of peracid radicals, rather than   to
  cross-terminations with other species.   Scheme II
- Formation of hydroperoxides by combination of ROO
  with HOO .

The justification for this model is the fact that, based on the kinetic equations derived from it, quite a satisfactory fit between experiment and model was obtained. Assuming steady state conditions for all transient species involved in the set of reactions according to Scheme II, the following equation for the formation of peracids results from this rather crude model:

$$\frac{1}{I_o} \frac{d(Peracids)}{dt} \sim \frac{A}{cI_o} (-1 + \sqrt{1 + cI_o} \ )$$

where β and c are constants.

β is the part of peracid-radicals resulting from the cage $[RCH_2OO· \cdot OOCH_2R)$.

c is of the form: $c = \dfrac{K_r}{(k_1[Isooctane])^2}$ const.

Two limiting cases are of interest:

a) for $cI_o \ll 1$: $\dfrac{1}{I_o} \dfrac{d(Peracids)}{dt} \longrightarrow \dfrac{\beta}{2}$

b) for $\sqrt{cI_o} \gg 1$: $\dfrac{1}{I_o} \dfrac{d(Peracids)}{dt} \longrightarrow \dfrac{\beta}{\sqrt{cI_o}}$

Adjusting the two constants β and c to our experimental results, the fit between model and experimental data obtained can be seen in Figure 4A and 4B. The best fit was obtained with $\beta = 0.8$ and $c = 2.4 \times 10^4$.

As Figure 4B shows, also the $>$N-H inhibited photooxidation is quite conveniently covered by our model: The calculation in this case is based on the assumption that peracid radicals would completely be scavenged by the amine. In fact, previous work, which has partly been published (1, 13), indicated that TMP derivatives are indeed very powerful scavengers of peracid radicals.

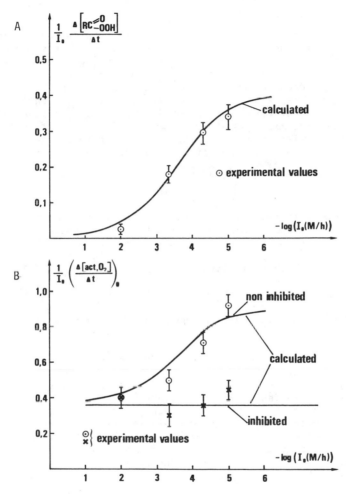

Figure 4.   Fit between kinetic model of isooctane photooxidation and experimental data.   (A)   Relative rate of peracid formation against $-\log I_o$.   (B)   Relative rate of iodometrically titratable total "active oxygen" formation against $-\log I_o$.

In the case of secondary amines, the following reaction takes place, with high yield of the corresponding $>$NO· radical:

$$RC{\overset{\nearrow O}{\underset{\searrow OO·}{}}} + >NH \quad\longrightarrow\quad RC{\overset{\nearrow O}{\underset{\searrow OH}{}}} + >NO·$$

This reaction actually takes place in the photooxidation of isooctane. This is shown in Figure 5 for the case of the secondary TMP derivative used in our photooxidation experiments.

The reaction of peracid radicals with $>$N-H seems, according to Figure 5, indeed to be quite fast. The rate of initiation applied in this experiment was comparatively high. It can't be excluded that under these rather extreme conditions, a certain amount of self termination of peracid radicals has, as a competitive reaction, also occurred. According to Figure 5, the relation between peracid radical production and amine conversion is

$$\left(\frac{d[NH]}{dt}\right)_{initial} \approx 0.7 \quad I_o'$$

Note: It has been observed that peracids derived from isooctane are also destroyed by $>$N-H. This reaction seems to be catalytic and proceeds with only partial conversion of $>$N-H to $>$NO· . It can therefore be excluded that this reaction plays a dominant role in the experiment shown in Figure 5.

Investigations on 2,4-Dimethylpentane

Investigations on this model compound have not been carried out in much detail. The results were, in the main, similar to those found for isooctane; i.e.,
- Low kinetic chain length of photooxidations, even at very low rates of tertiary BuO initiation.
- Formation of a species destroyable by addition of 7-tetradecene (presumably peracids) besides "ordinary" hydroperoxides.
- Inhibition of part of the photooxidation products (presumably peracids) in the presence of TMP-derivatives.

Quantitative assessments turned out to be much more difficult in 2,4-dimethylpentane than in isooctane, due to formation of relatively unstable photooxidation products.

Discussion

The photooxidation of isooctane turned out to be more complicated than originally expected. Of special interest was the rather surprising result that, during tertiary BuOOBu photolysis, practically statistical radical attack of the different C-H sites of isooctane was observed. At first glance, this result was not expected from literature data ([11], [12]). For example, according to investigations carried out by Niki and Kamiya ([12]) tertiary BuO· attack on primary, secondary and tertiary C-H sites occurs with a selectivity of 1:7:20. This selectivity ratio derives from experiments on hydro-

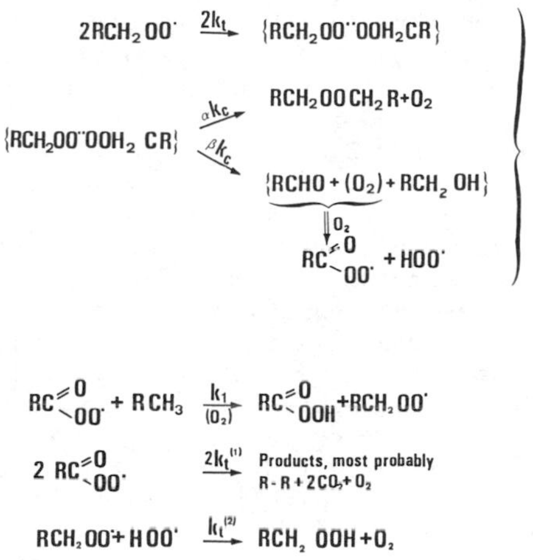

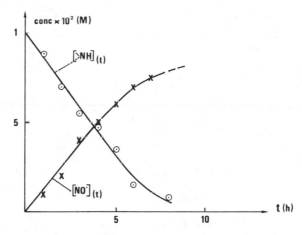

Scheme II.   Model for photooxidation of isooctane (dominating reactions).

Figure 5.   Consumption of $>$N-H and formation of $>$NO· in isooctane photooxidation $I_o \approx 5 \times 10^{-3}$ M isooctyl radicals/h, corresponding (according to Scheme II) to a rate of peracid formation of $I_o' = \beta I_o/2 \approx 2 \times 10^{-3}$ M/h (with $\beta \approx 0.8$).

carbons having different ratios of primary, secondary and tertiary
C-H. However, according to these authors, the overall reactivities
of model compounds for polypropylene, such as 2,4-dimethylpentane
and isooctane turned out to be much less than the reactivities
calculated from the ratio 1:7:20 mentioned above. The same also
applied for the reactivity for the polypropylene unit structure,
which has been measured on atactic polypropylene dissolved in benzene. Niki and Kamiya (15) attribute this behavior to "the preferred conformation" around the tertiary C-H group in these substances, presumably a steric crowding effect, obstruction tertiary
BuO· attack on this site and probably also on the immediately adjacent secondary CH_2-group. This means that, by steric hindrence,
H-abstraction selectivity of the different C-H sites in polypropylene and related low molecular compounds is expected to be
reduced. Nevertheless our results, which indicate almost statistical attack of all C-H groups in isooctane were however rather surprising. For the time being we did not go into details concerning
this question. It should, however, be pointed out that Niki's investigations have been carried out using thermal decomposition of
di-tert-butylperoxyoxalate in benzene solutions of hydrocarbons,
whereas our results refer to direct photolytic cleavage of tertiary
BuOOBu in undiluted isooctane.

At the present time, a discussion of the results of our model
investigations in terms of possible consequences for polypropylene
must be completely speculative. Apart from the differences expected between liquid phase and solid polymer photooxidation kinetics, differences in the chemical structure between our model substance, isooctane, and the structural unit of polypropylene have
to be also considered. With respect to the number of CH_3-groups
per structural unit, isooctane and polypropylene differ by a ratio
of 5:1.

Assuming, as our experiments in isooctane indicate, statistical
attack of all C-H sites by alkoxy and OH· radicals resulting from
photocleavage of hydroperoxides, it has to be recognized that in
polypropylene, H-abstraction would with 50% probability take place
at methyl-groups. This means that up to 25% of the macroperoxi-
radical-pairs produced in this initiation process might be pairs of
primary peroxy radicals and the remaining 75% would be pairs between
secondary, tertiary and mixed peroxy radicals.

Due to the extremely high radical concentration within such
pairs of immediately adjacent radicals, termination processes would
be favored over the slow H-abstraction reaction of these peroxy
radicals. This idea of pair recombination within cages of macroperoxiradicals has been put forward by Carlsson & Wiles in 1980
(2, 8, 9). According to these authors, this process is considered
to be the main difference between solid state and liquid state
photooxidation processes: In liquid state photooxidation of low
molecular species, diffusion quickly randomizes radical popula-
tions. In the solid state process, however, the macroperoxy radical
pair produced from each successful initiation will only separate by
slow segmental diffusion. This situation is exemplified in
Scheme III. As long as the recombination processes within peroxiradical pairs can be considered as true termination processes, only
a relatively small fraction of peroxy radicals produced by initia-

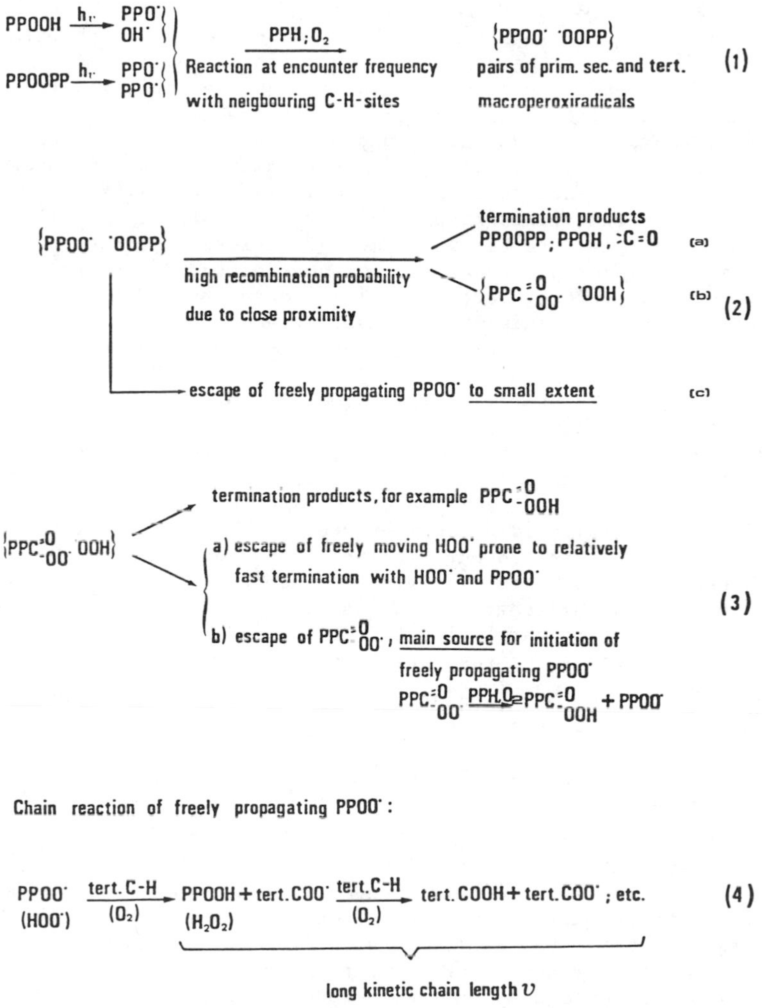

PPOOH $\xrightarrow{h\nu}$ $\left.\begin{array}{l} PPO^\cdot \\ OH^\cdot \end{array}\right\}$ $\left.\begin{array}{l} \\ \end{array}\right\}$ $\overline{\begin{array}{c} PPH;O_2 \\ \text{Reaction at encounter frequency} \\ \text{with neigbouring C-H-sites} \end{array}}$ $\{PPOO^\cdot \ ^\cdot OOPP\}$ pairs of prim. sec. and tert. macroperoxiradicals $\qquad$ (1)

PPOOPP $\xrightarrow{h\nu}$ $\left.\begin{array}{l} PPO^\cdot \\ PPO^\cdot \end{array}\right\}$

$\{PPOO^\cdot \ ^\cdot OOPP\}$ $\xrightarrow{\begin{array}{c} \text{high recombination probability} \\ \text{due to close proximity} \end{array}}$

termination products
PPOOPP ; PPOH , $>C=O$ (a)

$\{PPC\substack{:O \\ -OO^\cdot} \ ^\cdot OOH\}$ (b) (2)

escape of freely propagating PPOO⁻ to small extent (c)

$\{PPC\substack{:O \\ -OO^\cdot} \ OOH\}$

termination products, for example $PPC\substack{:O \\ -OOH}$

a) escape of freely moving HOO⁻ prone to relatively fast termination with HOO⁻ and PPOO⁻

(3)

b) escape of $PPC\substack{:O \\ OO^\cdot}$, main source for initiation of freely propagating PPOO⁻

$PPC\substack{:O \\ OO^\cdot} \xrightarrow{PPH,O} PPC\substack{:O \\ OOH} + PPOO^\cdot$

Chain reaction of freely propagating PPOO⁻ :

PPOO⁻ $\xrightarrow[(O_2)]{\text{tert. C-H}}$ PPOOH + tert. COO⁻ $\xrightarrow[(O_2)]{\text{tert. C-H}}$ tert. COOH + tert. COO⁻ ; etc. (4)
(HOO⁻)   (H₂O₂)

long kinetic chain length υ

Scheme III. Proposed model of polypropylene photooxidation.

tion would be expected to escape the radical cages as freely propagating macroperoxiradicals (process (2c) in Scheme III).

In the light of our investigations in isooctane, however, the
recombination process within pairs of primary peroxy radicals is not
a true termination (see Scheme II): Macroperacid radicals and
freely mobile HOO· radicals are expected to be formed from the primary tetroxide recombination cage by interaction with O_2 (processes (2b) and (3) in Scheme III).

The freely mobile HOO· radical is expected to escape relatively quickly from the cage, leaving a single macroperacid radical, which, if not inhibited, would undergo H-abstraction from
surrounding C-H sites (process (3b) in Scheme III). As a result,
isolated macroperoxy radicals would be produced, propagating for a
large number of steps before random termination occurs (process (4)
in Scheme III). The escaping HOO· radicals (process (3a)) are also
prone to hydrogen abstraction. However, these freely mobile species can be expected to terminate much faster than macroperoxy radicals, which can only migrate by segmental diffusion and by chain
reactions. Therefore, radical chains induced by HOO· seem less
probable than those initiated by macroperacid radicals.

Two conclusions result from the model presented in Scheme III:
- Inhibition of peracid radicals at the time of their formation
  would reduce the number of initiated photooxidation chains (see
  step (3b) in Scheme III).
- Scavenging of propagating macroperoxy radicals would reduce the
  kinetic chain length of photooxidation (see step (4) in Scheme
  III).

Tetramethylpiperidine derivatives are capable of acting in
both ways: according to our findings, these additives react very
efficiently with peracid radicals. In addition, they are expected
to accumulate at hydroperoxide sites by complex formation (1, 2).
This means they are partly located at the sites where photooxidation is initiated. The respective complex formation constants are
however low [5 to 20 M^{-1} in solution (1, 2)]. In the presence
of TMP-derivatives, the relatively large contribution of PPOO
chain reactions expected to be induced by peracid radicals would
most probably be extensively inhibited. TMP-derivatives are, on
the other hand, only low scavengers of alkylperoxy radicals (2,
7). The respective rate constant for secondary amines in solution
has been evaluated to roughly 3 Msec^{-1}. However, according to
Carlsson & Wiles, even weak scavenger interaction is already sufficient to dramatically reduce the kinetic chain length of the propagating oxidation steps (2). In addition, due to complex formation
with ROOH sites, reaction of TMP with macroperoxy radicals within
radical cages could also be considered to occur, reducing in this
way the number of initiated photooxidation chains. In conclusion,
the action of TMP-derivatives and their conversion products in
polypropylene seems to be at least twofold:

They seem to interfere, by slow reactions, with propagating
macroperoxy radicals; secondly, they might be capable of reducing
the number of initiated chain reactions by fast inhibition of peracid radicals produced by non-terminating recombination of primary
macroperoxy radicals within initiated radical pairs.

The contribution of the two stabilizing mechanisms proposed is
difficult to estimate at present time. Reliable values for the
rate constants of all species involved in PP photooxidation and
stabilization are lacking. Much depends on the ability of HOO· ,
escaping from radical cages, to initiate propagating macroperoxi-
radical chains. From kinetic considerations derived from Scheme
III, it follows that the higher this ability is, then the possible
stabilizing contribution due to peracid radical scavenging would
consequently be lower. The following consideration covers the most
unfavorable case with respect to this contribution: Each HOO· radical leaving a radical pair (equation (3a) in Scheme III) is assumed to initiate a propagating peroxiradical chain. The kinetic
equations corresponding to this case are given in Scheme IV. Although strongly simplified, they represent the essential features
and roughly apply to kinetic chain length >10.
     In Figure 6 the result of a model calculation is presented.
The set of parameter used (see Figure 6) is arbitrary and has only
a qualitative meaning. The results of the calculation cover three
cases:

Curve a: Non-inhibited photooxidation ($[>NX] = 0$; $p = 1$; $\nu_o = 200$)
Curve b: Inhibited photooxidation by exclusively $>NX/PPOO·$ interaction

$$(p = 1; \frac{k_p [PPH]}{k_{NX}[>NX]} = 20)$$

According to predictions made by Carlsson ($\underline{2}$) appreciable stabilization already occurs even in the case of relatively low PPOO·/NX
interaction. (For higher values of ν_o correspondingly lower stabilizer reactivity would be needed to achieve the same stabilizing
effect.) The condition for stabilization by PPOO·/ NX interaction
is, according to equation (1) simply given by

$$k_{NX}[>NX][PPOO·] \gg 2k_r[PPOO·]^2$$

As long as this relation holds, the kinetic chain length of photooxidation is greatly reduced.

Curve c: Inhibited photooxidation by $>NX/PPOO·$ interaction and
complete peracid radical destruction

$$(p = 0; \frac{k_p [PPH]}{k_{NX}[>NX]} = 20)$$

Under the conditions considered here, complete peracid radical
destruction approximately doubles the effect of exclusive PPOO·/
$>NX$ interaction. Therefore, partial peracid radical scavenging
would give an effect somewhere in between curves b and c.
     As already mentioned, the estimation presented here applies
to the rather unrealistic assumption, that each HOO· radical
leaving a pair of recombining primary peroxy radicals would be able
to initiate a propagating oxidative chain. In reality, this might

$$\frac{d[PPOO']}{dt} \simeq J(t) - 2k_t[PPOO']^2 - k_{NX}[\text{>}NX][PPOO'] = 0 \qquad (1)$$
(steady state approximation)

$$\frac{d[PPOOH]}{dt} \simeq k_p[PPH][PPOO'], \qquad (2)$$

where k_p: propagation const. $\left.\right|$ of PPOO'
       k_t: termination const. $\left.\right|$
       k_{NX}: rate const. of >NX with PPOO'
       $J(t)$: rate of initiation of propagating
               macroperoxiradical chains.

$$J(t) \simeq \phi[PPOOH] = J_{RCOOO'}^{(t)} + J_{HOO'}^{(t)}, \qquad (3)$$

where $J_{HOO'} \simeq \frac{\Phi}{2}[PPOOH]$ = rate of chain initiation
                             by HOO'.

       $J_{RCOOO'} = p\frac{\Phi}{2}[PPOOH]$ = rate of chain initi-
                             ation by RCOOO'.

with
$$0 \leq p \leq 1$$

p=1 refers to no interaction of peracid radicals with >NX.

p=0 refers to complete scavenging of peracid radicals
     by >NX.

Scheme IV. Kinetics for initiation of propagating peroxy
radical chains by HOO'.

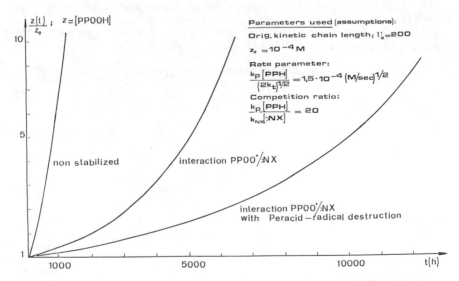

Figure 6. Model calculation according to Scheme III (for further
explanation see text).

not be the case, since freely mobile HOO· radicals are expected to
disappear quite quickly by recombination with other species before
chain initiation can take place. Accordingly, chain initiation by
peracid radicals might be of much more importance than assumed in
our model calculation. Therefore, peracid radical scavenging by
TMP derivatives is expected to be an important additional stabilizing process in polypropylene.

Experimental

a) Radical counting and photooxidation experiments

Irradiations were carried out on an optical bench in quartz cells
of 1 cm path length using light from a 200 W high pressure Hg lamp.
The light was filtered by a 315 nm interference filter and rendered
parallel by use of a quartz lens. In order to protect the interference filter from damage caused by long time exposure to short
wave length radiation, a pyrex plate (cut off <300 nm) was placed
in front of this filter.

Tertiary BuO· radicals were produced at constant rate by
photoylsis of tertiary BuOOBu dissolved in isooctane (conc. $\approx$
0.5 M). The production rate of isooctylradicals (t–BuO· + RH $\longrightarrow$
t–BuOH + R·) was measured by $>$NO· consumption ($>$NO· + R· $\longrightarrow >$NOR)
under complete absence of oxygen (N_2 flushing before and during
irradiation). The $>$NO· used (initial concentration: 2.10^{-2} M)
was Nitroxide I described in ($\underline{1}$). $>$NO· consumption was measured
spectroscopically (λ max. = $47\overline{0}$ nm, $\epsilon_{NO} \simeq 10(Mcm)^{-1}$). Photooxidations were carried out either in 10 ml quartz cells (in case of
high radical initiation) or in a special cell of approximately 100
ml and 1 cm path length (in case of low initiation). The cell surface was in all cases uniformly illuminated. The rate of radical
production I was controlled by uniform attenuation of light intensity, using a combination of calibrated copper grids placed in
the space between the lens and the reaction cell. The TMP derivatives used in photooxidation studies were amine II in ($\underline{1}$) and a
$>$N–octyl derivative.

b) Analytical procedures

Total concentration of hydroperoxides, peroxides and peracids: The
iodometric method described by Carlsson & Wiles ($\underline{16}$) was applied.
Note: tertiary BuOOBu is not measured under these conditions;
other peroxides, for example RCH_2OOCH_2R are however titratable.
Determination of "act. O_2" in the presence of $>$NO· : Since $>$NO·
also converts I^- to I_3^-, corrections for this contribution
have been made by gas chromatographic estimation of ($>$NO·).
(Standard used: Nitroxide derived from amine I).
Titration of peracids. Peracids have been determined by making use
of the quantitative and selective raction of peracids with olefins
[epoxidation ($\underline{10}$)]: To a given volume of irradiated isooctane an
equal volume of 7-tetradecene solution (2.10^{-2}M) was added.
After 15 hours standing in the dark, the remaining hydroperoxides
and peroxides were titrated ($C_{peracids} = C_{total} - C_{remaining}$).

c) Distribution of >NO-isooctyl isomers formed by trapping of
isooctyl radicals with >NO·.
Irradiation was carried out as for radical counting; the 100 ml
cell was used, and full light intensity was applied ($I \approx 10^{-2}$M
sec.$^{-1}$). The distribution of isomers was measured by gas chromatography using a Varian 2740 FID instrument (OV 101, Poropak
columns).

Identification: Separation of isomers by preparative thin layer
chromatography; identification by ^{13}C-NMR in the spectroscopic
laboratory of our Analytical Department.

Acknowledgement

The author wishes to thank Dr. G. Rist for recording and inter-
preting ^{13}C-NMR spectra. He is very grateful to Dr. H. Gysling
for very valuable encouragement and support. He also thanks the
management of CIBA-GEIGY Limited in Basle for permission to report
the results of this investigation.

Literature Cited

1. B. Felder, R. Schumacher and F. Sitek, ACS Symposium on Photodegradation and Photostabilization of Coatings, ACS Symposium
Series 151, 65 (1980).
2. D.J. Carlsson, K.H. Chan and D.M. Wiles, ACS Symposium on
Photodegradation and Photostabilization of Coatings, ACS Symposium
Series 151, 51 (1980).
3. D. Bellus, H. Lind and J.F. Wyatt, J. Chem. Soc. Chem. Commun.
1972, 1199.
4. B. Felder and R. Schumacher, Angew. Makromol. Chem. 31, 35
(1973).
5. J.B. Shilov and E.T. Denisov, Vysokomol. Soed. 16A 2313 (1974).
6. D.J. Carlsson, D.W. Grattan and D.M. Wiles, Coatings and Plastic Preprints 39, 628 (1978).
7. D.W. Grattan, D.J. Carlsson and D.M. Wiles, Polym. Degrad.
Stab. 1, 69 (1979).
8. D.J. Carlsson, K.H. Chan, A. Garton and D.M. Wiles, Pure Appl.
Chem. 52, 389 (1980).
9. A. Garton, D.J. Carlsson and D.M. Wiles, Die Makromol. Chem.
52, 389 (1980).
10. Houben Weyl, Bd. II, Analytische Methoden, 308 (1953).
11. C. Walling and B.B. Jacknow, J. Amer. Chem. Soc. 82, 6108
(1960).
12. E. Niki and Y. Kamiya, J. Org. Chem. 38 Nr. 7, 1403 (1973).
13. B. Felder, R. Schumacher and F. Sitek, Helv. Chim. Acta 63,
132 (1980).
14. D.J. Carlsson, K.H. Chan, J. Durmis, and D.M. Wiles,
J. Polymer Sci., Pol. Chem. Ed., 20, 575 (1982).
15. E. Niki and Y. Kamiya, Bull. Soc. Japan, 38, 3226 (1975).
16. D.J. Carlsson and D.M. Wiles, Macromolecules 2, 597 (1969).

RECEIVED December 7, 1984

Reactions of Aminyl Radicals and Mechanisms of Amine Regeneration as Inhibitors of Oxidation

E. T. DENISOV

Institute of Chemical Physics, USSR Academy of Sciences, 142 432 Chernogolovka, USSR

Diphenylaminyl radicals, In·, produced in tetraphenylhydrazine decomposition, react with the formation of
diphenyl amine and oligomeric semidienes. Diphenyl
amine is formed in the reaction of In· with labile
dimers, AmAm, with iminoquinone structure. The rate
constant of Am· recombination measured by flash
photolysis technique (FTP) was found to be
1.8×10^7 l.mole^{-1}s^{-1} (cyclohexane, 293°K). The combination of In· and RO_2· was studied by FPT using monitoring at two different wavelengths. It runs with the
rate constant of 6×10^8 l.mole^{-1}s^{-1} (cyclohexane,
293°K) with formation of iminoquinone as the main
product. Radicals Am· react with cumylhydroperoxide
very fast (k = 1.1×10^5 l.mole^{-1}s^{-1}, cyclohexane,
293°K, FTP). The mechanism of this reaction is
complicated at [ROOH] > 5×10^{-3} mole l^{-1} by parallel
consecutive reaction with proton transfer and InH·
formation and electron transfer from ROO$^-$ to InH·.
Quinone imine retards the oxidation of n-heptadecane
(393°K) with regeneration of 4-hydroxydiphenylamine in
the presence of hydroperoxide.

Each aromatic amine molecule, InH, terminates many free radical
chains in autooxidation of alcohols and amines due to the ability of
oxyperoxy and aminoperoxy radicals to oxidize InH as well as to
reduce In to InH (1). However, the coefficient of inhibition,
f > 2, can be very often observed in oxidizing hydrocarbons too (2).
Therefore, some reduction of aminyl radicals to InH proceeds in
oxidizing hydrocarbons. To elucidate the ways of such reduction we
have studied the products and kinetics of the reactions of diphenylaminyl radical In·.

Recombination of Aminyl Radicals. A convenient source of In· is
tetraphenylhydrazine, InIn. Diphenylamine and oligomeric semidienes
were found to be the products of InIn decomposition at T > 400°K
(3,4). The kinetics of their formation were studied by Varlamov (5).
He found that InH and o-semidienes were primary stable products of

InIn decomposition and p-semidienes were formed after any period of
induction. InH was proposed to be formed in the following sequence
of reactions ($\underline{4},\underline{5}$):

$$In^{\cdot} + In^{\cdot} \quad - \cdot \quad In-C_6H_5=N-C_6H_5 \xrightarrow{In^{\cdot}} InH + (C_6H_5)_2NC_6H_4\dot{N}C_6H_5$$

As the result of such reaction pathways, oligomeric semidienes were
formed. The yield of InH was found to be 80% on decomposed InIn
(CCl_4, 348°K). Such reactions may be treated as regeneration of InH
from In$^{\cdot}$. The kinetics of InIn decomposition are first order ($\underline{6}$); the
introduction of a free radical acceptor increases the rate constant.
For example, InIn decomposes with the rate constant 6.3×10^{-5} s^{-1} in
CCl_4 at 348°K and in the presence of N-phenyl-2-naphthylamine the
rate constant is 14.8×10^{-5} s^{-1}. This suggests that the decomposition
of InIn includes an equilibrium stage and may be described by the
following scheme:

$$InIn \underset{k_{-1}}{\overset{k_1}{\rightleftharpoons}} 2In^{\cdot} \xrightarrow{k_2} Products$$

From the data mentioned above, one can estimate the ratio
$k_2/k_{-1} = 0.74$.
    The kinetics of In$^{\cdot}$ recombination were studied by FPT in cyclo-
hexane solution ($\underline{7}$). The concentration of In$^{\cdot}$ was measured spectrophotometrically at $\lambda = 770$ nm. The extinction coefficient was found
to be 3.9×10^3 l.mole^{-1} cm^{-1}. In$\cdot$ radicals were found to disappear in
a bimolecular reaction with the overall rate constant of
1.8×10^7 l.mole^{-1} s^{-1}, which is the sum of rate constants on In$\cdot$
recombination into InIn and quinone imine formation equal to
0.75×10^7 l.mole^{-1} s^{-1} and 1.05×10^7 l.mole^{-1} s^{-1}, respectively (cyclohexane, 293°K).

<u>Reaction of In$\cdot$ With RO$_2^{\cdot}$.</u> The kinetics of In$\cdot$ reaction with RO$_2^{\cdot}$ from
cyclohexane were also studied by FPT ($\underline{7}$). Di-t-butylperoxide was decomposed photochemically, and the reaction of $(CH_3)_3CO^{\cdot}$ radicals with
cyclohexane produced cyclohexyl radicals. The latter were transformed into peroxy radicals RO$_2^{\cdot}$ after the addition of oxygen.
Radicals In$\cdot$ were formed by the photolysis of InIn. The kinetics of
RO$_2^{\cdot}$ and In$\cdot$ disappearance were monitored simultaneously at $\lambda = 270$
and 770 nm, respectively. Under the experimental conditions when
$[RO_2^{\cdot}] \gg [In^{\cdot}]$, In$\cdot$ disappeared mainly by the reaction with RO$_2$.
Because of the slow recombination of RO$_2^{\cdot}$, there is a nearly constant
RO$_2^{\cdot}$ concentration ($\sim 6 \times 10^{-6}$ mole.l^{-1}). The rate constant of the
reaction with In$_1^{\cdot}$ with RO$_2^{\cdot}$ was estimated to be equal to the
6×10^8 l.mole^{-1} s^{-1} in the temperature range of 283-303°K in cyclohexane. The decomposition of InIn in the presence of cumylhydroperoxide yielded diphenylamine, quinone imine, cumyl alcohol and
acetophenone ($\underline{8}$). Diphenylnitroxide was detected by EPR techniques.
The comparison of the concentrations of decomposed ROOH and InH and
ROH formed shows that ROOH is not the only donor of hydrogen atoms
but a part of InH and ROH was formed in the reactions of In$\cdot$ and RO$\cdot$
with labile intermediate products. The most probable products are
labile quinone imines $RO_2C_6H_5C_6H_4$ and $(C_6H_5)_2NC_6H_5NC_6H_5$, which react
with In$\cdot$ and Ro$\cdot$ forming InH and ROH. Quinone imine apparently is

formed from $RO_2C_6H_5NC_6H_5$. So $RO_2^{\cdot}$ reacts with a nitrogen atom and benzene ring of aminyl radical:

$$RO_2^{\cdot} + In^{\cdot} \longrightarrow RO^{\cdot} + InO^{\cdot}$$

$$RO_2^{\cdot} + In^{\cdot} \longrightarrow RO_2C_6H_5NC_6H_5$$

The Mechanisms of Aminyl Radical Reaction with ROOH. The kinetics of In$^{\cdot}$ reaction with cumylhydroperoxide were studied in cyclohexane at 393°K using FPT. The concentration of In$^{\cdot}$ was measured spectrophotometrically at $\lambda = 812$ nm. Aminyl radical disappeared in the presence of ROOH in two reactions, namely In$^{\cdot}$ + In$^{\cdot}$ and In$^{\cdot}$ + ROOH. The reaction rate constant of In$^{\cdot}$ with ROOH at [ROOH] = $1.4-5.1 \times 10^{-3}$ was calculated via computer from experimental kinetic curves of In$^{\cdot}$ disappearance and found to be 1.1×10^{5} l.mole^{-1}s^{-1}. A new absorption was observed at λ max = 690 nm in experiments with [ROOH] > 10^{-2} mole.l^{-1}. This is attributed to a radical-cation InH^{+}, which may be formed by protonation of aminyl radical:

$$In^{\cdot} + HOOR \rightleftharpoons InH^{+} + RO_2^{-}$$

The kinetics of In$^{\cdot}$ and InH^{+} consumption were measured at $\lambda = 712$ nm. It was found that In$^{\cdot}$ + InH^{+} disappear in the presence of ROOH with first order kinetics, the observed rate constant $k_{obs} = [ROOH]^{-1} \Delta \ln D_{712}/t$ decreases with increasing hydroperoxide concentration. The formation of InH$^{\cdot}$ and dependence of k_{obs} on [ROOH] may be explained by the following kinetic scheme:

$$In^{\cdot} + HOOR \xrightarrow{k_3} InH + RO_2^{\cdot}$$

$$In^{\cdot} + HOOR \xrightleftharpoons{K} InH^{+} + RO_2^{-}$$

$$InH^{+} + RO_2^{-} \xrightarrow{k_4} InH + RO_2^{\cdot}$$

The treatment of the experimental data according to this scheme gives the following values: $k_3 = 1.1 \times 10^{5}$ l.mole^{-1}s^{-1}, K = 42 l.mole^{-1}, $k_4 = 1.0 \times 10^{3}$ s^{-1} (293°K, cyclohexane). When ROOH concentration is high enough (0.1 mole.l^{-1} or more) the reaction runs via protonation of In$^{\cdot}$ and electron transfer.

Quinoneimine as an Inhibitor of Oxidation Reactions. Quinoneimine, $C_6H_5NC_6H_4O$, was found to retard oxidation for a long period of time. If quinoneimine is introduced into oxidizing n-heptadecane (393°K) containing hydroperoxide, it retards oxidation. The higher the hydroperoxide concentration the stronger the retarding action of quinoneimine. Aminophenol, $HOC_6H_4NC_6H_5$, was found in small concentration among the reaction products. This means that quinoneimine is reduced to aminophenol in oxidizing hydrocarbon in the presence of hydroperoxide. Since the aminophenol is oxidized by peroxy radicals to quinoneimine, the opposing cyclic oxidation-reduction reactions proceed in oxidizing hydrocarbon with participation of $RO_2^{\cdot}$ and ROOH.

Hence, the extra chain termination ($f > 2$) when aromatic amines are used as inhibitors of oxidation may be the result of the following reactions:

$$In^\cdot + In^\cdot \longrightarrow InC_6H_5NC_6H_5$$

$$RO_2^\cdot + InC_6H_5NC_6H_5 \longrightarrow ROOH + InC_6H_4\dot{N}C_6H_5$$

$$In^\cdot + RO_2^\cdot \longrightarrow RO_2C_6H_5NC_6H_5$$

$$In^\cdot + RO_2C_6H_5NC_6H_5 \longrightarrow InH + RO^\cdot + OC_6H_4NC_6H_5$$

$$ROOH + OC_6H_4NC_6H_5 \longrightarrow \ldots \longrightarrow HOC_6H_4NHC_6H_5$$

All the reactions mentioned above make the inhibiting coefficient f much more than 2, when amines are used as inhibitors. At present we do not know either the role of alkyl radicals in such reactions or the stages of quinoneimine reduction to aminophenol.

Literature Cited

1. Denisov, E.T. in "Developments in Polymer Stabilization 3"; Scott, G., Ed.; Appl. Sci. Publ. LTD: London, 1980; pp. 1-20.
2. Berger, H.; Bolsman T.A.B.M.; Brower, D.M. in "Developments in Polymer Stabilization 6"; Scott, G., Ed.; Appl. Sci. Publ. LTD: London, 1983; pp. 1-27.
3. Musso, H. Chem. Ber. 1959, 92, 2881.
4. Welzel, P. Chem. Ber. 1970, 103, 1318.
5. Varlamov, V.T. Izv. Akad. Nauk SSSR, ser.khim. 1982, 1481.
6. Varlamov, V.T. Izv. Akad. Nauk SSSR, ser.khim. 1982, 1629.
7. Varlamov, V.T. Safiullin, R.L.; Denisov, E.T. Khimicheskaya Fizika 1983, 408.
8. Varlamov, V.T.; Denisov, E.T. Kinetika i Kataliz 1983, 24, 547.

RECEIVED December 7, 1984

Hindered Diazacycloalkanones as Ultraviolet Stabilizers and Antioxidants

J. T. LAI, P. N. SON, and E. JENNINGS

The BFGoodrich Company, Brecksville, OH 44141

The properties of hindered diazacycloalkanones were
evaluated as light stabilizers and antioxidants for
polypropylene. A totally hindered piperazinone was
found to be excellent light stabilizer with many other
desirable properties as a polymer additive. Partially
hindered decahydroquinoxalinone derivatives were found
to be surprisingly good antioxidants. The relationship between structure parameters and activity was
investigated.

Since Hindered Amine Light Stabilizers (HALS) were first introduced
about a decade ago, their commercial applications have proliferated
to the stage where they may well dominate[1] the polymer light stabilizers market in the near future. Among the many commercial and
experimental HALS, only the BFGoodrich Company offers one that does
not bear a 2,2,6,6-tetramethylpiperidine (1) structure. In this
presentation, we would like to report our work on the chemistry and
properties of our developmental HALS 1,1'-(1,2-ethanediyl) bis(3,3,-
5,5-tetramethyl-2-piperazinone) (2), and the interesting antioxidant
(AO) and light stabilizing properties of some substituted decahydro-
quinoxalinones (3).

1   2   3

Dr. Layer[2] has given one plausible explanation as to how 3
performs as an AO. The structural parameters of 3 will also be
probed to locate the origins of its activity.

0097–6156/85/0280–0091$06.00/0

1,1'-(1,2-Ethanediyl)bis(3,3,5,5-tetramethyl-2-piperazinone)
(2) (piperazine) was prepared from a novel synthesis[3] using easily
available raw materials:

piperazine 2

Piperazine HALS 2 is a colorless crystalline solid which melts
at 134-6°C.   It is a powerful UV stabilizer for polyolefins as
illustrated in the outdoor aging data in isotactic polypropylene
tapes.

2x100 mil Slit Tapes - Arizona Aged
Failure = Months to 50% Loss of Tensile

		Piperazine	Piperidine	Polymeric Piperidine
0.1 phr HALS +	AO-1	9	5	7
	AO-2	10	6	3
	AO-3	12	10	-
	AO-4	8	3	-

AO's are added at 0.1 phr. Piperidine: Bis(2,2,6,6-tetramethyl-
4-piperidinyl) decanedioate; Polymeric piperidine: Poly[[6-
[(1,1,3,3-tetramethylbutyl)amino]-1,3,5-trizaine-2,4-diyl(2,2,6,6-
tetramethyl-4-piperidinyl)imino-1,6-hexanediyl-(2,2,6,6-tetramethyl-
4-piperidinyl)imino]
AO-1:   1,3,5-Tris(3,5-di-t-butyl-4-hydroxyphenyl)methyl-1,3,5-
        triazine-2,4,6-(1H,3H,5H)-trione.
AO-2:   Tris 2-[β-(3,5-di-t-butyl-4-hydroxyphenyl)propionoxy]ethyl
        isocyanurate.
AO-3:   Octadecyl 3,5-di-t-butyl-4-hydroxybenzenepropionate.
AO-4:   Pentaerythritol tetrakis-[β-(4-hydroxy-3,5-di-t-butylphenyl)
        propionate].

		Piperazine	Piperidine	Polymeric Piperidine
0.25 phr HALS +	AO-1	15	13	-
	AO-2	17	15	-
	AO-3	21	19	13
	AO-4	14	15	15

Piperazine 2 is also highly synergistic with commercial UV absorbers, as shown in 20 mil-thick compression-molded polypropylene samples aged in a xenon Weather-Ometer.

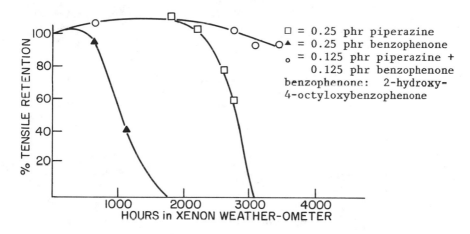

□ = 0.25 phr piperazine
▲ = 0.25 phr benzophenone
○ = 0.125 phr piperazine +
    0.125 phr benzophenone
benzophenone: 2-hydroxy-
4-octyloxybenzophenone

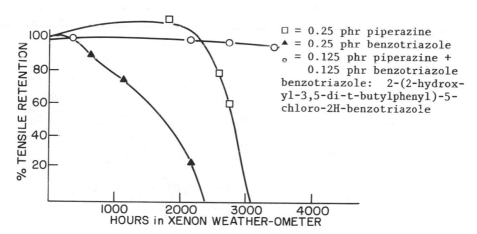

□ = 0.25 phr piperazine
▲ = 0.25 phr benzotriazole
○ = 0.125 phr piperazine +
    0.125 phr benzotriazole
benzotriazole: 2-(2-hydrox-
yl-3,5-di-t-butylphenyl)-5-
chloro-2H-benzotriazole

Piperazine HALS 2 has a positive effect on the processing stability of polypropylene when used in conjunction with a phenolic antioxidant:

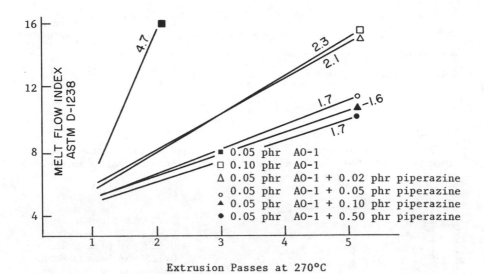

Extrusion Passes at 270°C

2 also has a very favorable water carryover behavior in polypropylene films:

Additive	Film Speed (ft/min)
1. Benzophenone	134
2. Piperidine HALS	78.5
3. Piperazine HALS	150.5

    PP containing 0.1 phr additive was extruded at 260°C into a
5 mil film that was quenched in a 37.8°C water bath. Tachometer
measurements of the film speed at the point of water carryover was
recorded. The higher the number, the faster the production would
be, which is preferred.

In the second part of this presentation, we would like to introduce
another family of amine stabilizers. Namely, decahydroquinoxalines
3 (R,R'=alkyl) which not only show light-stabilizing activity, but
also antioxidation properties.

3 is present as a mixture of cis- and trans- isomers. Cis-3,3-
dimethyl-decahydroquinoxalin-2-one (6) is a known compound[4] while

all trans-3 are novel. 3 (R'=H) can be synthesized from two different routes[5]: a) from the condensation of 1,2-diaminocyclohexane and
a ketone cyanohydrin in hot water, or b) from the reaction among
DCH, ketone, chloroform, and base.

Cis- and trans- 3 are easily separable by an acetone wash,
since the trans-isomer is insoluble in almost all solvents while the
cis-isomer is easily soluble. 3 (R'=H), either as the pure isomer
or as a mixture, was then alkylated with various alkylating agents
to give 3 (R' = alkyl, alkylene, etc.).

As a partially hindered amine, 3 showed light-stabilizing
activity lower than that of piperazine or piperidine HALS, but
comparable to commercial benzophenones. Like HALS, they are strongly synergistic with benzophenones and benzotriazoles in polyolefins.

Light-Stabilizing Activity of 3

Formulations[a] (phr)	Hrs. to Failure[b]
1. 3 (0.15)	1250
2. Benzophenone (0.15)	1260
3. Benzotriazole (0.15)	2100
4. 3 (0.075) + Benzophenone (0.075)	2540
5. 3 (0.075) + Benzotriazole (0.075)	2830

[a] 20 mil PP plaques containing 0.1 phr AO-2
[b] 50% tensile loss from aging in xenon Weather-Ometer®

More suprising to us is the antioxidation activity of 3 in
polypropylene. Not only strong primary antioxidants by themselves,
they are highly synergistic with other phenolic AO's.

Oven-aging of $\underset{\sim}{3}$ in Polypropylene at 125°C

Formulations (phr)	Days	
	10 mil	20 mil
1. AO-1 (0.1)	35	34
2. AO-1 (0.25)	87	91
3. AO-2 (0.1)	78	78
4. AO-2 (0.25)	134	210
5. 3a[a] (0.1)	59	184
6. 3a (0.25)	160	388
7. 3b[b] (0.1)	50	205
8. 3b (0.25)	201	438
9. AO-1 (0.125) + DSTDP[c] (0.125)	128	208
10. AO-1 (0.125) + DSTDP (0.25)	167	299
11. AO-1 (0.125) + 3a (0.125)	284	369
12. AO-1 (0.125) + 3b (0.125)	248	390

[a] $\underset{\sim}{3}$a: R = $-CH_2-\langle O \rangle-CH_2-$;   [b] $\underset{\sim}{3}$b: R = $-(CH_2)_6-$;

[c] distearyl thiodipropionate

To our knowledge, this is the best AO performance by any aliphatic amine. The magnitude of its activity is far greater than that of polymeric piperidine HALS. We did a structural probe by making various compounds similar to $\underset{\sim}{3}$ and tested their AO activities, trying to pinpoint which part of the molecule is responsible for its activity. Among the many compounds we tested, the following showed no AO activity at all.

The following compounds retain most or all the activity of $\underset{\sim}{3}$.

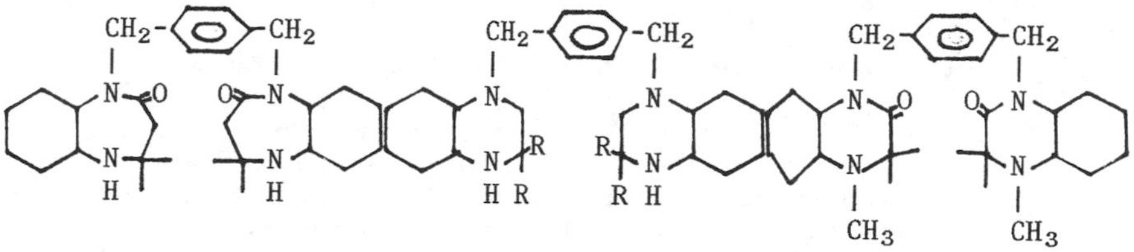

From this study, it is apparent that the most important features for the AO activity of 3 are the following:

a.   the fused bicylodiazacycloalkane ring system;
b.   diakyl substituents at the C-3 position;
c.   alkyl, instead of acyl substitution at the N-1 position.

Experimental Section

The polypropylene used in these experiments was Hercules Profax 6501. Small batches (48 gram resin) were fluxed at 190°C in the mixing cam heat of a Brabender Plasticorder with 1½ minutes addition time for resin and additives and a 3 minute mix at 30 rpm. Mixes were cooled in a room temperature hydraulic press. 20 mil thick plaques (8" x 8"), were pressed at 215°C and 20,000 lbs in a Pasadena hydraulic press.

2000 gram batches for outdoor aging were blended in a Henschel mixer for 3 minutes, extruded at 215°C from a Brabender extruder and pelletized. Slit polypropylene tape was produced by extruding 4 inch wide film from a temperature-programmed extruder, and putting a ½ inch wide strip at 14 ft/min through a 165°C oven where it is drawn 6 to 1 to produce a 23 mil thick oriented tape with tensile strength approximately 60,000 psi.

Samples were exposed in a Weather-Ometer Model 65-WR (Atlas Electric Device Company) which uses a filtered-xenon source of UV radiation with automatic regulation for constant irradiance, 50% relative humidity, 60°C black panel temperature, and 18 minutes of water spray every two hours.

Slit tape samples were mounted in aluminum frames and were exposed outdoors 45° south direct in Arizona (ASTM D1435-75) with no backing or cover. Samples were monitored at monthly intervals for tensile testing.

UV exposed samples were tested on an Instron tensile machine and tensile strength at yield was recorded. Failure time for exposed samples was loss of 50% of original tensile strength.

Oven aging of polypropylene plaques was done in an 125°C oven with 1" x 1" plaques in triplicate strung on aluminum wire. Degradation is the time for samples to crumble and fall from the wire.

Acknowledgment

We would like to thank Mr. George Kletecka for the measurement of melt flow index and water carry-over.

Literature Cited

1. "Stabilizers and Antioxidants for Plastics" by Skeist Labs, Inc., Dec. 1982.
2. "Electron Spin Resonance and Field Desorption - Mass Spectrometry Oxidation Studies of Partially Hindered Amines: 3,3-Dialkyldecahydroquinoxalin-2-ones." by R. W. Layer, J. T. Lai, R. P. Lattimer and J. C. Westfahl, this symposium.
3. J. T. Lai, J. Org. Chem. 45, 754(1980).
4. J Bindler and H. Schläpfer, U. S. Patent 2,920,077 to J. R. Geigy Corp. (1960).
5. J. T. Lai, Synthesis 71(1982).

RECEIVED December 7, 1984

3,3-Dialkyldecahydroquinoxalin-2-ones

Electron Spin Resonance and Field Desorption–Mass Spectrometry Oxidation Studies of Partially Hindered Amines

R. W. LAYER, J. T. LAI, R. P. LATTIMER, and J. C. WESTFAHL

The BFGoodrich Company, Brecksville, OH 44141

Fully hindered amines are excellent UV stabilizers, but poor thermal antioxidants, for polyolefins. We have found that partially hindered bicyclic amines, such as 3,3-dialkyldecahydroquinoxalin-2-ones, are excellent UV stabilizers and also excellent thermal antioxidants as well. The oxidation of these bicyclic amines with m-chloroperbenzoic acid was studied by electron spin resonance spectroscopy. They form stable, partially hindered nitroxyl radicals (6-line spectra). However, these primary radicals are easily oxidized to a new, fully hindered nitroxyl radical (3-line spectra) which are also very stable. Field desorption mass spectroscopic studies of the oxidation of these amines show that the primary nitroxyl radicals, which form first, lose two hydrogen atoms and add an oxygen. Thus, these bicyclic partially hindered amines are unique in being both stable and at the same time hydrogen atom donors.

SUMMARY

Fully hindered amines are known to be excellent UV stabilizers which retard the photodegradation of polyolefins. But they are not very effective thermal stabilizers. Their activity centers around their ability to form stable nitroxyl radicals which function as chain breaking electron acceptors but not as chain breaking hydrogen atom donors in the free radical oxidative process (1). Partially hindered amines, those that possess an alpha hydrogen atom, are generally much less effective as UV stabilizers. This has been related to the known instability of their nitroxyl radicals (2). For example, Coppinger has found that nitroxyl radicals of partially hindered amines are very unstable with life-times of only a few seconds (3). We have found that a class of partially hindered bicyclic amines, 3,3-dialkyldecahydroquinoxalin-2-ones, are excellent UV stabilizers and, more interestingly, excellent thermal stabilizers for polypropylene. This paper describes the results of an electron spin resonance and field desorption mass spectrometric studies of the oxidation of these partially hindered amines.

0097–6156/85/0280–0099$06.00/0

EXPERIMENTAL

The ESR spectra were obtained with a Varian E-3 spectrometer at room temperature. A saturated solution of amine in perchloroethylene was treated with a few drops of a saturated solution of m-chloroperbenzoic acid in perchloroethylene. The FD-Mass spectra were obtained on a Finnigan MAT 311A/Incos 2400 system using the same solutions as above.

RESULTS AND DISCUSSION

J. Lai, in a previous presentation, described the antioxidant activity of some partially hindered amines, such as α,α'-p-xylidine-bis-[1-(3,3-dialkyldecahydroquinoxalin-2-one)]. To learn how these compounds function, we studied their oxidation to nitroxyl radicals using m-chloroperbenzoic acid as the oxidant by electron spin resonance spectroscopy. A number of compounds were oxidized, and they gave spectra with anywhere from three to nine lines (Figure 1). Analysis of these spectra, by computer simulation, showed that the spectra could be described as a combination of two radicals. One of these radicals was the primary nitroxyl radical (I) expected from the amine. It had an ESR spectrum of six lines of equal intensity resulting from the hyperfine splitting of the nitrogen atom (~15 G) and the alpha hydrogen atom (~22 G). The other radical (II) was a fully hindered amine which gave a three-line ESR spectrum from the hyperfine splitting of the nitrogen atom (~14 G) (Figure 2). The observation that the three-line spectra were always obtained from amines of low solubility while the six-line spectra were obtained from highly soluble amines led us to investigate the role of oxidant level on the radical formed. We found that when very small amounts of oxidant were used, the six-line primary radical was obtained in every case. As progressively more oxidant was added, this primary radical was gradually oxidized to the new fully hindered nitroxyl radical (Figure 3 - experimental spectra). Similar results were obtained when the primary radical was allowed to stand with tert-butylhydroperoxide. The primary radical was slowly oxidized to fully hindered nitroxide (II) within 18 days. Air also slowly oxidized I to II.

Heretofore, the primary radicals of bicyclic nitroxyl radicals, with α-hydrogen atoms, had not been observed (4,5). This was due to the ease with which these radicals were oxidized rather than their intrinsic instabilities. For example, we found these radicals to be very stable. The ESR spectrum of a solution of 1-tetradecyl-3,3-dimethyldecahydroquinoxalin-2-one was monitored for 66 days with vitually no change in the ESR spectrum (Figure 4). On further standing in air, the primary radical was gradually oxidized to the secondary radical.

The secondary radical is also extremely stable and it was monitored for 231 days with only a small loss in the intensity of its ESR signal.

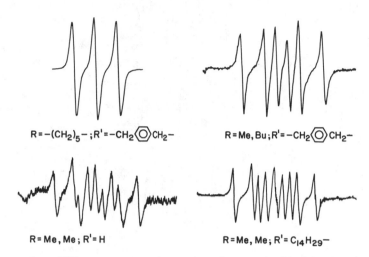

$R=-(CH_2)_5-$; $R'=-CH_2\langle O \rangle CH_2-$ $R=Me,Bu$; $R'=-CH_2\langle O \rangle CH_2-$

$R=Me,Me$; $R'=H$ $R=Me,Me$; $R'=C_{14}H_{29}-$

Figure 1. ESR spectra of oxidized 3,3-dialkyldecahydroquin-oxalin-2-ones.

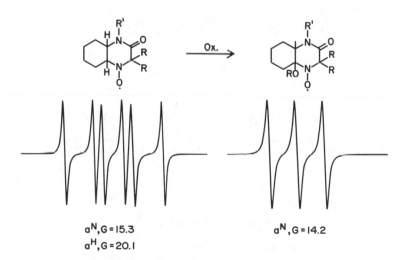

$a^N, G = 15.3$
$a^H, G = 20.1$ $a^N, G = 14.2$

Figure 2. Computed ESR spectra of the primary and secondary nitroxyl radicals of 1-tetradecyl-3,3-dimethylquinoxalin-2-one.

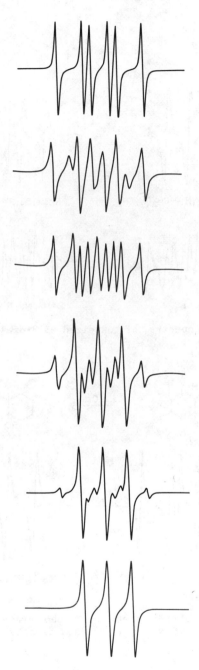

Figure 3.   ESR spectra obtained during the stepwise oxidation of 1-tetradecyl-3,3-dimethyldecahydroquinoxalin-2-one.

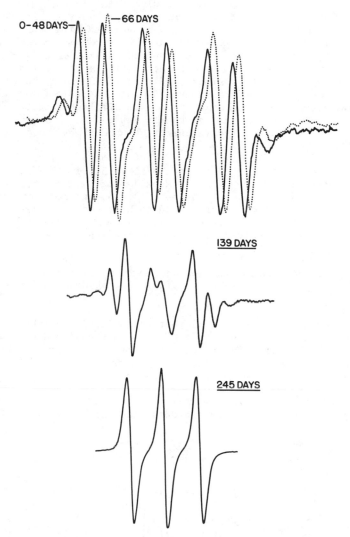

Figure 4. Stability of the primary nitroxyl radical from 1-tetradecyl-3,3-dimethyldecahydroquinoxalin-2-one.

A stable nitroxyl radical is also obtained when polypropylene containing a decahydroquinoxalin-2-one is irradiated in a Weathero-meter (Figure 5).

The structure of this new radical was investigated by FD-MS oxidation studies. We found that the first product formed was the primary nitroxyl radical (loss of one hydrogen atom and gain of one oxygen atom). This was oxidized further to a new radical which had lost two additional hydrogen atoms and added another oxygen atom. No other major chemical species were noted (Table I). Attesting to the ease of oxidation of the primary radical, it should be noted that the primary radical was oxidized even though a considerable amount of unoxidized amine was still present. The simplest structural assign-ment for this radical is given in Table I. Although a radical obtained by an "ene" rearrangement described by Moad is another possibility (6).

FD-MS oxidation studies of N-alkyl-3,3-dialkyldecahydroquin-oxalin-2-ones show that only the N-oxide form and no hydrogen atoms are lost. Similarly, the oxidation of 2,2,6,6-tetramethylpiperidines gives only the nitroxyl radical and no hydrogen atoms are lost. The facts prove that both a nitroxyl radical and hydrogen atoms on the bicyclic bridge are necessary for the oxidation of the hydrogen atoms to occur.

Similar results were obtained with another class of bicyclic partially hindered amines--2,2,4-trialkyldecahydroquinolines (Figure 6).

In summary, we have found that the oxidation of these partially hindered amines occurs to give two stable nitroxyl radicals which can function as chain breaking acceptors. The primary radical formed is easily oxidized and thus could also act as a hydrogen atom donor in route to the second stable nitroxyl racical.

These reactions might account for the stabilizing activity of these amines in polymers and are summarized in Scheme 1.

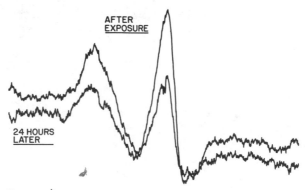

AFTER
EXPOSURE

24 HOURS
LATER

[a]O.I phr   [b]75 mil PLAQUES

Figure 5. ESR spectra of polypropylene containing a 3,3-dialkyl-decahydroquinoxalin-2-one[a] after two days in a Weatherometer[b].

Table I.   FD–MS Oxidation Studies of Spiro[cyclohexane–1,3'–1'–benzyldecahydroquinoxalin–2'–one]

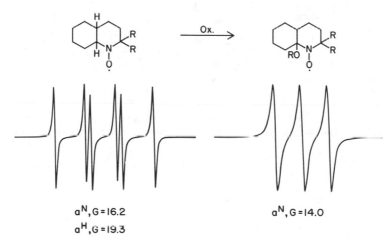

MW 312   MW 327   MW 341

Peroxide Added (μl)	Relative Intensities of Principal Ions*		
	312	327	341
100	10	2	0
200	10	4	0.5
400	10	33	3
800	10	600	2000
1200	10	10	2000

*Principal Ions are Observed as $MH^{+ \cdot}$

$a^N, G = 16.2$

$a^H, G = 19.3$

$a^N, G = 14.0$

Figure 6.   Primary nitroxyl radical from 2,2-dialkyldecahydroquinoline.

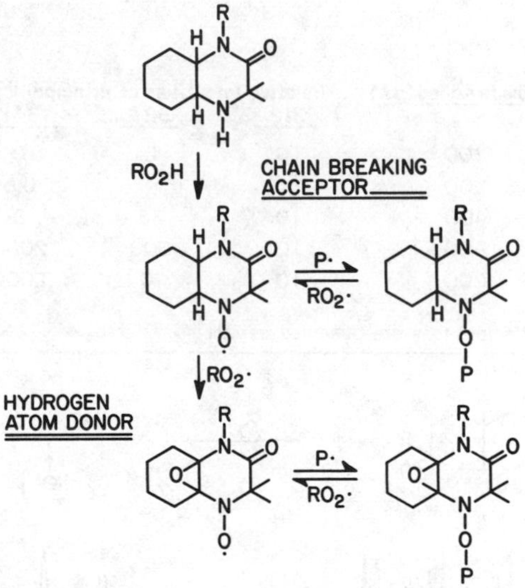

Scheme 1.   Possible mechanism for the antioxidant and light stabilization activity of 3,3-dialkyldecahydroquinoxalin-2-ones.

REFERENCES

1.   R. Bagheri, K. B. Chakraborty and G. Scott, Poly. Degrad. and
      Stab., 4, 1 (1982).
2.   N. S. Allen and A. Parkinson, Polym. Degrad. and Stab., 4, 161
      (1982).
3.   G. M. Coppinger and J. D. Swalen, J. Am. Chem. Soc., 83, 4900
      (1961).
4.   P. N. Son, Poly. Degrad. and Stab., 2, 295 (1980).
5.   Yu. A. Ivanov, A. I. Kokorin, A. V. Shapiro and E. G. Rozantsev,
      Izv. Akad. Nauk. SSSR Ser Khim, 10, 2217 (1976).
6.   G. Moad, E. Rizzardo and D. H. Solomon, Tetrahedron Lett., 1165
      (1981).

RECEIVED December 7, 1984

Electron Spin Resonance Study of Hindered Piperidine Derivatives in Polypropylene Matrix in the Presence of Other Additives

T. KELEN, F. TÜDŐS, G. BÁLINT, and A. ROCKENBAUER

Central Research Institute for Chemistry of the Hungarian Academy of Sciences, Budapest, Hungary

The intensive photostabilizing activity of hindered amine light stabilizers (HALS) is connected with the presence of N-oxyl radicals formed from them during photooxidation. Their ESR signal in polymers, however, is a superposition of the symmetrical triplet of the mobile N-oxyl radical and an asymmetrical signal of this radical being in a restrictive environment. In oxygen-poor atmosphere, the ESR signal of a tertiary hindered amine is somewhat different from that of a secondary one. Concentration changes of N-oxyl radicals and interaction between a HALS compound and a hindered phenol were investigated. It has been stated that the HALS exert an antioxidant effect, but the phenolic antioxidant hinders the photostabilizing effect of the HALS compound. This is presumably due to a direct reaction between the N-oxyl radical and the hindered phenol.

It is well known that HALS compounds (sterically hindered amine light stabilizers) exert excellent photostabilizing effect on polyolefins (1-3). Most of the authors agree that N-oxyl radicals formed from HALS compounds play an important role in the photostabilizing process (4-6).

Polymer systems, however, contain some other additives, e.g. antioxidants to avoid the thermal oxidation during processing, and it would be very interesting to know whether the concentration of the N-oxyl radicals can be influenced by other additives present in the polymer. The formation and nature of N-oxyl radicals, and the influence of a hindered phenolic antioxidant on the concentration of N-oxyl radicals, as well as on the photostabilizing activity of the HALS compound, were investigated.

Experimental

Materials. Experiments were carried out on a polypropylene composite filled with chalk and containing the following additives: a hindered phenolic antioxidant; IRGANOX 1010 (CIBA-GEIGY product) 0.028-0.113

0097–6156/85/0280–0109$06.00/0

mass % of the composite, a secondary hindered amine photostabilizer:
TINUVIN 770 (CIBA-GEIGY product) 0.028-0.337 mass % of the
composite, or a tertiary hindered amine photostabilizer: TINUVIN
622 (CIBA-GEIGY product) 0.05-0.02 mass % of the composite. To study
the effect of the nitroxyl radical which forms from the HALS
compounds in PP during photooxidation, we directly added 1.0 mass %
tetra-methyl-piperidine N-oxyl radical (TMP-NO·) into the PP
(Figure 1). The compound was mixed for 10-40 minutes in a Rheomix
600 chamber of a Rheocord (Haake) equipment at 170-185°C. A film of
about 100 um thickness was pressed from each composition in nitrogen
atmosphere.

Measurements. Carbonyl concentration was measured by means of IR
spectroscopy. N-oxyl radicals were investigated with a JEOL JES-FE
3X ESR spectrometer.

Thermooxidation. Thermooxidation of the films compression molded
from the composite was performed at 165°C by measuring oxygen
absorption.

Photooxidation. Photooxidation was followed by measuring the
concentration of carbonyl groups formed during irradiation of films
with a 500 W high-pressure mercury lamp.

Results and Discussion

Nature of the Radicals Formed from HALS Compounds. Earlier we have
shown (7) that the N-oxyl radicals formed from a secondary amine
(e.g. from TINUVIN 770) give both in powder form and in polypropylene
an asymmetric ESR signal (Figure 2a), which is a superposition of the
symmetrical triplet of the mobile and of the asymmetrical signal
characteristic of the immobile N-oxyl radicals. Increasing the
temperature of the sample - or dissolving the stabilizer in benzene -
causes transformation of the asymmetric component to the symmetric
triplet (Figure 2b). On the contrary, we have proven that the
decrease of the temperature of the solution containing N-oxyl
radicals results in the continuous change of the symmetric signal to
the asymmetric one (7).
     These changes are the consequences of the mobility change of the
free N-oxyl radicals. The same phenomenon has been observed with the
N-oxyl radicals formed from a tertiary amine, e.g. from TINUVIN 622.
     The amount of the N-oxyl radicals in tertiary amine, however, is
much less than that in the secondary one. Another difference between
the N-oxyl radicals formed from the different kind of amines is that
by decreasing the amount of the oxygen above the sample, the signal
of the radicals formed from the tertiary amine splits into secondary
lines (Figure 3), while the ESR spectrum of the radicals formed from
the TINUVIN 770 remains unchanged.
     This phenomenon cannot be interpreted on the basis of the
chemical structures of the different kind of HALS compounds. A
comparison of the structures (Figure 1) shows that the N-C bond
breaks in the tertiary amine type TINUVIN 622 when an N-oxyl radical
is formed. Thus, similar groups can be found in the vicinity of the
N-atoms in both compounds, which practically do not differ in respect
to their influence on the shape of the ESR signal.

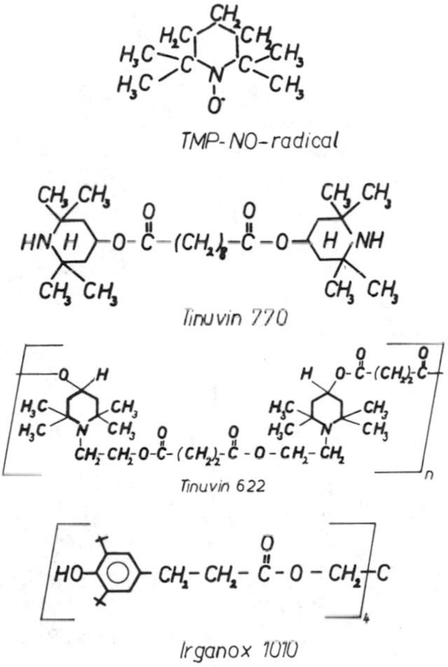

Figure 1. Structure of TMP–NO· radicals, Tinuvin 770, Tinuvin 622, and Irganox 1010.

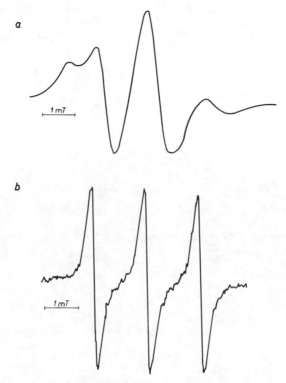

Figure 2. (a) ESR spectrum of N—oxyl radicals formed under
oxygen from Tinuvin 770 or 622 powder. (b) ESR spectrum
of N—oxyl radicals formed under oxygen from Tinuvin 770 or 622 in
benzene solution.

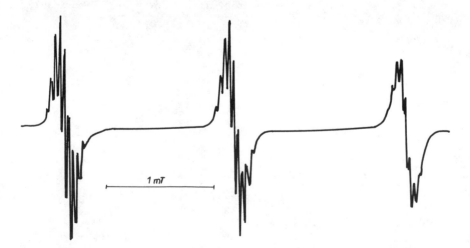

Figure 3. ESR spectrum of the N—oxyl radicals formed from Tinuvin
622 in benzene solution after decreasing the amount of oxygen above
the sample.

Concentration Change of Radicals in the Course of Irradiation. The radical concentration of PP containing 1.0 m % TMP-NO· radical shows a rapid decrease during irradiation with a Xenon lamp (7). This means that the radicals are consumed in the course of irradiation, presumably in reaction with some of the oxidation chain propagating radicals of the polymer.

The initial TINUVIN 770 concentration of the polymer decreases exponentially in the course of irradiation, while the concentration of radicals formed from TINUVIN 770 shows a maximum at the initial period of irradiation and then the curve gradually levels down.

The spontaneous rupture of the samples containing 1.0 m % TINUVIN 770 takes place at about 210 Mlxh (megaluxhours), and during this time no carbonyl formation can be detected (7-8). The concentration of the stabilizing transformation product of TINUVIN 770 (the N-oxyl radical) decreases slowly, even at very high dose: it can be assumed that the N-oxyl radicals may be regenerated from the product formed in the course of the stabilization (9-10).

The pattern of the ESR spectrum taken at room temperature does not alter during irradiation; i.e. the ratio of the two types of radicals (mobile and immobile) does not change, while their concentration changes. This may be attributed to the fact that irradiation is performed at about 70°C, and the distribution of the radicals is practically homogeneous at this temperature (7).

Interaction Between the HALS Compound and the Antioxidant Present in the Polymer System. Thermal oxidation of the films containing different amounts of IRGANOX and TINUVIN 770 as a function of TINUVIN 770 concentration is shown in Figure 4. As expected, the induction period of the oxygen absorption increases with the antioxidant content, but also with the increase of the TINUVIN 770 concentration. That means that the hindered amine can operate during the radical process of the thermal oxidation as a radical scavenger.

The maximal rate of the oxygen absorption is also highly affected by the hindered amine (Figure 5). As is evident from the figure, IRGANOX practically does not influence the curve probably because it is consumed by the end of the induction period. Thus, the decrease in the rate of oxidation must be due to the radical scavenger ability of the hindered amine or its product.

In photooxidation, however, a different effect can be observed which can be demonstrated by plotting the induction period of carbonyl formation as a function of TINUVIN 770 content (Figure 6). The three curves refer to the different amounts of phenolic antioxidant in the polymer. As shown in the figure, the increase of the amine content does not cause a linear increase in the photostability. On the other hand, in the presence of higher IRGANOX concentration, the stability curve runs lower. This points to an antagonistic effect between the amine and the phenolic antioxidant in the course of the photooxidation.

As it is mentioned, N-oxyl radical concentration changes through a maximum as a function of the irradiation time. Higher amounts of phenolic antioxidant result in lower concentration of the N-oxyl radicals (Figure 7). This is in accordance with the above results, namely with the lower photostability of the polymer in the presence of higher amount of antioxidant. This can be explained by presuming an interaction between the HALS, or rather its oxidation product, the N-oxyl radical and the phenolic antioxidant.

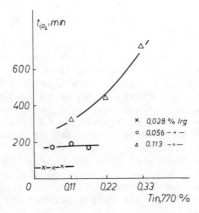

Figure 4. Induction period of oxygen absorption as a function of Tinuvin 770 concentration.

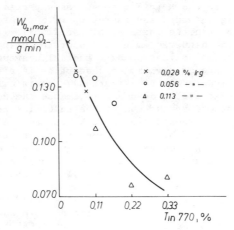

Figure 5. Maximal rate of oxygen absorption as a function of Tinuvin 770 concentration.

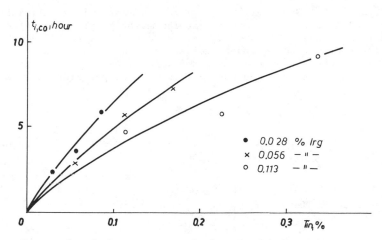

Figure 6. Induction period of carbonyl formation in photooxidation as a function of stabilizer content.

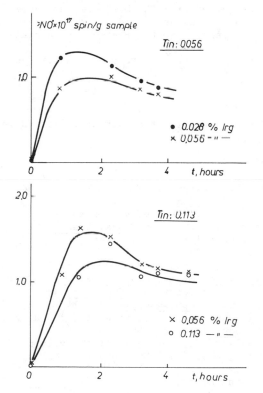

Figure 7. Time dependence of concentration of the N-oxyl radicals in polypropylene samples containing different amounts of Tinuvin 770 and Irganox 1010 during irradiation.

According to Ganem (11), N-oxyl radical can oxidize aliphatic
alcohols to ketones. A similar reaction might be presumed between
N-oxyl radical and IRGANOX 1010 (assumed also by Allen (12)), giving
a resonance-stabilized quinone radical and a hydroxyl amine (Equation
1). This quinone is photoactive, and sensitizes the photooxidation
of the polymer via hydrogen abstraction or hydroperoxide formation.

$$\text{(1)}$$

This direct reaction taking place between the hindered phenol
and N-oxyl radical formed from the HALS compound during processing
might be the reason for the decreased N-oxyl radical concentration
and the antagonistic effect in the presence of a hindered phenolic
antioxidant.

Further work is required to study the validity of the above
presumed reaction.

Acknowledgment

The authors thank M. Iring for the thermal oxidation measurements
and evaluation.

Literature Cited

1.   Shilov, Yu. B.; Denisov, E. T., Vysokomol. Soed. 1974, A 16,
     2313.
2.   Shlyapintokh, V. Ya.; Ivanov, V. B.; Khvostach, O. M.; Shapiro,
     A. B.; Rozantsev, E. G., Dokl. Akad. Nauk SSSR 1975, 225, 1132.
3.   Allen, N. S.; McKellar, J. F.; Wilson, D., Polym. Degr. Stab.
     1979, 1, 205.
4.   Grattan, D. W.; Reddoch, A. H.; Carlsson, D. J.; Wiles, D. M.,
     J. Polym. Sci., Polym. Letters Ed. 1978, 16, 143.
5.   Chakraborty, K. B.; Scott, G., Chem. and Ind. 1978, 237.
6.   Carlsson, D. J., J. Appl. Polym. Sci. 1978, 22, 2217.
7.   Balint, G.; Rockenbauer, A.; Kelen, T.; Tudos, F.; Jokay, L.,
     Polym. Photochem. 1981, 1, 139.
8.   Tudos, F.; Balint, G.; Kelen, T., In "Developments in Polymer
     Stabilization"; Scott, G., Ed.; Appl. Sci. Publ.: London, 1983;
     Vol. 6, p. 121.
9.   Felder, B.; Schumacher, R.; Sitek, F., Chem. and Ind. 1980, 155.
10. Carlsson, D. J.; Garton, A.; Wiles, D. M., In "Developments in
     Polymer Stabilization"; Scott, G., Ed.; Appl. Sci. Publ.:
     London, 1979; Vol. 1, p. 219.

11.   Ganem, B., J. Org. Chem. 1975, 40, 1998.
12.   Allen, N. S., Polym. Photochem. 1981, 1, 234.

RECEIVED December 7, 1984

Electron Spin Resonance Determination of Nitroxide Kinetics in Acrylic/Melamine Coatings

Relationship to Photodegradation and Photostabilization Kinetics

JOHN L. GERLOCK, DAVID R. BAUER, and LINDA M. BRIGGS

Research Staff, Ford Motor Company, Dearborn, MI 48121

Electron spin resonance (ESR) has been used to monitor the kinetics of nitroxide formation and decay during the photodegradation of acrylic/melamine coatings doped with either hindered amine light stabilizer or hindered amine based nitroxide. In the coatings doped with nitroxide, the nitroxide decreases as the nitroxide scavenges radicals produced in the coating. Measurements of the nitroxide decay rate provide a rapid means to determine the photoinitiation rate of radicals by the coating. For these coatings, the photodegradation rate has been found to be proportional to the square root of the initiation rate. The data suggests that measurements of the photoinitiation rate will provide useful information concerning the durability of polymers as long as the photooxidation rate is roughly constant over the service life of the material. The nitroxide kinetics of hindered amine doped coatings are more complex. The nitroxide concentration rises rapidly to a maximum then slowly decays. The slow decay of the nitroxide concentration in the hindered amine doped coatings together with the high rate of initiation demonstrates the importance of nitroxide regeneration as a stabilization mechanism. The effectiveness of the hindered amine stabilizers is a function of the kinetic chain length and of the lifetime of the stabilizer in the coating. The results found here are contrasted with degradation and stabilization studies in other polymers such as polypropylene and polyethylene.

Photodegradation in polymers is a complex chemical process that ultimately leads to a loss of physical properties. Most studies of photodegradation in polymeric coatings have involved measurements of the rates of loss of physical properties such as gloss. Outdoor "real time" exposure is generally regarded as being the most reliable indicator of coating durability (1). Outdoor exposure

0097–6156/85/0280–0119$06.00/0

suffers from the fact that long times can be required to
differentiate between good and very good coatings. Accelerated
tests shorten test times by employing harsher than ambient exposure
conditions (2). In such tests it is impossible to guarantee that
all photodegradation and photostabilization processes are
accelerated to the same extent. Thus it is not surprising that
gloss loss results in accelerated tests do not always correlate with
outdoor exposure results (3). Development of reliable rapid tests
of durability require improved understanding of the chemistry of
degradation and stabilization.

The chemistry of photodegradation in polymers has been the
subject of extensive study (4-9). Most of these studies are based
on studies of free radical oxidation of model compounds in solution
(10-13). The degradation and stabilization of polymers such as
polypropylene and polyethylene have been studied in particular
detail (14-33). In contrast, relatively few studies have been made
of the degradation chemistry of organic coatings (34-35). The
degradation chemistry of acrylic/melamine coatings which are
commonly used as automotive topcoats has only begun to be studied
(36-43). The interest in these materials is due both to the desire
to develop more durable coatings and to regulations lowering solvent
emissions in painting operations which has led to the development of
high solids coatings. The study of degradation chemistry in
crosslinked coatings has required the use of new techniques since
conventional solution techniques cannot be employed. It has been
found that both free radical oxidation and hydrolytic degradation
are important in acrylic/melamine coatings and that the two
processes can be interactive. Recently a technique has been
developed (41,42,44,45) which employs ESR to measure nitroxide decay
rates in coatings doped with nitroxide and photolyzed. From these
experiments it is possible to obtain a rapid and quantitative
measure of the photoinitiation rate in fully crosslinked coatings.
It was found for a series of acrylic/melamine and acrylic/urethane
coatings that the measured photoinitiation rate correlated well with
measures of the durability of these coatings. In this paper further
studies of acrylic/melamine photodegradation and stabilization are
reported. ESR studies of nitroxide kinetics in coatings doped with
persistent nitroxides or hindered amine light stabilizer (HALS)
additives are compared with infrared studies of coating chemical
degradation rates. The goals of this work are to understand the
factors that determine when the nitroxide based measurement of
photoinitiation rate is applicable and to study the mechanisms of
stabilization of coatings by HALS additives. The results obtained
on acrylic/melamine coatings are contrasted with the previous work
on other polymers such as polypropylene and polyethylene.

Experimental

The acrylic copolymers used in this study have been described
elsewhere (40). Number average molecular weights and glass
transition temperatures are listed in Table I. Number average
molecular weights were determined by gel permeation chromatography
(GPC). The GPC was calibrated with a series of fractionated
styrene-acrylic copolymers whose number average molecular weights

were determined by vapor phase osmometry. Glass transition
temperatures were determined by differential scanning calorimetry.
Acrylic/melamine coatings were formulated using a partially
alkylated melamine (Cymel 325, American Cyanamid Co.) in the ratio
70:30 acrylic copolymer:crosslinker. Coating $\underline{H}$ contained a
tetrafunctional oligoester rather than an acrylic copolymer (46).
Coatings were doped with either a hindered amine light stabilizer
(HALS $\underline{I}$) or a persistent nitroxide $\underline{I}$.

HALS $\underline{I}$ Nitroxide $\underline{I}$

HALS $\underline{I}$ and the benzotriazole ultraviolet (UV) light absorber CGL-900
were obtained from the Ciba-Geigy Corporation and recrystallized
before use. The synthesis of nitroxide $\underline{I}$ has been described (44).
Nonpigmented coatings were cast on quartz slides (for the ESR
studies) or on KRS-5 salt plates (for the IR studies) and cured at
130 C for 30 minutes. The coatings were then exposed in a modified
Atlas UV-2 UV weathering chamber. The chamber was modified to allow
air temperature and dew point to be controlled simultaneously.
Unless otherwise noted, the air temperature was held at 60 $\pm$1 C and
the dew point was 25 $\pm$2 C. No condensing humidity cycle was
employed. Samples doped with HALS $\underline{I}$ and the IR samples were mounted
approximately 6 cm away from two FS20 UV-B sunlamps. At various
times, the samples were removed from the chamber and analyzed. IR
spectra were obtained with a Nicolet Fourier transform IR. Samples
were mounted such that the same spot on the plate was always
measured. In this way it was possible to obtain information on
changes in film thickness as well and chemical composition on
degradation. The ESR samples were placed in a special sample holder
and the nitroxide concentration determined using an IBM-Bruker ESR
spectrometer equipped with an Aspect 2000 data system (42).
Nitroxide concentrations could be determined to better than $\pm$ 5%.
Nitroxide concentrations in the samples doped with nitroxide $\underline{I}$ had
to be monitored after short exposure times (of the order of
minutes). This required a further modification to the weathering
chamber which allowed the insertion and removal of samples without
disturbing the chamber environment. All modifications to the
chamber have been described in detail (41). Due to the placement of
the samples (approximately 9 cm below the sunlamps), the light
intensity in the nitroxide doping experiments was roughly one-half
that employed in the HALS doped experiments. Gloss loss studies
were made on TiO_2 pigmented versions of the acrylic/melamine
coatings exposed in a conventional Q-Panel Company QUV weathering
chamber using an exposure cycle consisting of 4 hours UV light at 60
°C followed by 4 hours of condensing humidity at 50 °C.

Results and discussion

Coating Photodegradation: Gloss Loss and Chemistry. On exposure to
UV light, the resin in a coating will begin to photooxidize. In

clear coats, the coating oxidizes throughout the bulk of the resin
ultimately leading to film failure due to cracking or delamination.
In pigmented coatings, resin degradation is limited to the surface
of the coating. As the coating oxidizes resin is lost from the
surface resulting in an increase in surface roughness $<\delta>$ and a
loss of gloss (47,48),

$$\text{Gloss} = 100 \times \exp -(4\pi \cos\Theta < \delta>/\lambda)^2 \tag{1}$$

where λ is the wavelength of light used and Θ is the angle of
reflection. The reduced surface roughness, $<S>$, can be written as

$$<S> = A<\delta> = (\ln(\text{gloss}/100))^{1/2} \tag{2}$$

where A is an instrument constant given by equation 1. Values of
gloss at $\Theta = 20^\circ$ were determined as a function of exposure time for
a series of acrylic/melamine coatings containing 35% by weight of a
coated rutile TiO_2 pigment (41). This pigment has been found to be
photochemically inactive (42). As shown in Figure 1, the reduced
surface roughness increases linearly with time out to a value of
around 1.0 (which corresponds to a gloss of around 35). The
physical rate of photodegradation is given by the slopes in Figure
1. Equivalently, the gloss data can be fit to the following
expression:

$$\text{Gloss}(t) = \text{Gloss}(0) \exp(-K_g t) \tag{3}$$

where K_g is a rate constant for gloss loss which depends on the
composition of the coating, the level of pigment, and the harshness
of the exposure. K_g is proportional to the slope in Figure 1.
Values of K_g are given in Table I.

The relationship between gloss loss and resin loss can be
demonstrated by comparing gloss loss rate constants to rates of
resin loss. Resin loss has been estimated in thin (5 - 10 micron)
unpigmented coating films by following the rate of decrease in
absorption of the C-H stretching band in the IR. As shown in Figure
2 the change in absorbance with time can be given by

$$A(t) = A(0) \exp(-K_e t) \tag{4}$$

where K_e is a rate constant for resin loss (erosion) which depends
on resin composition and exposure harshness. A comparision of K_g
determined for pigmented coatings in a conventional Q-panel QUV
weathering chamber and K_e determined for unpigmented coatings in our
modified weathering chamber are given in Table I. Within
experimental error the rate constants are proportional to each
other.

In these coatings the rate of change of surface roughness is
more or less constant with time over the useful life of the coating.
As shown in Figure 3, the increase in total C=O functionality
relative to C-H functionality is also linear with time implying that
the photooxidation rate is also more or less constant in time.
Other data suggests that linear photooxidation may be the rule

TABLE I. COATING KINETIC DATA

COATING	A	B	C	D	E	F	G	H
M_n	1700	6400	3900	3600	4500	2500	2700	840
T_g	-27	9	-26	18	-9	-11	-13	---
K_g x 10^3 hr^{-1}	5.1	1.4	3.0	2.1	2.8	3.6	3.5	2.2
K_e x 10^3 hr^{-1}	1.5	0.6	0.88	0.63	0.83	1.1	1.25	---
W_i x 10^8 mol/g,min	7.3	0.6	2.0	1.0	1.3	5.1	5.5	0.3
[$NO \cdot$]$_{max}$ x 10^8 mol/g	148	160	134	51	175	146	148	20
t_{max} hrs	25	400	170	800	180	180	200	---
d[$NO \cdot$]/dt$_i$ x10^8mol/g,min	0.31	0.026	0.070	0.0067	0.10	0.054	0.13	<0.001
K_{decay} x 10^4 hr^{-1}	8.0	5.2	5.3	----	7.0	6.0	7.5	---

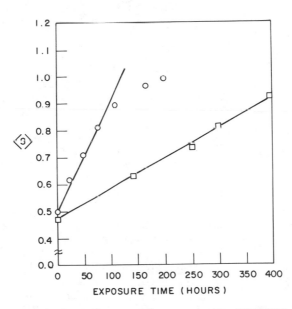

Figure 1. Reduced surface roughness versus exposure time for coating A (◯) and coating B (▢). Data from Ref. 40.

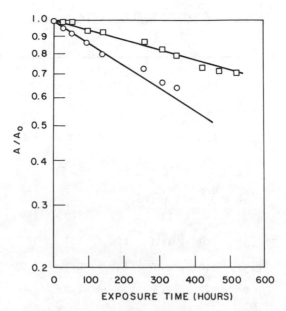

Figure 2.   Absorption of C-H stretching band versus exposure time for coating A (○) and coating B (□).   Samples were exposed in the modified weathering chamber at a dewpoint of 50 C.

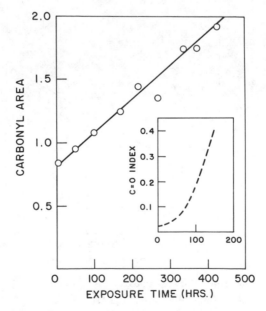

Figure 3.   Area of C=O band relative to C-H band for coating G exposed under the conditions of Figure 2.   Also shown is a similiar plot based on data from Ref. 35 for polypropylene.

rather than the exception in coatings. For example, oxygen uptake measurements have been found to be linear in time (49) as have weight loss data (50,51) on a variety of coatings. Not all polymers oxidize linearly with time. Polypropylene, for example undergoes rapid photooxidation after an induction period, Figure 3 (35). It cannot be expected that any polymer will continue to photooxidise at a constant rate indefinitely. Eventually the polymer will become so different from the starting material that its photodegradation rate must change. It is somewhat surprising that the photodegradation rate in the acrylic/melamine coatings remains roughly constant over the service life of the coating considering the large change in coating composition implied by Figure 3. By the time a pigmented coating has reached a gloss of 20, the carbonyl content on the surface of the coating can be expected to have almost doubled. By contrast less than 1% of the polypropylene groups have to be converted to hydroperoxide groups to cause mechanical failure in this polymer (21).

Photooxidation kinetics are generally interpreted in terms of free radical reactions. A typical scheme is given below:

$$A + h\nu \underset{k_{-i}}{\overset{k_i}{\rightleftharpoons}} A^* \left.\right\} \quad \text{(5)}$$

$$\text{Initiation}$$

$$A^* \xrightarrow{k_1} 2Y\cdot \quad \text{(6)}$$

$$Y\cdot + O_2 \xrightarrow{k_{O_2}} YOO\cdot \left.\right\} \quad \text{(7)}$$

$$\text{Propagation}$$

$$YOO\cdot + YH \xrightarrow{k_p} Y\cdot + YOOH \quad \text{(8)}$$

$$2\ YOO\cdot \xrightarrow{k_t} \text{Products} \qquad \text{Termination} \quad \text{(9)}$$

$$YOOH + h\nu \xrightarrow{k_{cb}} YO\cdot + HO\cdot \qquad \text{Chain branching} \quad \text{(10)}$$

$$YOOH + X \xrightarrow{k_{ncb}} ? \qquad \begin{array}{l}\text{Non-Chain branching} \quad \text{(11)}\\ \text{decomposition of YOOH.}\end{array}$$

The non-chain branching decompositon of YOOH is in accord with Ingold (11) and Denisov (26) who have described reactions of YOOH with Y· and YOO· which yield YOH and YO· In the treatments that follow, it is assumed that only a fraction, ς , of the YOOH species decompose by chain branching. The rate of primary photoinitiation of free radicals, W_i, is given by,

$$W_i = \frac{2k_1 \, k_i \, h\nu \, [A]}{k_{-i} + k_1} \qquad (12)$$

In the scheme above, linear photooxidation requires that $d[YOOH]/dt = 0$. With this assumption, and assuming steady state kinetics in both $Y\cdot$ and $YOO\cdot$ it can be shown (42) that the overall rate of photooxidation, W, is given by,

$$W = W_i + k_p \, [YH] \; \frac{\epsilon\, k_p \, [YH] + \sqrt{\epsilon^2 k_p^2 [YH]^2 + 2k_t \, W_i}}{2k_t} \qquad (13)$$

Equation 13 assumes that photooxidation is the sum of the primary photoinitiation event and the rate of hydrogen atom abstraction from the resin. If   is small and if the kinetic chain length (given by W/W_i) is at least 3, then to a good approximation, the rate of photooxidation is given by,

$$W = k_p \, [YH] \sqrt{(\, W_i/2k_t \,)} \qquad (14)$$

Resin erosion rates as measured by IR have been found to be proportional to the square root of the light intensity, Figure 4. A possible mechanism explaining the increase in photooxidation rate with humidity in acrylic/melamine coatings will be presented below.

Nitroxide I Decay Kinetics: Measurement of W_i. Since the rate of photooxidation exhibited by acrylic/melamine coatings is nearly constant, it should be possible to develop a rapid test of coating durability based on measurements of the initial photooxidation rate. However, it has been found that the rate of photooxidation under ambient conditions is slow and that long exposure times are required to obtain significant spectral changes. It has been shown that it is possible to measure the rate of photoinitiation rapidly and accurately under near ambient exposure conditions by measuring the decay rate of nitroxide concentration in coatings doped with a persistent nitroxide such as nitroxide I. The details of the technique have been described elsewhere (41,42,44,45) and are only briefly summarized here. The technique takes advantage of the fact that nitroxides react rapidly with alkyl, macroalkyl, and acyl radicals to yield amino ethers. The kinetics are complicated by the fact that nitroxides are photolabile. The excited state of nitroxide I can abstract a hydrogen to yield a hydroxyl amine and an alkyl radical (52,53). Assuming that the nitroxide concentration is sufficiently high to insure that the amount of chain branching is much less than W_i and that the reaction rate of nitroxide with radicals is much greater than the self termination rate, it can be shown that the initial rate of consumption of nitroxide I is given by (42,45)

$$-d[\rangle NO\cdot]/dt = C[\rangle NO\cdot] + W_i \qquad (15)$$

where C is a measure of the rate of hydrogen atom abstraction by

excited state nitroxide. Initial rates of nitroxide consumption are
measured to minimize complications from reactions of amino ethers
and hydroxyl amines to regenerate nitroxide. Photoinitiation rates
are determined by extrapolating the nitroxide decay rate measured as
a function of initial nitroxide concentration to zero nitroxide
concentration. Values of W_i for the series of acrylic/melamine
coatings studied here are given in Table $\underline{I}$ (41). The values of W_i
vary widely. The trends with polymer composition and molecular
weight are discussed elsewhere (41).

Within experimental error, it is found that the gloss loss
rates for these coatings are proportional to the square root of the
photoinitiation rate in agreement with equation 14. This, together
with the dependence of W_i on light intensity suggests that the
coating with the largest value of W_i (i.e., the one with the
shortest chain length) has a chain length of at least 3. It is
possible to estimate the chain length by comparing the
photoinitiation rate to measures of the photooxidation rate. For
example, using the measured increase in C=O and assuming that W_i is
constant over the course of the exposure yields a value for the
chain length for coating $\underline{G}$ to be around 5. Such an estimate is at
best qualitative but is does suggest that though the kinetic chain
lengths in thses coatings are small, they are long enough to make
the use of equation 14 valid. By contrast, typical chain lengths
for polyethylene are around 10 (16) and for polypropylene are around
100 (21).

A key assumption in using the nitroxide doping scheme to
measure W_i is that the nitroxide is sufficiently high in
concentration to control chain branching and termination. This
assumption must fail at low nitroxide concentration. When the
nitroxide concentration becomes insufficient to control chain
branching and termination, the nitroxide decay rate differs from
that predicted by equation 15. If a significant fraction of the
radicals formed undergo self termination, the nitroxide decay rate
would be reduced from that predicted by equation 15. If, on the
other hand, chain branching occurred, the nitroxide decay rate would
be higher than that predicted by equation 15 since more radicals
would be formed. To determine over what range in concentration
equation 15 is valid, the nitroxide decay rate was determined for
coating $\underline{G}$ at low light intensity over the concentration range of
nitroxide $\underline{I}$ of $15 - 600 \times 10^{-8}$ moles/g (45). No evidence for either
chain branching or self termination was observed in these coatings
at nitroxide concentrations of greater than 50×10^{-8} moles/g (45).
At concentrations of 50×10^{-8} moles/g and below, it was found that
the nitroxide decay rate was greater than that predicted by equation
15 indicative of the onset of chain branching. Wiles and coworkers
(21) have calculated that the minimum concentration of nitroxide
necessary to control chain branching in polypropylene is $150 - 500 \times 10^{-8}$ moles/g. A general expression for the nitroxide decay rate can
be derived by adding the reaction of nitroxide $\underline{I}$ with Y$\cdot$ to
reactions 6 - 11 (42). This expression was found to give a good fit
to the experimental decay rates at low nitroxide concentration.
From the parameters derived from the fit and estimates of the
relative reactivity of oxygen and nitroxide with Y$\cdot$ (12) it can be

concluded that for acrylic/melamine coatings, the fraction of
hydroperoxides decomposing by chain branching (ϵ) is small (0.01 -
0.05). Over this range of values of ϵ it can also be concluded from
equation 14 that the photooxidation rate will be roughly
proportional to the square root of the photoinitiation rate as long
as the chain length is between 3 - 30. The low value of ϵ together
with the relatively short chain lengths implies that photooxidation
due to chain branching by hydroperoxides is relatively unimportant
in acrylic/melamine coatings. These studies suggest that a
measurement of the photoinitiation rate will provide useful
information about the durability of a polymer as long as the
photooxidation rate is more or less constant over the service life
of the material. One requirement for this to be true is that the
hydroperoxide chain branching be relatively unimportant to the
degradation kinetics.

HALS Based Nitroxide Kinetics and Photostabilization. Both UV
absorbers and HALS additives are commonly added to coatings to
reduce the rate of photooxidation. As is shown in Figure 4, the
addition of 2% by weight of CGL-900 or HALS-I to coating G reduces
the rate of resin loss by roughly a factor of 2. The rate of resin
loss is reduced by even a greater amount when 1% by weight of both
CGL-900 and HALS-I are added to the coating. In both cases the rate
of photooxidation is roughly constant in time. In some cases in
HALS-I doped coatings it has been observed that after extensive
exposure, the rate of degradation abruptly increases, possibly due
to depletion of stablilizer. Some but not necessarily all of the
reduction of degradation rate by the addition of UV absorber is due
to the reduction of the initiation rate by screening. Studies of
other possible mechanisms of stabilization such as energy transfer
are in progress. The effect of HALS-I on the photooxidation
kinetics of coating G is similiar to that observed in polyethylene
(20) and a butadiene nitrile rubber (54). The effect of HALS-I on
the photooxidation kinetics in polypropylene is much different.
Instead of simply reducing the rate of photooxidation, the addition
of stabilizers to polypropylene has the effect of prolonging the
induction period, Figure 5, as well as reducing the rate of
oxidation after the induction period. The greater degree of
effectiveness of HALS-I in polypropylene is due to the much longer
chain length in polypropylene than in either polyethylene or the
acrylic/melamine coatings (21).

HALS additives function by interfering with the propagation of
free radicals via the following general reactions:

$$\text{>NH} + \text{YOO·} \longrightarrow \text{>NO·} \tag{16}$$

$$\text{>NO·} + \text{Y·} \longrightarrow \text{>NOY} \tag{17}$$

$$\text{>NOY} + \text{YOO·} \longrightarrow \text{>NO·} + \text{YOOY} \tag{18}$$

Through these reactions the HALS additive and it byproducts reduce
the steady-state concentration of Y· and YOO· thus reducing kinetic
chain length. The longer the initial chain length, the larger is
the effect of adding HALS.

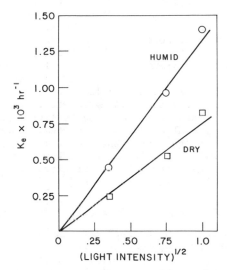

Figure 4. K_e versus the square root of the light intensity for coating G. The dewpoints were 50 C ($\bigcirc$) and −40 C ($\square$).

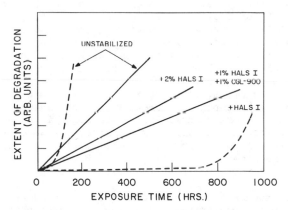

Figure 5. Extent of photodegradation versus exposure time for stabilized and unstabilized coating G (——) and polypropylene (---) showing the effect of hindered amine and UV absorbers on the photodegradation kinetics.

One method to follow the stabilization kinetics of HALS
additives is to follow the nitroxide concentration as a function of
time of exposure. As shown in Figure 6, the nitroxide concentration
in HALS-I doped coatings rises more or less rapidly to a maximum and
then slowly decays. Despite significant differences in
photooxidation chemistry, similiar behavior is observed in HALS
doped polypropylene samples (31) though the maximum nitroxide
concentration observed is roughly 10 times lower than observed for
the acrylic/melamine coatings. In the initial part of the exposure,
up to about 30% of the maximum nitroxide level reached in the
coating, the increase in nitroxide with time is found to be roughly
linear (Figure 7). It is also found that there is some nitroxide
formation during cure of these coatings suggesting that there is a
small amount of free radical thermal degradation during cure. The
amount of nitroxide formation during cure seems to be greatest for
the least durable coatings. From the slopes in Figure 7, the
initial rate of formation of nitroxide, $(d[\text{>NO·}]/dt)_i$, can be
measured. At long times, the decrease in nitroxide concentration
has been found to fit the following expression:

$$[\text{>NO·}](t) = [\text{>NO·}]_{max} \exp(-K_{decay}(t - t_{max})) \qquad (19)$$

The nitroxide kinetics of HALS doped coatings are a complex function
of formation, decay, and regeneration and to date no analytical
expression has been derived which completely predicts the nitroxide
behavior. It is possible to describe the kinetic behavior through
parameters such as the maximum nitroxide concentration ($[\text{>NO·}]_{max}$),
the time to maximum (t_{max}), and the above mentioned nitroxide
formation and decay rates. Values of these parameters are given in
Table I. Except for Coatings D and H, all values of $[\text{>NO·}]_{max}$ are
around 150 x 10^{-8} moles/g and no obvious trends with coating
composition are observed. Values of K_{decay} are also nearly
independent of coating composition. Higher rates of nitroxide
formation and shorter values of t_{max} generally correspond to
coatings that have higher rates of gloss loss though there are
exceptions. For example, coatings E and F have the same value of
t_{max} but have significantly different gloss loss rates. Also the
initial rate of nitroxide formation in Coating F is lower than in
Coating E even though the rate of gloss loss is greater in Coating
F. The time to maximum nitroxide and the initial rate of formation
of nitroxide are slowest in Coatings D and H even though the rate of
gloss loss in Coating B is lower than either D or H.

The exact mechanisms for the conversion of hindered amine to
nitroxide are not known in these coatings. According to the
treatment above, the initial rate of formation of nitroxide should
be proportional to [YOO·]. From the scheme of equation 5, the
overall rate of photooxidation should also be proportional to
[YOO·]. Even though there appears to be a qualitative correlation
between the initial nitroxide formation rate and the rate of
degradation, the initial formation rate is hardly proportional to
either K_g or K_e. Apparently nitroxide formation from HALS-I is more
complex than indicated by the simple scheme above. Compared to the
excellent correlation found between photodegradation rates and
photoinitiation rates in these coatings, no single parameter in the

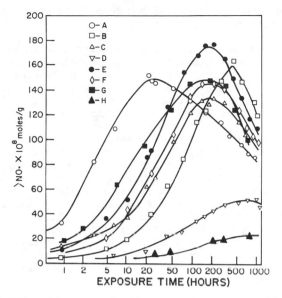

Figure 6. Nitroxide concentration versus exposure time in acrylic/melamine coatings A-H. Coatings were doped with 1% by weight HALS-I and exposed in the modified weather chamber at a dewpoint of 25 C.

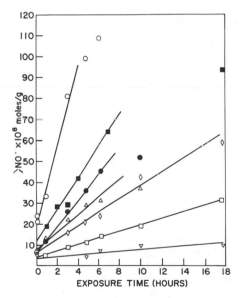

Figure 7. Initial buildup of nitroxide on exposure of HALS-I doped coatings. Samples and exposure conditions are given in Figure 6.

HALS-$\underline{I}$ based nitroxide kinetics has been found which correlates with photodegradation rate.

Despite the lack of direct correlation between the hindered amine based nitroxide kinetics and the photodegradation rates of the acrylic/melamine coatings, the hindered amine based nitroxide kinetics have been found to be very sensitive to exposure conditions (38). In particular, it has been found that humidity increases the rate for formation of nitroxide from HALS-$\underline{I}$ doped coatings (38) even though it does not affect the photoinitiation rate (42). It has been suggested that hydrolysis is primarily responsible for this effect (38). During hydrolysis of acrylic/melamine coatings, formaldehyde is released into the coating. Formaldehyde is easily oxidised to a peracid which can increase the rate of photooxidation at constant initiation rate as observed in Figure $\underline{3}$. Peracids are also known to be able to rapidly convert hindered amines to nitroxide (55). The role of formaldehyde in HALS based nitroxide kinetics is currently being investigated by comparing acrylic/melamine coatings to acrylic/urethane coatings which release no formaldehyde on cure or degradation. It should also be noted that HALS-$\underline{I}$, in addiiton to reducing the photooxidation rate, also reduces the rate of hydrolysis observed in acrylic/melamine coatings.

The high rate of photoinitiation of radicals in acrylic/melamine coatings together with the long lifetime of nitroxide in the HALS-$\underline{I}$ doped coatings confirms the key role that nitroxide regeneration (reaction 18) plays in the effectiveness of HALS additives. Without nitroxide regeneration, the hindered amine would be rapidly consumed and the nitroxide would be completely converted to amino ether. The initial rate of formation of nitroxide in the HALS-$\underline{I}$ doped coatings is much smaller than W_i indicating that either few of the radicals formed convert hindered amine to nitroxide or that some of the nitroxide immediately reacts with alkyl radicals to form amino ethers. Preliminary measurements of the loss of HALS-$\underline{I}$ itself from the coating during exposure (43) suggests that both processes are occurring. The initial rate of loss of HALS-$\underline{I}$ from coating $\underline{G}$ is smaller that W_i but is greater than the initial formation rate of nitroxide. The rate of conversion of HALS-$\underline{I}$ to nitroxide does not seem large enough relative to W_i to greatly reduce the photodegradation rate. The average concentration of nitroxide is large enough in most of the coatings to reduce the photodegradation rate. However, it should be noted that the addition of HALS-$\underline{I}$ to coating $\underline{H}$ reduces the photodegradation rate even though the maximum nitroxide concentration is only 20 x 10^{-8} moles/g. In polypropylene it has also been suggested that the nitroxide concentration observed is insufficient to account for the degree of stabilization observed (21). To account for the apparent high efficiency of HALS based additives in this polymer it has been suggested that HALS preferentially associates with hydroperoxides, the known major chromophore in polypropylene (30). It is not known at this point whether or not similiar associations exist in the very polar acrylic/melamine coatings studied here.

The measurements of the concentration of HALS-$\underline{I}$ in the coating as a function of exposure time show that the hindered amine is essentially consumed long before the nitroxide concentration has decreased significantly from its maximum. This means that for most of the exposure time stabilization by HALS-I involves reactions 17 and 18 but not 16. The length of time that the additive is effective thus depends not on the lifetime of the HALS itself but rather depends on the value of K_{decay}. The value of K_{decay} does not depend greatly on coating composition. It is sensitive to exposure conditions, however. It has been found that K_{decay} is proportional to light intensity. The mechanism whereby nitroxide or amino ethers are lost from the coating is unknown at present. Understanding the factors that control K_{decay} could lead to substantial improvements in effectiveness of HALS additives. This behavior may also explain the synergism that is often observed between HALS additives and UV absorbers. UV absorbers reduce the photoinitiation rate by reducing the light intensity. Reducing the photoinitiation rate increases the kinetic chain length thus increasing the effectiveness of the HALS additive. Reducing the light intensity also reduces the value of K_{decay} thus increasing the period of time that the photostabilizer is effective.

Conclusion

The kinetics of nitroxide formation and decay have been studied by ESR in acrylic/melamine coatings doped with persistent nitroxides or hindered amine light stabilizers. These kinetics together with measurements of the rates of chemical and physical changes occurring in the coating have been used to study the photodegradation and photostabilization chemistry of these coatings. It has been found that photodegradation rate as measured by gloss loss of pigmented coatings is constant with time but depends strongly on the nature of the acrylic copolymer used in the coating. Measurement of the the nitroxide decay rate in coatings doped with a hindered amine based nitroxide provide a convenient and rapid means to determine the photoinitiation rate in these coatings. It has been found that the photodegradation rate is simply proportional to the square root of the photoinitiation rate. Comparison of infrared spectroscopic measures of the photooxidation rate with the photoinitiation rate yield relatively short kinetic chain lengths (5 - 15). The data also suggest that hydroperoxide chain branching is unimportant in acrylic/melamine coatings. Use of the nitroxide based assay of photoinitiation rate is predicted to correlate with long term measurements of durability as long as the photooxidation rates are roughly constant over the service life of the material and the kinetic chain lengths are reasonably short (< 30). The addition of hindered amine light stabilizers were found to reduce the rate of photodegradation in these coatings. Comparision of the photoinitiation rate and various nitroxide kinetic parameters in the hindered amine doped coatings confirmed the importance of nitroxide recycling to the stabilization of these coatings. The exact mechanism of conversion of hindered amine to nitroxide and of nitroxide regeneration are not yet known in these coatings. Further measurements of the kinetics of HALS doped coatings as a function of exposure conditions are in progress.

Literature Cited

1.   Ellinger, M. L.   Prog. Org. Coat. 1977, 5, 21.
2.   Grossman, G. W.   J. Coat. Techno. 1977, 49 (633), 45.
3.   Papenroth, W.   DEFAZET 1974, 28, 284.
4.   Ranby, B.; Rabek, J. F.   "Photodegradation, Photooxidation,
     and Photostabilization of Polymers"; John Wiley and Sons: New
     York, 1975.
5.   Scott, G.   "Developments in Polymer Stabilization" Vols. 1-5;
     Applied Science: London.
6.   Labana, S. S.   "Ultraviolet Light Induced Reactions in
     Polymers"; ACS Symposium Series No. 25, American Chemical
     Society: Washington D.C., 1976.
7.   Emanuel, N. M.   Polym. Eng. Sci. 1980, 20, 662.
8.   Carlsson, D. J.; Wiles, D. M.   Rubb. Chem. Tech. 1974, 47,
     991.
9.   Scott, G.   J. Photochem. 1984, 25, 83.
10. Emanuel, N. M.; Denisov, E. T.; Maizus, Z. K. "Liquid Phase
     Oxidation of Hydrocarbons"; Plenum Press: New York, 1967.
11. Ingold, K. U.   Chem. Rev. 1961, 61, 563.
12. Brownlie, I. T.; Ingold, K. U.   Can. J. Chem. 1967, 45,
     2427.
13. Ingold, K. U.   Accts. Chem. Res. 1969, 2, 1.
14. Carlsson, D. J.; Wiles, D. M.   Macromolecules 1969, 2, 597.
15. Niki, E.; Decker, C.; Mayo, F. R.   J. Polym. Sci.
     Polym. Chem. Ed. 1973, 11, 2813.
16. Decker, C.; Mayo, F. R.; Richardson, H. R.   ibid 1973, 11,
     2879.
17. Shilov, Y. B.; Denisov, E. T.   Polym. Sci. USSR 1974, 16,
     2009.
18. Shilov, Y. B.; Denisov, E. T.   Polym. Sci. USSR 1978, 20,
     2079.
19. Chakraborty, K. B.; Scott, G.   Euro. Polym. J. 1977, 13,
     1007.
20. Chakraborty, K. B.; Scott, G.   Chem. Ind. 1978, 237.
21. Carlsson, D. J.; Garton, A.; Wiles, D. M. In "Developments
     in Polymer Stabilization - 1"; Scott, G. Ed.; Applied Sci.:
     London, 1979; p. 219.
22. Carlsson, D. J.; Grattan, D. W.; Wiles, D. M.   Org. Coat.
     Plast. Chem. 1978, 39, 628.
23. Allen, N. S; McKellar, J. F.   J. App. Polym. Sci. 1978, 22,
     3277.
24. Carlsson, D. J.; Grattan, D. W.; Suprunchuk, T.; Wiles,
     D. M.   J. App. Polym. Sci. 1978, 22, 2217.
25. Kiryushkin, S. G.; Shlyapnikov, Y. A.   Polym. Sci. USSR
     1980, 22, 1310.
26. Denisov, E. T. In "Developments in Polymer Stabilization - 5";
     Scott, G. Ed.; Applied Science: London, 1982; p. 23.
27. Krisyuk, B. E.; Popov, A. A.; Griva, A. P.; Denisov, E. T.
     Dokl. Phys. Chem. 1983, 269, 156.
28. Allen, N. S.   Polym. Deg. Stab. 1980, 2, 129.
29. Carlsson, D. J.; Chan, K. H.; Garton, A.; Wiles, D. M. Pure
     and Appl. Chem. 1980 52, 389.
30. Chan, K. H.; Carlsson, D. J.; Wiles, D. M.   J. Polym. Sci.
     Polym. Lett. 1980, 18, 607.

31. Carlsson, D. J.; Chan, K. H.; Wiles, D. M.   J. Polym. Sci. Polym. Lett. 1981, 19, 549.
32. Carlsson, D. J.; Chan, K. H.; Durmis, J.; Wiles, D. M. J. Polym. Sci. Polym. Chem. Ed. 1982, 20, 575.
33. Kurumada, T.; Ohsawa, H.; Fujita, T.; Toda, T. J. Polym. Sci. Polym. Chem. Ed. 1984, 22, 277.
34. Krejcar, E.; Kolar, O.   Prog. Org. Coat. 1972, 1, 249.
35. Morimoto, K.; Lida, T.   ibid 1973, 2, 35.
36. Killgoar, P. C., Jr.; van Oene, H. In "Ultraviolet Light Induced Reactions of Polymers"; Labana, S. S. Ed.; ACS Symposium Series No. 25 American Chemical Society: Washington D.C., 1976; p. 407.
37. Bauer, D. R.   J. Appl. Polym. Sci. 1982, 27, 3651.
38. Gerlock, J. L.; van Oene, H.; Bauer, D. R.   Euro. Polym. J. 1983, 19, 11.
39. English, A. D.; Spinelli, H. J. In "Characterization of Highly Crosslinked Polymers"; Labana, S. S.; Dickie, R. A., Eds.; ACS Symposium Series 243, American Chemical Society: Washington DC, 1984; p. 257.
40. Bauer, D. R.; Briggs, L. M. ibid., p. 271.
41. Gerlock, J. L.; Bauer, D. R.; Briggs, L. M. ibid., p. 285.
42. Gerlock, J. L.; Bauer, D. R.; Briggs, L. M.; Dickie, R. A. J. Coat. Techno. Submitted.
43. Gerlock, J. L.; Bauer, D. R.; Briggs, L. M.   Polymer Preprints 1984, 25, 30.
44. Gerlock, J. L.   Anal. Chem. 1983, 54, 1529.
45. Gerlock, J. L.; Bauer, D. R.   J. Polym. Sci. Polym. Lett. Ed. 1984, 22, 477.
46. Chattha, M. S.; Cassatta, J. C.   J. Coat. Technol. 1983, 55 (700), 39.
47. Beckman, P.; Spizzichino, A. "The Scattering of Electromagnetic Waves from Rough Surfaces"; Pergamon Press Ltd.: Oxford, 1963; Chapter 5.
48. Simpson, L. A.   Progr. Org. Coat. 1978, 6, 1.
49. Berner, G.; Kreibach, U. T. In "Sixth International Conference in Organic Coatings Science and Technology"; Parfitt, G. D.; Patsis, A. V. Eds.; Technomic Pub.: Westport, CT, 1982; p. 334.
50. Kaiser, W. D.   Korrosion (Dresden) 1976, 7, 33.
51. Finzel, W. A.   J. Coat. Techno. 1980, 52 (660), 55.
52. Bogatryeva A. I.; Buchachenko, A. L.   Kinetics and Catalysis 1971, 12, 1226.
53. Keana, J. F. W.; Dinerstein, R.; Baitis, F.   J. Org. Chem. 1971, 36, 209.
54. Shlypintokh, V. Y.; Ivanov V. B. In "Developments in Polymer Stabilization -5"; Scott, G. Ed.; Applied Sci.: London, 1982; p. 41.
55. Rozantzev, E. G. "Free Nitroxyl Radicals"; Plenum Press: New York, 1970; Chap. 9.

RECEIVED October 26, 1984

Stabilization of Polypropylene Multifilaments
Utility of Oligomeric Hindered Amine Light Stabilizers

ROBERT J. TUCKER and PETER V. SUSI

Polymer Products Division, American Cyanamid Company, Bridgewater, NJ 08807

The light stabilization of polypropylene (PP) fibers places special requirements on the stabilization system due to the high surface to volume ratio, the severe processing conditions, and the post spinning treatments. Earlier systems provided only moderate performance lifetimes thus limiting the markets for PP fibers. The development of oligomeric hindered amine light stabilizers (HALS) has led to PP fibers with greatly increased service lifetimes, thus opening up new applications for this fiber. Some of the factors leading to the enhanced performance obtainable with certain newer oligomeric HALS are described and evaluated. Also, the importance of processing conditions and interactions of HALS with other additives on the light stability of the resulting fiber are discussed.

Because of polypropylene's unusual properties, such as light weight and ease of fabrication, it has been used to make a variety of fabrics and is expected to be the growth fiber of of the eighties (1,2). Adding impetus to this expectation was the development of hindered amine light stabilizers (HALS) (3) that enable polypropylene (PP) fibers to penetrate new markets. The processes for producing these fibers range from conventional melt spinning for continuous filament and staple through heavy denier monofilaments produced by extrusion into water, with additional large volumes produced from film by slitting or splitting. Even direct production of fabrics from polymer by spun-bonded processes is possible with polypropylene. The deniers of fibers produced by the varied techniques cover a wide range varying from micro deniers produced by melt blowing processes to the heavy deniers used in carpet backings, sacks, bags and rope or cordage (2). Essential to the continued success for this polymer in the fiber market is the ongoing effort to continually enhance the stability of the products produced especially towards oxygen, heat and light.

0097–6156/85/0280–0137$06.00/0

While the early HALS gave considerably greater light stability
to PP fibers than classical light stabilizers, they still failed to
perform well in low denier fibers. These place special requirements
on the stabilizer system due to the high surface area, the severe
processing conditions used, and the use of various colorants and
post spinning treatments such as tentering, laundering, and dry
cleaning. More recently, oligomeric type HALS have been found to
provide the best balance of properties for most applications. The
products attain their improved structural properties without a
substantial reduction in hindered amine content, thus retaining a
high specific activity. Data are presented showing the superior
performance of certain oligomeric HALS in PP multifilaments.

Experimental

As used herein, yarn denier is the number of grams per 9000 meters.
Yarn tenacity is the tensile stress expressed as force per unit
linear density of the unstrained specimen, in grams-force per denier
(gf/den.). The majority of tests were carried out on multifilament
yarns prepared from Hercules PRO-FAX 6401 polypropylene powder. To
the base polymer was added 0.05% calcium stearate and 0.1%
1,3,5-tris(3,5-di-t-butyl-4-hydroxybenzyl)isocyanurate (processing
stabilizer). Additives were dry blended into the powder and the
resulting blends extruded at 227°C and pelletized. The pellets were
spun into yarns, using a NRM extruder at 280°C with a 30 hole die,
and the yarns were then drawn at a 6:1 ratio in two stages and given
a 2Z twist. The yarns (240/30 denier) were woven into test strips
(40 yarns per inch) and used for exposure studies. Accelerated
light stability studies were carried out in an Atlas Xenon Arc
Weather-Ometer (WOM) with a 6500 Watt burner. Operating conditions
were 30% relative humidity, 44°C ambient temperature and a black
panel temperature of 65 ± 3°C. For the GM-WOM test, an Atlas twin
globe enclosed carbon arc unit was used with a 3.8 hour light cycle
and 1 hour water mist cycle. The ambient temperature was 72°C with
a black panel temperature of 89° ± 3°C during the light cycle.
Tentering was simulated by heating the yarns at 120°C for 20 minutes
in a circulating air oven. In the laundering test, yarns were
machine washed with detergent and dried three times. For the dry
cleaning tests, yarns were commercially dry cleaned three times. In
all studies, the failure point was a 50% loss in original breaking
strength of the yarns as measured by Instron tensile property
measurements. HALS structures are shown in Figure 1.

Results and Discussion

Most hindered amine light stabilizers have evolved from the
discovery (4) that compounds containing a 2,2,6,6-
tetramethylpiperidine moiety can stabilize polymers against
photodegradation and this moiety has been incorporated into HALS of
various types (Figure 1). Much has been published on the mechanism
of action of HALS and the literature in this area has recently been
critically reviewed (5,6). While the complete mechanism of action
has not been fully elucidated, the high performance of HALS is
generally attributed to the ability of their oxidation products to

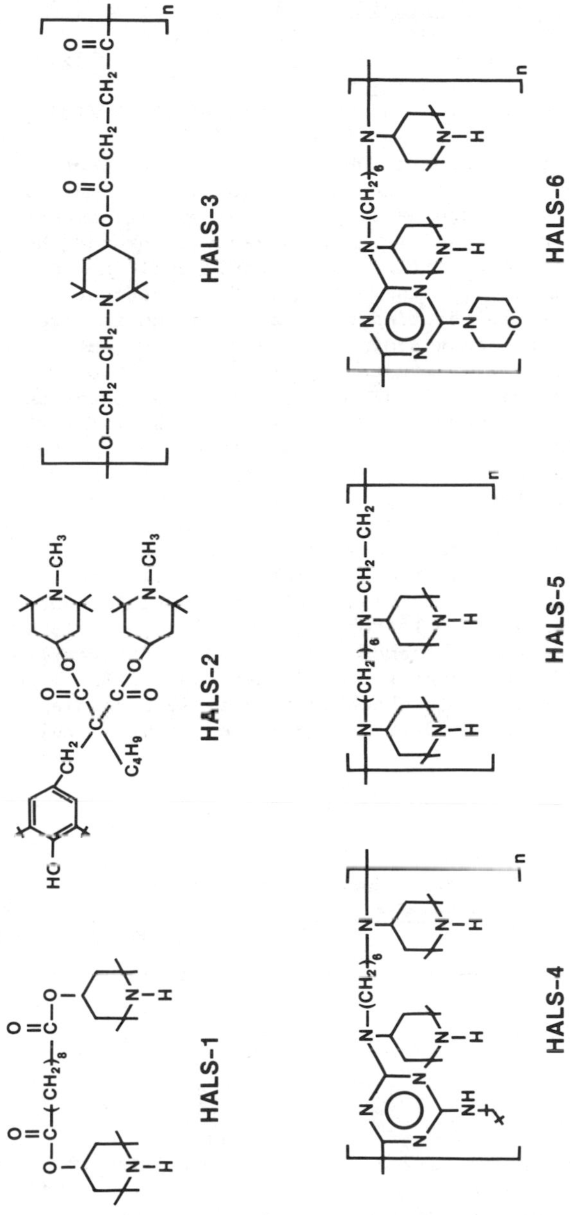

Figure 1. Structures of HALS.

act as radical scavengers in a cyclic self-perpetuating fashion as shown in Equations 1 through 3 (7,8).

$$\overset{\diagdown}{\underset{\diagup}{N}}-H \quad \underrightarrow{\text{oxidation}} \quad \overset{\diagdown}{\underset{\diagup}{N}}-O\cdot \qquad (1)$$

$$\overset{\diagdown}{\underset{\diagup}{N}}-O\cdot \ + \ R\cdot \ \longrightarrow \ \overset{\diagdown}{\underset{\diagup}{N}}-O-R \qquad (2)$$

$$\overset{\diagdown}{\underset{\diagup}{N}}-O-R + \ ROO\cdot \ \longrightarrow \ \overset{\diagdown}{\underset{\diagup}{N}}-O\cdot + \ ROOR \quad (3)$$

Recently, most efforts in the HALS area have centered around obtaining the best stabilization at the lowest cost in a variety of demanding applications such as polypropylene multifilaments. While the stabilizing activity of HALS is centered around the hindered piperidine nitrogen, the rest of the molecule still has an influence on overall performance. It is felt that HALS concentrate in the amorphous area of polyolefins where degradation is more likely to occur due to increased oxygen diffusion and a lack of crystalline order. Some chemical structures may be more suitable or sterically more favorable than others in allowing closer association of the HALS with potential damage sites. This "compatibility" can be a very important HALS attribute and an important factor in the effectiveness of a HALS over the life of a polymer substrate.

The stabilization of polypropylene yarns is a demanding application because of the extremely high surface area (9) and intimate exposure of the filaments to oxygen and light throughout their very thin cross section. Table I shows the effect of thickness on the thermal stability of polypropylene containing, as the antioxidant system, 0.1% tetrakis[methylene(3,5-di-tert-butyl-4-hydroxyhydrocinnamate)]methane and 0.3% distearylthiodipropionate. At 150°C, an 8 denier per filament yarn fails much sooner than a 0.4 mil thick film which in turn is less stable than a 4.0 mil film demonstrating the dramatic effect of sample thickness on thermal stability.

Table I. PP Thermal Stability versus Thickness

Sample Thickness	Hours to Embrittlement at 150°C
4.0 mil Film	>500
0.4 mil Film	200
240/30 Denier Yarn (8 dpf)	25

In Table II, the effect of filament diameter on the light stability of polypropylene yarn, containing 0.1% octadecyl 3,5-di-tert-butyl-4-hydroxyhydrocinnamate as the the antioxidant, is shown. Failure was the time to 50% original breaking strength. The lower denier fiber showed a greatly reduced lifetime due to its much smaller cross section, and thus increased exposure to light and oxygen.

Table II.   PP Light Stability versus Thickness

Denier per Filament	Filament Diameter	Hours to Failure Xenon WOM
146	6.0 mil	750
8	1.4 mils	200

The severe processing conditions used to produce low denier fibers require a stabilizer with very good thermal stability and low volatility. Volatility of the stabilizer can also be an important factor in the effectiveness of a HALS under end-use conditions. The volatility of several commercial HALS, as measured by thermogravimetric analysis (TGA), is shown in Table III. The oligomeric HALS (3-6) have the best thermogravimetric profile, showing low product loss at high temperatures, and should survive polymer processing better than HALS 1 and 2, as well as remain in the high surface area fibers over time.

Table III.   Relative Volatility of HALS by TGA[a]

	Temperature ($^{\circ}$C) at X% Weight Loss		
	T_5	T_{10}	T_{20}
HALS 1	236	251	267
HALS 2	275	291	305
HALS 3	318	329	339
HALS 4	331	351	380
HALS 5	277	301	325
HALS 6	344	371	401

[a]Heating Rate 10°C/minute in Air

Another important attribute of a HALS is its effect on the thermo-oxidative stability of polypropylene multifilaments. This property is important in certain end-use applications where elevated temperature over a period of time is experienced, such as in automobile rear shelf fabrics. The excellent performance of the oligomeric HALS 4 and 6, as determined by 120°C oven aging, is shown in Figure 2. The superior activity of these products may be due not only to their low volatility but also to the presence of a triazine moiety in the structure, which appears to have a positive effect on the thermo-oxidative stability of polypropylene.

The GM-WOM is a high temperature (72 $^{\circ}$C), high humidity accelerated weathering test, specified by General Motors, for fibers and plastics for automotive interior applications. In this unit, the oligomeric HALS 4 and 6 gave the best performance, although HALS 2 was also very effective (Figure 3).

Oligomeric HALS generally outperform other types when a thermal treatment, such as a tentering or a latexing operation, is performed on the yarns. As shown in Figure 4, in simulated tentered yarns (heated at 120 $^{\circ}$C for 20 minutes), the oligomeric HALS 3,4 and 6 showed the best performance in the Xenon WOM.

Another important consideration in selecting a light stabilizer system for polypropylene yarns is the resistance to activity loss after laundering or dry cleaning. In Figure 5 data are presented

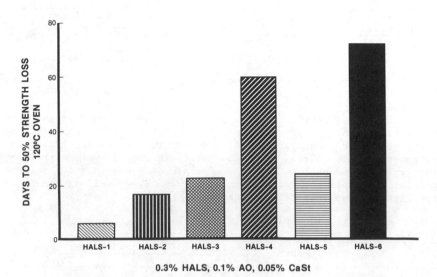

Figure 2. Thermal stabilizing activity in 240/30 denier polypropylene yarn (yarn tenacities 5.2 ± 4%).

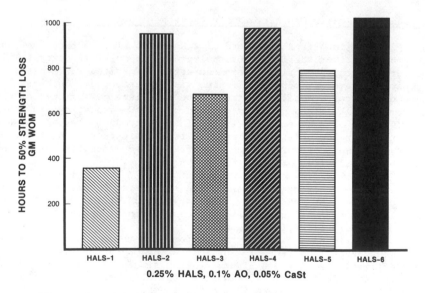

Figure 3. Light stabilizing activity in 240/30 denier polypropylene yarn (yarn tenacities 4.7 ± 4%).

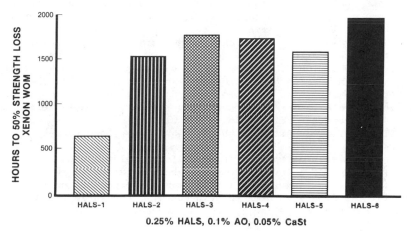

Figure 4. Light stabilizing activity in 240/30 denier polypropylene yarn (simulated tentering).

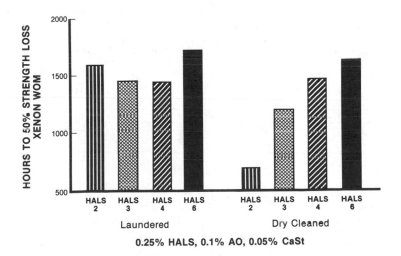

Figure 5. Laundering and dry cleaning effects on 240/30 denier polypropylene yarns.

showing the effect of these operations on yarns containing various HALS as measured by Xenon-WOM exposure. While HALS 2 showed good activity after laundering, it lost most of its activity after the dry cleaning treatment. The oligomeric HALS 6 outperformed all the other HALS evaluated, demonstrating good resistance to extraction from the yarns by a hot aqueous detergent solution or by organic solvents used in dry cleaning.

The results of a concentration versus light stabilizing activity study with two oligomeric HALS (4 and 6) are shown in Figure 6. In the GM-WOM unit, HALS 6 shows a better activity response to increasing concentration than does HALS 4. At the lower concentration levels of 0.15% to 0.3%, HALS 6 shows an almost linear activity increase. The differences observed are probably related to subtle structural differences resulting in altered polymer compatibility. This effect with HALS 4 has also been seen by other workers (10).

Another area of concern in stabilizing polypropylene fibers is the development of color due to the stabilizer system. While most HALS are colorless and impart little or no color on production, "gas yellowing" of fibers in use can be a concern. In Table IV, data are shown on the gas yellowing resistance of several HALS at 0.5% concentration in natural polypropylene multifilament. The yarns were exposed using a modified AATCC 23-1972 test for 1 cycle and then evaluated using a gray scale comparator with a 5.0 rating indicating no change, and a 1.0 rating denoting severe color change. With 0.1% phenolic antioxidant present, HALS 2 and HALS 6 showed a barely perceptible color development, while HALS 4 discolored to a greater extent. In the absence of the phenolic antioxidant, HALS 4 still showed a noticeable discoloration, while the yarns containing HALS 6 showed no color development, indicating that discoloration in the presence of the phenolic antioxidant was due to the latter and not to HALS 6 itself.

Table IV.   Gas Yellowing Resistance in 240/30 Denier PP Yarns[a]

	% Phenolic A.O.	Rating[b]
HALS 2	0.1	4.5
HALS 4	0.1	4.0
HALS 6	0.1	4.5
HALS 4	0	4.0
HALS 6	0	5.0

[a]0.5% HALS
[b]Modified AATCC 23-1972 test - 1 cycle; 5.0 = no color change; 1.0 = severe color change

Polypropylene fibers are often pigmented and the pigments used can influence the light stability of the system (11). Some improve stability, some are neutral, while others are deterimenal due to their prodegradative tendencies or to pigment-stabilizer interactions (12). A comparison of two oligomeric HALS (4 and 6) in blue and red pigmented yarns is shown in Table V. As can be seen, even structurally similar products show different relative stabilization effectiveness with different pigments.

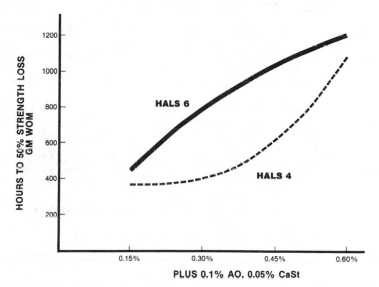

Figure 6. Concentration effects in 240/30 denier polypropylene yarns.

Table V. Stabilization of Pigmented PP Multiflament[a]

| Stabilizer | % Concentration | Hours to Failure GM WOM | |
		Blue	Red
HALS 4	0.2	390	---
	0.4	620	---
	0.5	---	370
	0.8	775	---
HALS 6	0.2	565	---
	0.4	855	---
	0.5	---	375
	0.8	1230	---

[a]600/34 Denier blue yarn; 300/70 denier red yarn

Finally, the importance of using optimum processing conditions is illustrated in Table VI. Here similar formulations were examined but the spin pack temperature used in the preparation of the yarns was incresed. Both the oligomeric HALS 4 and 6 gave yarns with improved light stability when processed at the higher temperature.

Table VI. Effect of Processing Conditions on Yarn Stability [a]

| Stabilizer | Processing Temperature | Hours to Failure | | |
		GM WOM	Xenon WOM NT	ST
HALS 4	265°C	390	785	675
	280°C	980	1878	1772
HALS 6	265°C	475	840	710
	280°C	1067	2440	1950

[a]240/30 Denier yarns (0.25% HALS); NT = non-tentered; ST = simulated

Summary

In polypropylene multifilaments, oligomeric hindered amine light stabilizers have been found to offer superior light stabilizing activity. Properties such as good thermal stability, low volatility, compatibility, and extraction resistance have been shown to be important factors favoring the use of oligomeric HALS in polypropylene fiber applications. Differences in molecular structure among the oligomeric HALS are probably responsible for the best overall stabilizing activity observed with HALS 4 and 6.

Acknowledgements

The authors wish to acknowledge the contribution made to these studies by co-workers and thank the management of American Cyanamid Company for permission to publish the work and Miss J. C. Williams for typing the manuscript.

Literature Cited

1. Blore, J. H. Knitting Times November 6 1978, p. 24.
2. Polypropylene Fiber Symposium, New York, N.Y., 1981
3. Patel, A. R.; Usilton, J. J. In "Stabilization and Degradation of Polymers"; Allara, D. L.; Hawkins, W. L., Eds.; ADVANCES IN CHEMISTRY SERIES No. 169, American Chemical Society: Washington, D. C., 1978; p. 116.
4. Sankyo Co. Ltd. British Patent 1 196 224, 1970.
5. Sedlar, J.; Marchal, J.; Petruj, J. Polymer Photochemistry 1982, 2, 175.
6. Dagonneau, M. et al. Rev. Macromol. Chem. Phys. 1982, C22 (2), 169.
7. Carlsson, D. J.; Grattan, D. W.; Wiles, D. M. Organic Coatings and Plastics Chemistry 1980, 39, 628.
8. Durmis, J. et al. J. Polym. Sci., Polym. Lett. Ed. 1981, 19, 549.
9. Carlsson, D. J.; Wiles, D. M. J. Macromol. Sci., Rev. Macromol. Chem. 1976, C14, 65.
10. Gugumus, F.; Linhart, H. Chemicke Vlakna 1982, 32, 94.
11. Klemchuk, P. P. Polymer Photochemistry 1963, 3, 1.
12. Uzlmeier, C. SPE Journal 1970, 26, 69.

RECEIVED October 26, 1984

Hexahydropyrimidines as Hindered Amine Light Stabilizers

C. E. RAMEY and C. J. ROSTEK

Chemical Division, Ferro Corporation, Bedford, OH 44146

Novel hindered amine light stabilizers (HALS) derived
from 2,2,4,4,6-pentamethylhexahydropyrimidine were
shown to exhibit excellent performance in exposed polypropylene films. Test data for other HALS compounds
prepared from the previously described 2,2,5,5-tetra-
methyl-4-imidazolidinone and 4,4-dimethyloxazolidine
ring systems are provided for comparison. Inferior
performance was generally observed for those additives
which would be expected to form low molecular weight
nitroxide radicals upon oxidation. A relatively large
extention of film lifetime was produced by formulations
containing a combination of HALS and commercial hydroxybenzoate stabilizer. This effect was not evident when
HALS compounds containing an intramolecualr hydroxybenzoate group were tested.

Hindered amine light stabilizers are at least partially converted to
their corresponding nitroxides during the processing (1) and exposure
of stabilized polymers. Allen (2) showed that the hydroperoxides
present in thermally oxidized polypropylene were capable of effecting
this transformation under compression molding conditions. Hindered
amine nitroxides have been recognized as key intermediates in the
stabilization mechanism of HALS compounds (1, 3-5). Although the
parent hindered amines may contribute to photostabilization (6), in
our development work the stability and nature of the nitroxide radicals resulting from HALS oxidation were assumed to be critical to
viable stabilizer activity. At the outset of our HALS program a
number of chemically stable nitroxide free radicals had been identified from work in spin labelling (7), but only derivatives of tetramethylpiperidine had appeared in the additive marketplace.

0097-6156/85/0280-0149$06.00/0
© 1985 American Chemical Society

We wanted to investigate how the effectiveness of a given light stabilizer might depend upon the particular heterocyclic amine used in its synthesis. The polarity, volatility and compatibility of the original additive were also known to be important factors in determining stabilizer performance (8). Three heterocyclic ring systems were chosen primarily for their ease of synthesis. The compounds 2,2,4,4,6-pentamethylhexahydropyrimidine (HHP) (9,10) and 2,2,5,5,-tetramethyl-4-imidazolidinone (IMZ) (11-13) were chosen as starting materials to evaluate the effect of derivativization and substitution on the activity of the resulting light stabilizers. A few 4,4,-dimethyloxazolidines were made from 2-amino-2-methyl-1-propanol and ketones to examine this structurally limited ring system (see Results and Discussion).

HHP   IMZ   4,4-Dimethyloxazolidine

Experimental

The additives were tested in polypropylene (Profax 6501, Hercules) films containing 0.1% Goodrite 3114 (phenolic antioxidant of B.F. Goodrich Co.) and .05% calcium stearate. The candidate additives were added at 0.25% or 0.5% final concentrations as methylene chloride solutions (100 ml methylene chloride/100g polypropylene) and the mixture stirred 15-20 mins. while the solvent was allowed to evaporate. The resultant powder was then dried and extruded into 3/32 inch strand, which was cut into pellets. After drying, the pellets were extruded into a broad (8") band. A 1/4" section was slit from the band and oriented by drawing at 175°F at a 7:1 draw ratio. The dimensions of the oriented film are about 1 x 80 mils.

The oriented film specimens were mounted on aluminum frames and exposed on an Atlas Weather-Ometer, Model 65WR. An 18 minute spray cycle together with an 102 minute cycle at 55% relative humidity and approximately 65°C was used. At regular intervals, the test specimens were removed from exposure and their tensile strength measured on an Instron Model 1102. A decrease in tensile strength, expressed as tenacity, over the tensile strength of the same formulation before exposure, is a measure of the deterioration of the physical properties of the polymer. "Failure" in this test is defined as a loss of 50% or more of the initial sample tenacity.

Photomicrographs to observe additive compatibility in the oriented film were taken at 400X using a transmission light microscope. Blooming was measured at 70°C using unoriented film samples 7 mil in thickness. Thermal oxidative stability was measured in a circulating air oven at 140°C on polypropylene films 5" x 1" x .025".

Results & Discussion

The compounds in Table I represent some early trials arising from screening IMZ derivatives and oxazolidines. The lower activity shown by compounds (1) and (2) is consistent with the oxazolidine group being lost as a low molecular weight fragment after nitroxide formation occurs. This illustrates a drawback of the oxazolidine compounds compared to other systems, that only the amine nitrogen and hindering group (derived from the starting ketone) are available to build the molecular weight of the stabilizer. A similar effect is apparently shown by compound (5) which is bridged through the hindered amine nitrogens, allowing the imidazolidinone group form low molecular weight nitroxide radicals. Compound (3) is a low molecular weight IMZ derivative which showed a long film lifetime at 0.5% concentration. However, this activity was not reflected in the higher molecular weight analog (4) which was tested at a lower concentration. Here the methyleneamino group is probably being lost from the 3-position of the IMZ group due to thermal instability. This is confirmed by the high thermogravimetric analysis (TGA) weight loss of the ethylenediamine derived model compound (16).

In Table II are shown some intermediate candidates where better activity is beginning to become apparent. The substitution of the hydroxybenzoate group in compound (8) does not seem to make much improvement in the activity of the long side-chain IMZ derivatives. The polymer lifetime of the HHP based additive (9) was encouraging, as the earlier failure of diisocyanate adduct (10) was attributed to its lower compatibility in the polypropylene film.

In Table III, the positive interaction between IMZ stearate (6) and the hydroxybenzoate stabilizer UV-Chek AM-340 is shown. This degree of lifetime enhancement was not achieved in any of the HALS compounds synthesized in which the hydroxybenzoate group was present as an intramolecular substituent. The proximity of the hindered phenolic group may in some way affect generation of the nitroxide radicals from the amine nitrogen.

The substitution of a hydroxybenzoate group for a stearate group seemed to have a detrimental result in comparing compounds (14) and (15) in Table IV. Incorporation of a more conventional phenolic antioxidant group in (13) may have suppressed activity even more. However, the effect of physical factors on the stabilizing activity of these additives should not be underestimated. The IMZ hydroxybenzoate compound (12) has good activity possibly due to its physical properties. Also in Table IV, bis HHP sebacamide (11) shows commercially viable stabilizing capabilities. The difunctional type of structure of compound (11) and Tinuvin 770 (bis(2,2,6,6-tetramethyl-4-piperidinyl)decanedioate) may be a factor in optimizing the activity of these additives.

Table I. Initial Compounds

Conc %	Additive	Compatibility	w/o Lifetime
0.25	(1)	G	850 hours
0.25	(2)	G	900
0.5	(3)	G	2820
0.25	(4)	G	890
0.5	(5)	G	1320

Compatibility: G = Good, F = Fair

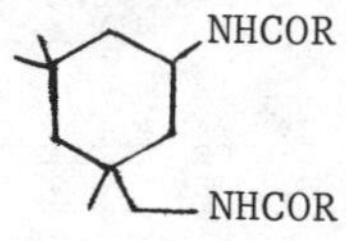

(1)

$R-C(=O)-NHC_{18}H_{37}$

(2)

R =

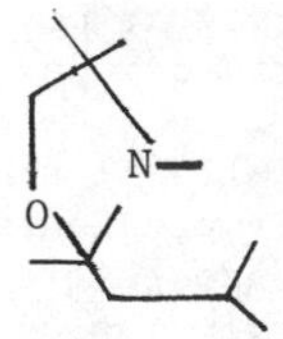

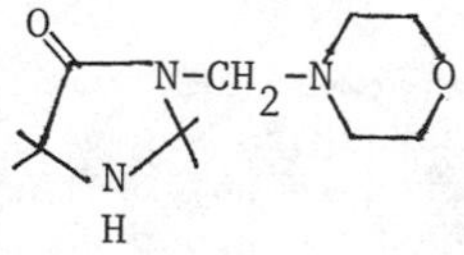

(3)

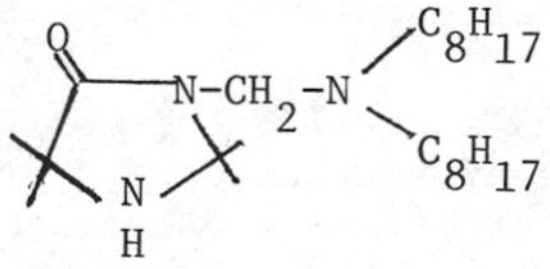

(4)

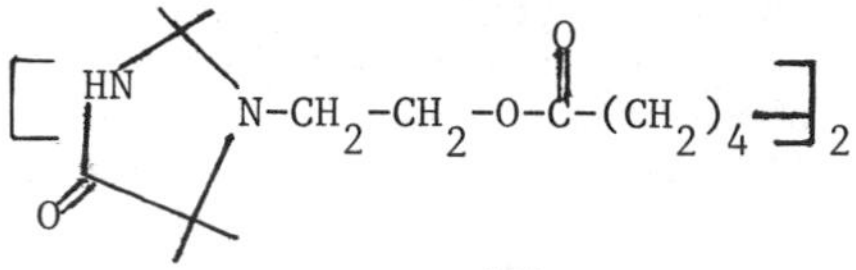

(5)

Table II. Intermediate Compounds

Conc %	Additive	Compatibility	w/o Lifetime
0.25	(6)	G-	1950 hours
0.25	(7)	G	1970
0.25	(8)	G	1980
0.25	(9)	G	2210
0.25	(10)	F	1790

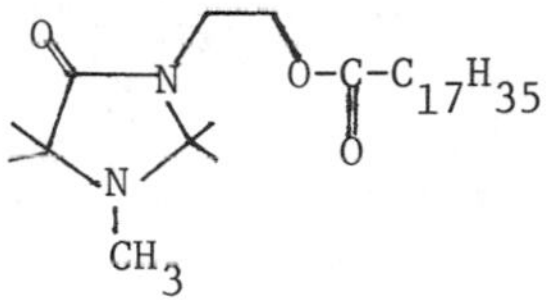

(6)

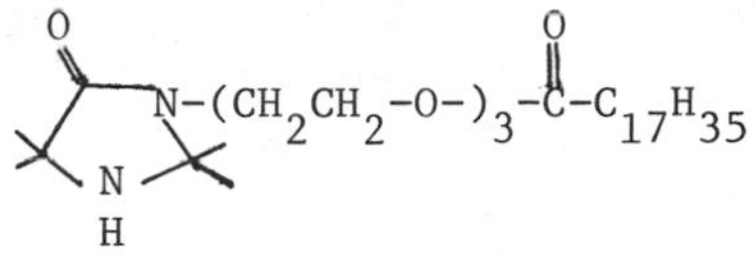

(7)

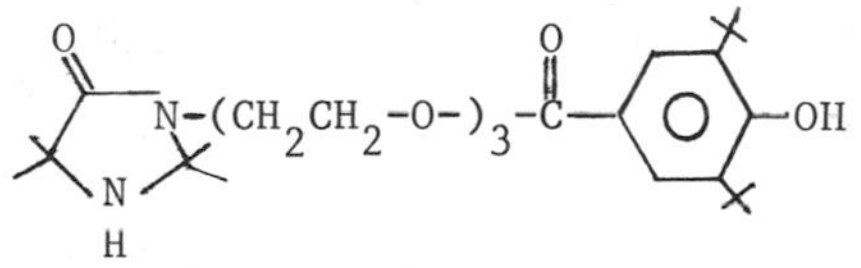

(8)

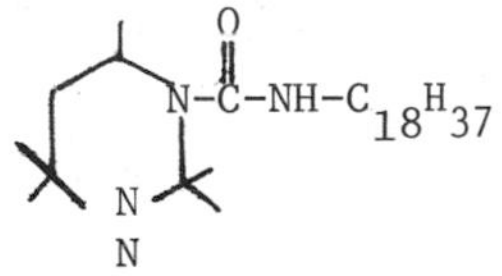

(9)

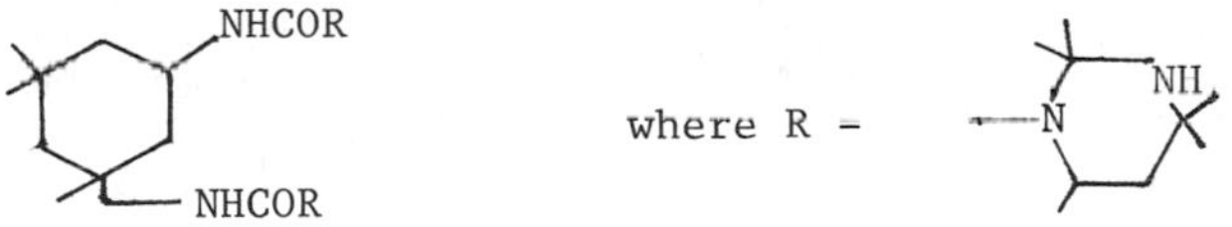

(10)

where R –

Table III. Hydroxybenzoate-HALS Interaction

Conc %	Compound	w/o Lifetime
0.25	Chimassorb 994*	1970 hours
0.25	Tinuvin 622*	1750
0.25	IMZ (6)	1950
0.25	AM-340	1350
0.25	AM-340 + 0.25 IMZ (6)	2550
0.25	Tinuvin 770*	2460

*All are commercial HALS of Ciba-Geigy Corporation
UV-Chek AM-340 = Hydroxybenzoate stabilizer of Ferro Corporation

Table IV. Final Compounds

Conc %	Additive	Compatibility	w/o Lifetime
0.25	(11)	G	2920 hours
0.25	(12)	G	2410
0.25	(13)	G	1450
0.25	(14)	G	2750
0.25	(15)	G	1950

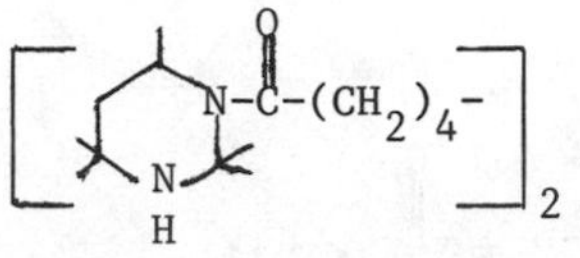

(11)

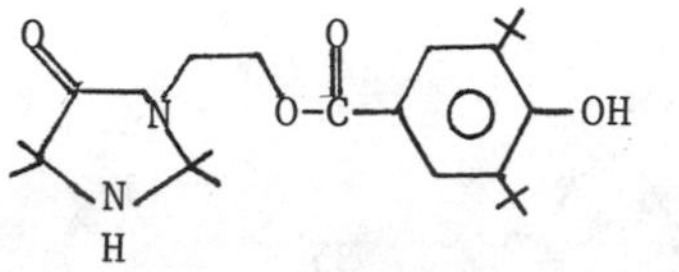

(12)

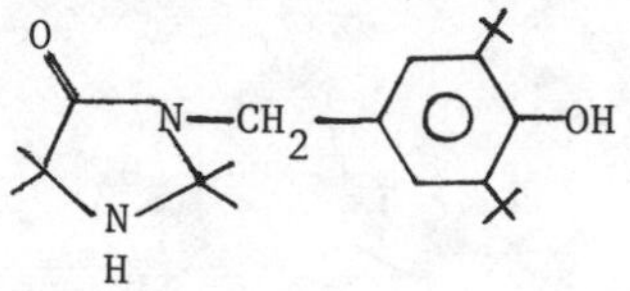

(13)

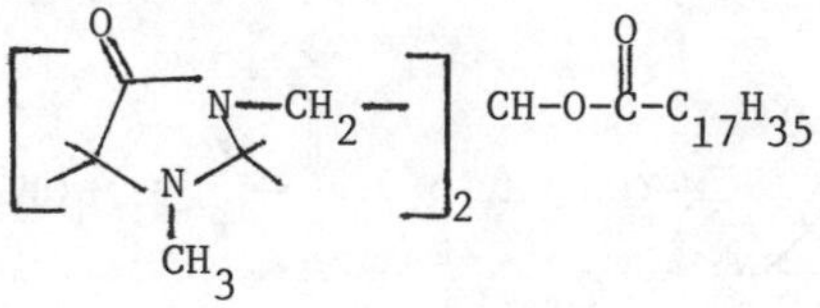

(14)

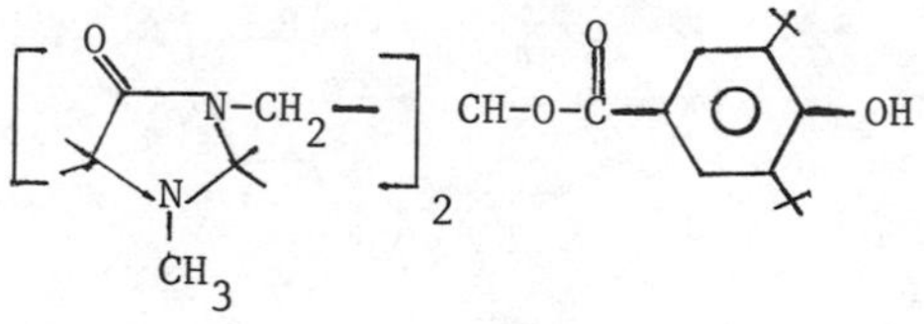

(15)

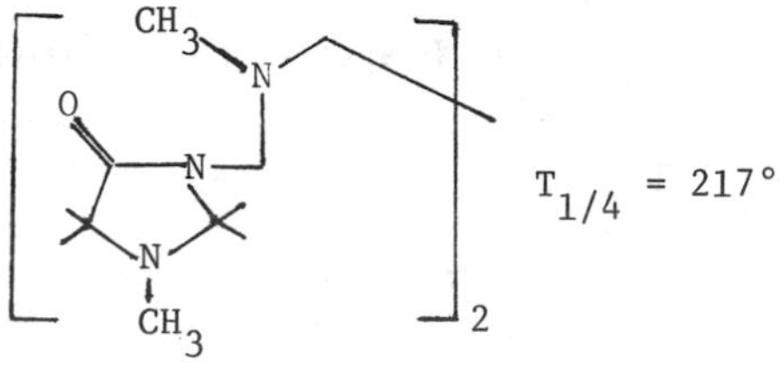

$T_{1/4} = 217°$

(16)

Acknowledgments

The authors wish to acknowledge the work of Dr. Goutam Gupta, who was the first to prepare HHP sebacamide and who also prepared the oxazolidines. We wish to thank, also, Dr. Ronald E. Thompson, who prepared the higher IMZ compounds, and Walter J. Wawro, Sr., who worked out the preparation of HHP by catalytic reduction.

Literature Cited

1. Bagheri, R; Chakraborty, K.B.; Scott, G. Polym. Degrad. and Stab. 1982, 4, 1.

2. Allen, N.S. Polym. Degrad. and Stab. 1980, 2, 179.

3. Grattan, D.W.; Reddoch, A.H.; Carlsson, D.J.; Wiles, D.M. J. Polym. Sci., Polym. Lett. Ed. 1978, 16, 143.

4. Hodgeman, D.K.C. J. Polym. Sci., Polym. Chem. Ed. 1980, 18, 533.

5. Carlsson, D.J.; Chan, K.H.; Wiles, D.M. J. Polym. Sci., Polym. Lett. Ed. 1981, 19, 549.

6. Carlsson, D.J.; Chan, K.H.; Dumas, J.; Wiles, D.M. J.Polm. Sci., Polym. Chem. Ed. 1982, 20, 575.

7. Keana, J.D.W. Chem. Reviews 1978, 78(1), 37.

8. Bailey, D; Vogl, O. J. Macromol. Sci-Rev. Macromol. Chem. 1976, C14(2), 267.

9. Gupta, G.; Ramey, C.E. U.S. Patent 4, 404, 302, 1983.

10. Matter, E. Helv. Chim. Acta 1947, 30, 1114-23.

11. Ramey, C.E.; Rostek, C.J. U.S. Patent 4, 448, 969, 1984.

12. Murayama, K.; Morimura, S.; Toda, T.; Akagi, S; Kurumada, T.; Watanabe, I. Ger. Offen. 1, 817, 703, 1969.

13. Ferrini, P.; Marxer, A. Helv. Chim Acta 1963, Vol. XLVI, Fasc IV, 1207-1212.

RECEIVED October 26, 1984

Mechanisms of Aromatic Amine Antidegradants

JAN POSPISIL

Institute of Macromolecular Chemistry, Czechoslovak Academy of Sciences, 162 06 Prague 6, Czechoslovakia

Nitroxides and benzoquinonediimines are formed from aromatic amines and diamines respectively as a consequence of amine involvement in antioxidant and/or antiozonant processes. Their participation in antioxidant regenerative mechanisms is suggested. Features of phenylenediamine involvement in antiozonant processes are discussed in relation to contemporary theories.

Secondary aromatic amines are effective antioxidants in the protection of saturated hydrocarbon polymers (polyolefins) against autooxidation. Their role in the stabilization of unsaturated hydrocarbon polymers (rubbers) is more complex: depending on their structure, they impart protection against autooxidation, metal catalyzed oxidation, flex-cracking, and ozonation. The understanding of antioxidant, antiflex-cracking and antiozonant processes together with involved mechanistic relations are of both scientific and economic interest.

Rubber stabilizers are generally very reactive organic compounds. Research of their reactivity under the influence of deteriogens involved in rubber weathering indicates stabilizer transformation and formation of some different classes of products. Data on their structure and reactivity under the influence of various deteriogens and/or reactive intermediates arising in oxidized and ozonized rubber are of importance in the elucidation of the individual pathways of amine protection mechanisms. Because of the extreme reactivity of many of amine transformation products observed even during their analyses, all mechanistic conclusions have to be made very carefully: the specific property of a product may be incorrectly ascribed to another compound formed by consecutive transformations instead of to the originally arising structure. A specific role may be played by the acidity of some deteriogens or impurities and processing additives.

Antioxidant Properties. Participation of secondary amines in autooxidation processes involves, in particular, reactions with free-

0097–6156/85/0280–0157$06.00/0

radicals and hydroperoxides generated in the autooxidized substrate
and with ground and singlet state molecular oxugen (1). Intermediate
formation of N-centered radicals (aminyls) is operative in these
reactions. Aminyls may react in mesomeric iminocyclohexadienyl forms
(C-centered radicals) and are transformed into products created via
recombination, coupling, disproportionation, and oxidation processes.
It is of importance to distinguish between the reactivity of
secondary monoamines (only diarylamines are considered as effective
antioxidants) and that of bifunctional N,N'-disubstituted 1,4-
phenylenediamine (PD). Two principal types of products are formed
important from the point of view of antioxidant mechanisms:
Nitroxides (characteristic of monoamines) and benzoquinone diimines
(BQDI, characteristic of PD).

Involvement of Nitroxides. Diarylnitroxides are formed with high
efficiency from diarylamines (2). Free electron delocalization is
characteristic of them. Steric and polar effects govern their
reactivity in dimerization, disproportionation and fragmentation
reactions (1). The reactivity of diarylnitroxides is higher than
that of nitroxides derived from sterically hindered piperidines
(HALSes) (3). It may be, therefore, misleading to transfer generally
the results of mechanistic studies obtained with the latter into the
aromatic series. The ability of diarylnitroxides to oxidize phenols
or thiols or to abstract hydrogen atom from C-H bonds increases the
complexity of processes taking place in amine stabilized rubber
vulcanizates. The reactivity with a sterically hindered phenoxyl was
confirmed by the isolation of a 2,5-cyclohexadiene-1-onyl derivative
(4). The recombination with radicals R· and $RO_2^·$ accounts for
diarylnitroxide antioxidant activity (5), which is however weaker
than that of the corresponding amine. The reaction with R· results
in O-alkyl-N,N-diarylhydroxylamine formation. This compound is
considered to be involved in an important mechanistic pathway: its
thermolysis creates the corresponding N-hydroxylamine and an olefin.
The former is an antioxidant species regenerating diarylnitroxide in
reactions with $RO_2^·$ and ROOH (6). A regenerative cyclical process
was proposed (7) (Scheme 1, R^1 = phenyl).

This process is very important in the HALS stabilization
mechanism. It should, however, be considered only as a minority
pathway in diarylamines just because of the lower stability of the
corresponding diarylnitroxides. This results in the participation of
the latter in side reactions leading to antioxidant ineffective
species, e.g. benzoquinone (BQ) and nitrobenzene (7). Transformation
of diarylnitroxide into a mixture of diarylamine and N-aryl-1,4-
benzoquinone monoimine-N-oxide (4, 8) seems therefore to be a more
probable pathway regenerating partly amine antioxidant than the
hydroxylamine/nitroxide cyclical process (Scheme 2).

The latter pathway is very vague in the 1,4-PD series. No
nitroxide corresponding to N,N'-diphenyl-1,4-PD (DPPD) was detected
in the oxidized substrate by Adamic and co-workers (2). It has been
explained as a consequence of the quick oxidation of the initially
formed monoaminyl into BQDI (4) without nitroxide formation. An ESR
signal characteristic of nitroxide corresponding to N-isopropyl-N'-
phenyl-1,4-phenylenediamine (IPPD) was detected in an oxygen
deficient system (9), i.e. under rubber fatiguing conditions. It is
therefore possible that the mononitroxide is an intermediate formed

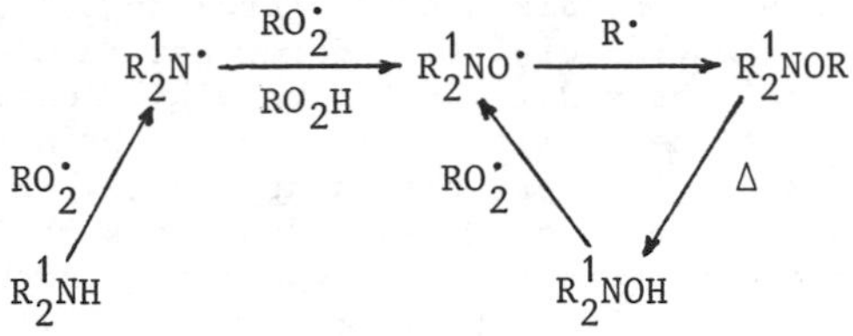

Scheme 1.   Regenerative cyclical process.

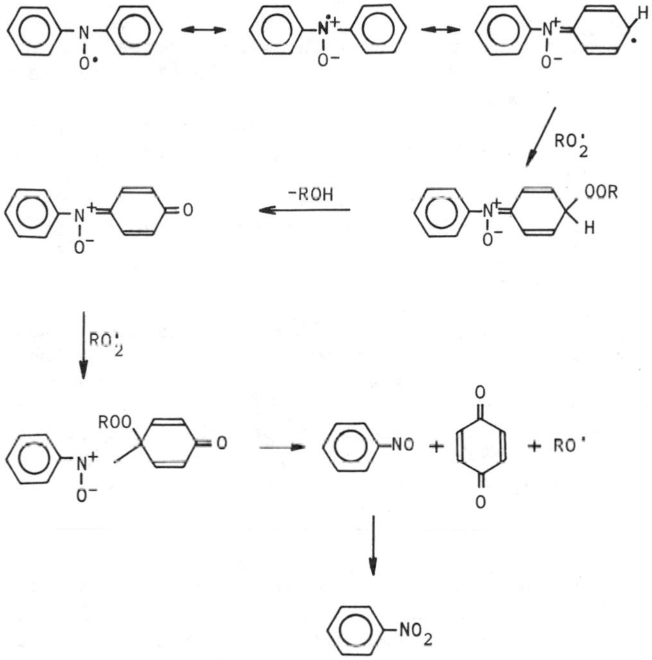

Scheme 2.   Participation of diarylnitroxides in side reactions leading to antioxidant ineffective species.

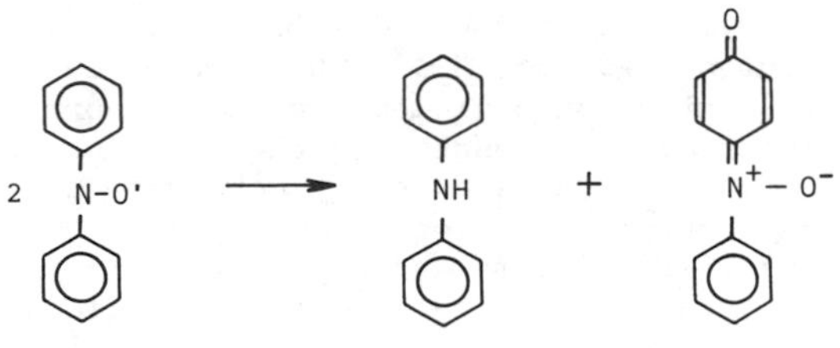

in PD under these conditions - although in a very low concentration
only - and is immediately ozidized to dinitroxide. The latter exists
in the more stable mesomeric dinitrone$_2$form, not involved in the
regenerative process (Scheme 3: R^1, R^2 are alkyls or aryls).

The importance of the cyclical regeneration of nitroxide in the
aromatic series seems to be therefore questionable: N,N'-di-
substituted PD, most probably not involved in this cycle, are
generally more efficient chain-breaking antioxidants than both
diphenylamine and N-phenyl-1-naphthylamine, potentially partly
involved in the nitoxide cycle. It may, therefore, be supposed that
the high antioxidant activity of PD more probably accounts for the
positive cooperative effects of PD with its principal oxidative
transformation product, BQDI.

<u>Involvement of Benzoquinonediimines</u>. BQDI's are easily formed from
PD by various oxidizing agents (<u>1</u>) and were detected in PD-stabilized
hydrocarbons during the thermal oxidation (10, 11) and photooxidation (<u>12</u>). PD and BQDI create a reqox system. An equilibrium
influenced by the acidity of the medium and oxidation potential of
the participating compounds in the reaction mixture is established.
Simultaneously, the chemical stability of BQDI, influenced by the
N,N'-substitution, is reflected in the final product composition.
Important differences in the reactivity of BQDI substituted by
N-sec.alkyls and N-aryls have been observed. The hydrolysis pro-
ceeds more easily on the C=N-sec.alkyl bond than on the
C=N-aryl bond (<u>13</u>).

N-Isopropyl-N'-phenyl-1,4-BQDI (IP-BQDI) is hydrolyzed in the
presence of organic carboxylic acids into N-phenyl-1,4-benzoquinone
monoimine (BQMI) (<u>13,14</u>). Decomposition of IP-BQDI takes place also
in the absence of water: approximately one half of IP-BQDI is reduced to the corresponding IPPD while the other half yields a complicated mixture of products (<u>14</u>). Because of the higher hydrolytic
stability of N,N'-diphenyl-1,4-BQDI (DP-BQDI), a series of characteristics transformation products was isolated and pathways of
their formation, involving hydrolytic, condensation, and redox reactions, were estabilished (<u>13</u>). These processes result in the
formation of BQMI, BQDI of the Bandrowski base type (I, II) and
nitrogen heterocyclic compounds, derivatives of phenazine (III) and
flourindine (IV). 2-Substituted and 2,5-disubstituted DPPD,
corresponding to the isolated Bandrowski BQDI I and II are formed
only as intermediates. Because of their low redox potentials, all PD
derivatives formed (with the only exception of DPPD) are present in
the mixture exclusively in their oxidized forms.

The antioxidant property on N,N'-disubstituted 1,4-DBDI has
been evidenced in polyunsaturated hydrocarbons, i.e., in
squalene (<u>11</u>) and vulcanized NR (<u>15</u>). BQDIs are of comparable
anti-fatigue and anti-abrasion efficiency with the corresponding PD
and are slightly inferior antiozonants in the NR vulcanizate. In
oxidized and photo-oxidized tetralin or cylcohexane, they have only
a concentration dependent retardation effect (<u>11</u>). Both the
efficient scavenging of R· radicals and regeneration of corresponding
PD may be responsible for the antifatigue and antioxidant properties of BQDI (<u>13-15</u>). The latter, formed via scavenging of RO$_2$·
or O$_2$ in oxidized substrate stabilized with PD, contributes thus
to the intergrally observed efficiency ascribed to PD. An efficient

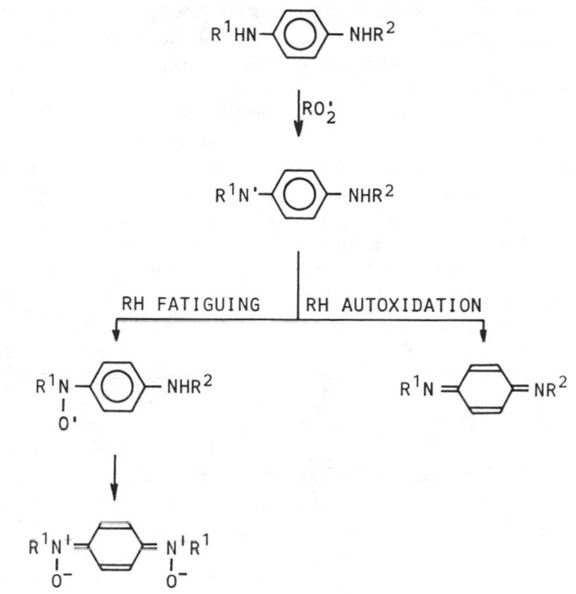

Scheme 3. Pathway for the 1,4–PD series.

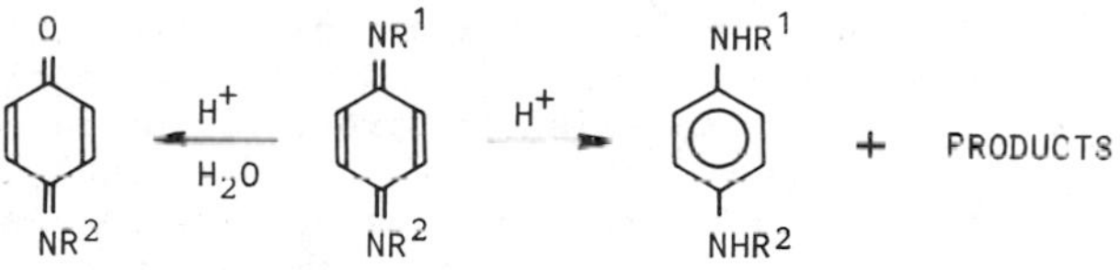

R^1 = iso PROPYL, R^2 = PHENYL

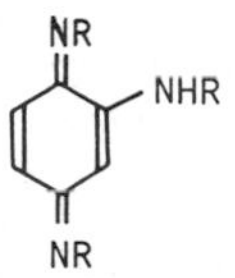

I

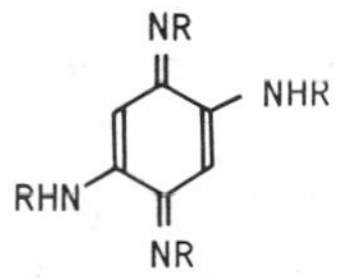

II

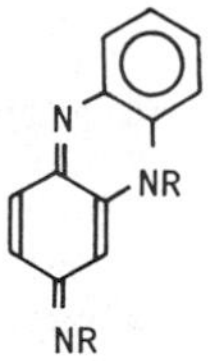

III

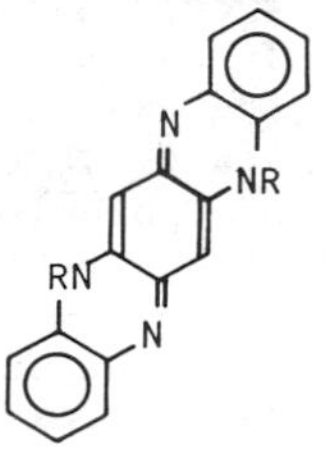

IV

antioxidant cooperative system is formed in the PD/BQDI mixture (16) as a result of alternative scavenging of both the propagating radicals $RO_2^{\cdot}$ and R: This cooperation may be expressed in particular in oxygen deficient fatiguing processes.

The regeneration of N,N'-disubstituted PD from the corresponding BQDI is dependent on reaction conditions and reactants. No conversion of BQDI into PD was observed during the low-temperature oxidation (65°C) of various hydrocarbons (11), perhaps because of the steadily high rate of $RO_2^{\cdot}$ generation. About 50% of BQDI was converted into PD in a weakly acid medium (14), and 60-75% in raw NR or NR compounded with sulphur, N-cyclohexyl-2-benzothiazolesulphenamide, HAF black e.t.c. after treatment at the vulcanization temperature (140°C) (15). The reduction of BQDI into PD observed in NR seems to be due to thermal processes. It is not influenced by vulcanization ingredients. In analogy to BQDIs, N-aryl-4-BQMIs were converted into the corresponding N-aryl-4-aminophenyls with a yield of 30-39%. The remaining part of BQDI or BQMI respectively was not recovered from NR by extraction and is considered to be polymer bound. The character of the polymeric species and the mode of the linkage into NR were not established. It is important that both products, i.e., regenerated PD and rubber-bound species, possess antioxidant properties.

N,N'-Disubstituted PD react with the same chain-breaking mechanism as phenols (17). A mixture of aminic and phenolic $RO_2^{\cdot}$ scavenger is able to be involved in homosynergism (1). The general mechanism of the latter accounts for a regeneration of the more efficient chain-breaker (i.e., amine) via hydrogen transfer from the less efficient one (i.e., phenol) to the primarily formed aminyl. An equilibrium between O- and N-centered radical is suggested. Complication in this simple mechanism is caused by the participation of the respective radicals in coupling, disproportionation, oxidation and recombination reactions. It is connected with the formation of products different from those in the original mixture and not involved in the regeneration cycle. A stepwise depletion of antioxidant active species depedent on both the amine and phenol structures should therefore be considered.

PD applied as a component in the amine/phenol homosynergistic system is converted during $RO_2^{\cdot}$ scavenging into the corresponding BQDI. To obtain more information about the regeneration of PD from BQDI in the presence of phenol, a product study has been performed in benzene solution in a weak acid medium using a mixture of DP-BQDI or IP-BQDI with 2,6-ditert-butylpheonol (14). The reactivity of BQDI plays a specific role: DP-BQDI was converted into DPPD and tetra-tert-butylbiphenyldiol was formed. An additional reaction was observed with IP-BQDI as a consequence of the presence of the reactive C=N-isopropyl moiety. In the absence of oxygen (i.e., under conditions simulating fatiquing of a hydrocarbon polymer), IPPD was created in about 75% yield,i.e., the PD regeneration was enhanced by 50% in comparison with the phenol-free process. About 15% of IP-BQDI was converted into 2,6-ditert.butyl-4-(4-phenylamino-4-phenylimino)-2,5-cyclohexadiene-1-one (imino-CHD). We suggest that a reaction of 2,6-ditert.butylphenoxyl and Wurster's cation radical combined with isopropyl group elimination participates in the imino-CHD formation. 5,5,3',5'-Tetratert.butyl-4,4'-biphenyldiol was formed 2,6-ditert.-butylphenol via the respective phenoxyl. 2,6-Ditert.butyl-1,4-BQ and

3,5,3',5'-tertratert.butyl-4,4'-diphenoquinone form, if oxygen is present in the reaction mixture (Scheme 4, R- is Propyl).

The product study revealed the regeneration of the strong PD antioxidant as a result of the cooperation of the weak antioxidant, 2,6-ditert.butylphenol, with BQDI originating in the first step of the antioxidant regenerative cycle from PD. Another strong antioxidant, tetratert.butyldiphenyldiol, is simultaneously generated in the process. 2,4-Dialkylphenols do not contribute to the PD regeneration.

The imino-CHD created in the regeneration cycle also possesses antioxidant properties (18). It is transformed during the stabilization of oxidized squalene into an aminyl able to dimerize. The dimer is decomposed quickly in oxidized squalene and is reduced into the starting imino-CHD. At the same time, however, the stabilizing effect of imino-CHD in squalene is stepwise lost without the title compound being destroyed. It seems that the squalene autooxidation is accelerated because of the effect of some intermediates formed from squalene during regeneration of imino-CHD from the aminyl. This regeneration should be therefore considered as an undesirable one.

Factors in Ozone Weathering of Rubbers and Stabilizers.

Ozone is a specific atmospheric pollutant characteristic of urban and some industrial areas of the troposphere. Its concentration is very variable. Because of both the photochemical character of the origin of ozone and its high reactivity with organic atmospheric pollutants as well as some organic materials on the Earth's surface, its night concentration drops to zero. Only ground state molecular oxygen attacks rubber in this period of the day, but mechanically initiated processes and reactive ozonation intermediates and products remain involved.

Due to the high reactivity of ozone with unsaturated hydrocarbons moieties, surface cracking of stressed or flexed NR, BR, NBR, and SBR vulcanizates arises. Rubber goods designed for outdoor applications must therefore be stabilized against both O_2 and O_3 attacks. Antioxidant protection mechanisms have been discussed in detail (1). Discussions dealing with antiozonant mechanism involve some contradictory experimental observations.

Any approach to the formulation of an antiozonant mechanism should reflect all possible interactions of a stabilizer molecule with ozone and active species formed via rubber ozonation. The latter involves an ionic mechanism, and a variety of active oxygen containing products having structures of ozonides, polymeric ozonides, and peroxides is formed (19). A partial occurrence of free radical processes during hydrocarbon ozonation may be ascribed to the simultaneous attack of the O_2 and O_3 mixture. Under realistic outdoor service conditions, rubber weathering is caused not only by ozonation. Autooxidation and flex-cracking are equivalent deterioration processes (surpassing ozonation in the service period in the dark). Moreover, singlet oxygen has to be considered as another deteriogen, although present in trace concentration only. Its occurrence in the ozonation of polymer C-H bonds has been reported (20). Thus, a mixture of radical, ionic and molecular species arising in simultaneously proceeding ozonation, autooxidation and fatiguing of vulcanizates under dynamic conditions creates a fairly complicated reactive system. Any oversimplification in the rating

of particular weathering factors separately from the other ones may
be a source of serious misunderstandings in the interpretation of the
integral role of rubber oxidation processes and their involvement in
stabilizer mechanisms and transformation. Many experiments have to
be done in simplified model systems in which some important factors
are neglected. This sometimes distorts the interpretation very
gravely.

Rubber oxidation products able to undergo redox and/or
condensation reactions must be considered in particular as reactive
partners with antiozonant species. Analogies in the reactivity with
low molecular weight organics are mostly considered. Limits
controlling polymer-analogous reactions cannot however, be omitted in
particular restrictions of the reactivity by physical environmental
factors in the solid matrix.

Antiozonant Properties. Aromatic secondary diamines are the only
class of organic chemicals able to reduce efficiently the ozone crack
growth of vulcanizates under dynamic conditions and be acceptable at
the same time from both the technical and toxicological points of
view. The presence of a secondary aromatic amine moiety itself in a
molecule is not a sufficient condition to attain antiozonants
efficiency. (E.g., secondary monomaines are only antiozidants and
flex-crack inhibitors without appreciable antiozonant activity. On
the other hand, all N,N'-disubstituted PD antiozonants are also
efficient antoxidants and most of them also act as flex-crack
inhibitors (1). Both these stabilization activities have to be
considered in the complex antiozonant mechanism, together with some
metal deactivating activity.

An extensive screening of structure-activity relations revealed
(1) the outstanding properties of N,N'-disubstituted PD. It is
generally accepted that the presence of N-sec.alkyls accounts for
better antiozonant protection than that of N-prim. and N-tert.alkyls
or N-aryls (21). This may be one of the clues to decipher the
chemical pathways of the antiozonant mechanism. The final effect is
moreover fully dependent on the composition of the vulcanizate. The
structure of commercially used antiozonants is an optimum compromise
of efficiency, physical properties and toxicity. N, N'-Disec.alkyl-
1-4-PD are used in the U.S.A., N-sec.alkyl-N'-aryl-1,4-PD are
preferred in Europe. N,N'-Diaryl derivaties are not applied as antiozonants in NR, BR, IR, or SBR. One of the reasons may be their low
solubility in rubber vulcanizates (22). It does not allow them to
reach a concentration level in the rubber bulk which is able to act
as a long-term operative store of a stabilizer ready to supply the
rubber surface slowly but continuously with active compounds by
migration and to maintain the protective effect without inefficient
quick blooming of an incompatible PD. A chemical reason accounting
for the minority antiozonant role of N,N'-diaryl PD is discussed
later.

Antiozonant Mechanism. No simple model approach to the explanation
of the antiozonant activity of PD is applicable. Most ideas were
influenced by the fact that the rubber ozonation is a surface
process, not exceeding a thickness of about 40 molecular diameters
(23). The reaction of ozone with an antiozonant in the vulcanizate
surface layer and replenishment of the consumed stabilizer by means

of the migration of the fresh one to the rubber surface from the
rubber bulk was therefore considered as the most probable antiozonant
activity explanation in the earliest theories. It has been admitted
successively that more processes or reactants may be involved. A
high surface concentration of PD has been reported also in modern
mechanistic ozonation studies (24) as a necessary condition to
achieve antiozonant efficiency. But even this condition cannot be
the sole one valid as it may be extrapolated from some experimental
proofs. The high surface antiozonant concentration is also difficult
to be maintained during the long service time of vulcanizates: antiozonants are depleted on the surface not only be chemical (i.e.,
oxidation/ozonation reactions) but also by physical processes
(volatilization, leaching, abrasion). Only one part of the chemical
depletion processes contributes to the antiozonant efficiency. Moreover, the decisive role of antiozonant migration has been made doubtful by the results of Scott (25) obtained with the polymer bound
4-(mercaptoacetamido)diphenylamine (V). The high rubber protection
was achieved even in extracted samples in spite of the restricted
migration of structurally bound antiozonant species.

Disharmonies in the Conception of the Direct O₃/Antiozonant Reaction
Importance. Four antiozonant theories have been formulated within
the last 25 years. Ozone scavenging theory suggests a preferential
direct reaction of an antiozonant with ozone on the rubber surface as
a decisive process (26-27). As the antiozonant is depleted via
direct ozonation on the surface, fresh antiozonant diffuses rapidly
from the rubber bulk to reestablish the equilibrium surface concentration. At a comparable additive concentration and migration rate,
the antiozonant efficiency of an additive should be therefore
dependent on its ozonation rate and the vulcanizate will be protected
until the antiozonant is depleted below the lowest critical concentration. From this point of view, the ozonation rate seems to be a
more important factor than the total amount of ozone scavenged by one
mole of an antiozonant (this latter phenomenon may be called
ozonation factor). Relations between antiozonant efficiency in
vulcanizate and antiozonant ozonation rate or antiozonant surface
concentration have been indeed reported in some papers and an
appreciable higher ozonation rate of PD in comparison with rubber
unsaturation, a preferential consumption of an antiozonant in model
olefin solution or in rubber were observed. The rubber surface was
not attacked by ozone until the antiozonant was almost completely
consumed (28).
     Theory of protective film formation supposes creation of a
rubber surface film from PD oxidation and/or ozonation products.
Ozone attack on the oxidatively violated rubber surface is thus
prevented (21, 27, 29, 30). Creation of the surface layer was confirmed using microscopy. The theory is in agreement with the
observation that the initial rapid ozone consumption is stabilized
rubber drops and may be renewed after mechanical break of the formed
film (27). There is a chemical proof of the theory. Ozonation
products of N,N'-bis(1-methylheptyl)-1,4-PD (DOPPD) form a surface
film on ozonized and DOPPD doped vulcanized NR (31) and carbon-black
loaded NR (24, 32). In addition to unreacted DOPPD, many of the low-
molecular weight compounds observed in the film were found also in
the ozonation of pure DOPPD (24); the only difference was a

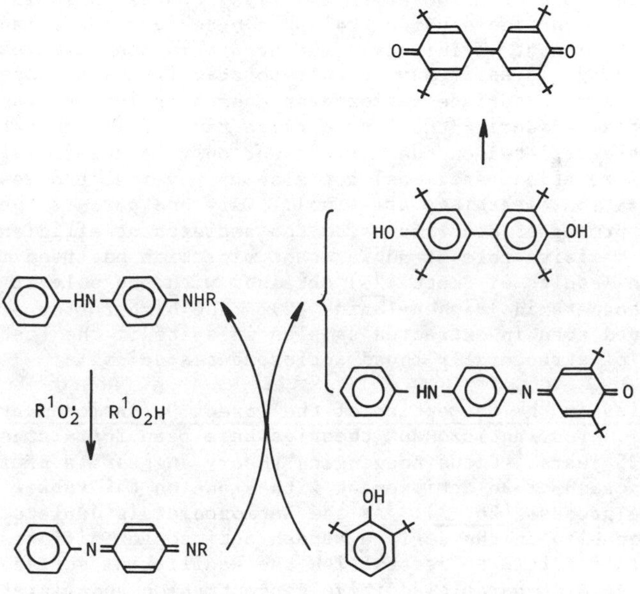

Scheme 4.   Mechanism for imino-CHD formation.

V

restricted formation of Bandrowski bases in the surface film. No
rubber ozonation products were found using ATR-IR in the degraded
surface layer of stabilized NR (31, 32), although ozonides and
carbonyl compounds were identified on the surface of degraded but
unstabilized vulcanized unloaded NR (31), clay (31) or carbon black
loaded NR (32). The diffusion of DOPPD to the rubber surface is
supposed to be sufficiently rapid to account for the observed antiozonant effect (33). The report experimental data have been
interpreted as a dual scavenger-protective film formation theory of
antiozonant mechanism of N,N'-disubstituted 1,4-PD (24, 32).
     There exist numbers of experimental observations and
contradictory results making vague the univocal validity of both the
simple ozone scavenger and the protective film theories. Products
formed via ozonation of N-(1,3-dimethylbutyl)-N'-phenyl-1,4-PD (HPPD)
and DOPPD are acetone soluble. But another part of the products
formed in PD doped NR and estimated by Lorenz and Parks (29) to about
23-37% are "unextractable nitrogen compounds" and should be believed
to be polymer bound moieties formed via a reaction of PD with
ozonized rubber. There are other experimental objections too: the
diffusion rate of PD to the surface is not quick enough to ensure a
high antiozonant concentration to be attained within a short time
period after the production of rubber goods (34). Antiozonant
efficiency is not unlimitedly proportional to the antiozonant concentration (30). The rate constants of direct amine ozonation do not
vary so expressively as their antiozonant efficiencies (35) and no
reasonable direct relation between these two phenomena could be
found. Moreover, an easy reactivity with O_3 is not a sufficient
condition for a compound to be an antiozonant as may be exemplified
on various reactive organic compounds having no antiozonant properties. There is also no clean-cut difference between the overall
ozone consumption (ozone equivalents) by various N,N'-disubsituted
1,4-PD (36). Therefore, a comparison of the antiozonant activity of
a particular compound in a vulcanizate with the compound reactivity
with ozone is a vague mechanistic explanation and is of minor value
from the point of view of generalization of the process. The
chemistry of PD ozonation is more important.
     In spite of many objections, ozone scavenger and protective film
theories cannot be neglected because of serious experimental evidence. They should be considered as an important part of the overall
antiozonant mechanism. Their role prevails in rubber solution or in
very thin rubber films. It is very probable that a part of an antiozonant is wastefully depleted just because of direct ozonation.

Involvement of the Ozonized Rubber Moieties in Antiozonant Mechanism.
The rubber chain relinking theory (30) is consistent in part with the
self-healing film formation theory (37): a reaction between an antiozonant or some of its transformation products and ozonized elastomer
is considered. Either scission of ozonized rubber is prevented in
this way or severed parts of the rubber chain are recombined (i.e.,
relinked). A "self-healing" film resistant to ozonation is formed on
the rubber surface. Such a film formed by the contribution of non-
volatile and flexible fragments of the rubber matrix should be more
persistent than any film suggested in the protective film theory.
Creation of an unextractable polymer bound part of originally added
PD has been reported (29) and considered as a piece of evidence of

the reactivity of PD with ozonized rubber. To explain the mode of formation of a link between antiozonant or its transformation products and the rubber network, reactions of antiozonant with rubber zwitterions, ozonides, and aldehydes have been considered. E.g., the formation of an addition product from zwitterion (37) or from ozonide (29) and PD accounting for the polymer bound moiety was suggested (Scheme 5).

A nitroxide and Wurster's cation-radical were reported to be intermediates in the reaction between IPPD and a low molecular weight ozonide (38). Some experimental proof bringing serious objections against relinking theory must also be considered: the reaction of rubber ozonides with IPPD and DPPD has been reported to be very slow (slower than with ozone) (28, 29, 38, 39). No products arising from the interaction between N,N'-disubstituted PD and vulcanized NR during ozonation were evidenced in (24, 31, 32). The low-rate reactivity of PD with ozonides in comparison with that with ozone is rather opposed to the chain-linking theory. Both reactions should, however, be considered also from the point of view of their relative importance: O_3 is present in the troposphere only during sunshine hours. On the contrary, the accumulation of ozonides (and of other ozonation products) in the rubber surface increases stepwise and permanently and the oxygenated species react with PD also in the dark period. The importance of the reaction with ozonide is thus augmented. Moreover, the reaction of PD with an ozonide may be appreciably accelerated in a weak acid media, i.e., by the presence of aliphatic carboxylic acids (39). Ozonide is reduced into a mix- ture of aldehydes in this process (Scheme 6). This may be one of the pathways leading to polymer bound species.

Using extrapolation from organic chemistry mechanisms, link and network formation via reaction between PD and aldehydic (29, 30, 33) or zwitterionic (37) and groups formed in ozonized rubber or its fragments with aldehydes formed by reduction of rubber gozonides (39) or with aldehydes arising from the ozonation of N-sec.alkyls in PD molecule (24, 40) seems to be operative. All these model conceptions indicate possibilities of rubber chain re-linking or extension or network formation in ozonized rubber. Whatever the mode of linkage formation may be, it has been supposed that this chemical transforma- tion causes relaxation of stressed rubber surface and increases the critical energy necessary for the surface ozone cracks formation (41). At the same time, a chemically modified, relaxed, flexible rubber surface layer less sensitive to the further O_3 attack is created. This chemical surface restructuring needs some time and may be one of the causes of differences between antiozonant efficiency observed in natural and accelerated ozonation tests. At the same time, it may explain the failure in monitoring rubber/antiozonant interactions in short ozonation experiments.

Products Studies in Antiozonant Mechanisms. Experimental proof of the individual antiozonant theories based on product studies are scarce. Using a very sophisticated instrumental analytical approach, the composition of a very complicated mixture of ozonation products of two technically important antiozonants, i.e., DOPPD and HPPD has been revealed and the reactivity pathways with ozone have been established (24, 40). Influence of the character of N-substituents on the ozonation mechanism has been evidenced. Some important mechanistic differences between N,N'-disec-alkyl-substituted PD and between those N-sec.alkyl-N'phenyl substituted are thus clarified. Two principal pathways govern the ozonation of DOPPD (24): (i) amine oxide pathway; (ii) N-sec.alkyl (i.e. side chain) oxidation pathway. The third mechanistic feature is a minority nitroxide radical pathway leading to a stable dinitrone.

In the HPPD ozonation, the free radical chemistry seems to be more important than in DOPPD, due to the stabilizing effect provided by the diarylamine moiety (40). The nitroxide pathway is therefore of relevant importance in HPPH ozonation to amine oxide and side chain oxidation pathways. Because of the influence of N-substituent effects the ozonation of HPPH occurs only on the aliphatic side of the molecule and a nitrone is the most abundant ozonation product. The formation of a dinitrone - in contrast to the ozonation of DOPPD - is inhibited most probably just by the stabilizing effect of the N-phenyl group. The same structural moiety stabilizes aromatic nitro and nitroso compounds formed via the amine oxide pathway: no Bandrowski bases were detected in the ozonation product of HPPD. The authors (40) favor the N-alkyl side reactivity in HPPD also in the interpretation of the formation of N-phenyl-N'-acyl-1,4-PD, i.e. in the condensation with aliphatic aldehydes.

Formation of an aminyl radical on the aromatic side of HPPD is a result of the NH-bond oxidation. The N-N coupling of aminyls accounts for the formation of interesting dimers, e.g. VI, able to be oxidized into nitroxide and nitrone. Formation of a C centered radical, mesomeric to the originally formed aminyl, should be expected. Thus, also other oligomeric products may be formed via C-N and C-C couplings.

By extrapolating data of Lattimer and coworkers (24, 40), it may be anticipated that the ozonation of DPPD proceeds most probably via nitroxide and aminyl formation pathways. Dinitrone and N-N or N-C and C-C coupling products should therefore be considered as ozonation products formed via free radical intermediates. The reactivity of DPPD with aldehydes will be very limited. All these mechanistic features together with the low ozonation rate of DPPD in comparison with N,N'-disec.alkyl PD (26) may be responsible for the generally reported poor antiozonant efficiency of N,N'-diaryl PD in comparison with N-sec.alkyl-N'-aryl and N,N'-disec.alkyl PD.

There is a common mechanistic feature in the direct ozonation of various N,N'-disubstituted 1,4-PD: Formation or accumulation of any simple BQDI, BQMI or BQ derivatives was not observed (12, 24, 40), although BQDI themselves were proven to be ozone resistant (18). But it has been evidenced in experiments aimed to clarify aspects of the chain-relinking theory that IPPD and DPPD are oxidized by 1-hexadecane ozonide in the presence of aliphatic carboxylic acids (39) to corresponding BQDI (aliphatic acids which catalyze the reaction may arise either during oxidation of rubber or are used as rubber compounding ingredients). The basicity of PD and stability of BQDI play an important role.

Nitroxide or Wurster's cation-radical may be involved in the formation of BQDI (38, 39). The latter is transformed in the weakly acidic medium used for the reaction of PD with ozonide into a rather complicated mixture of products mentioned earlier.

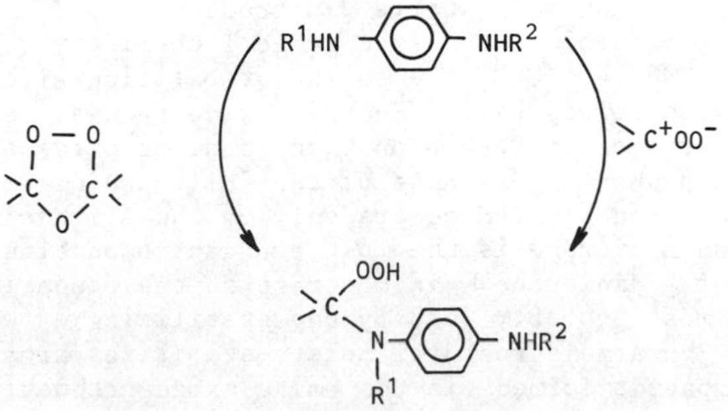

Scheme 5. Mechanism of the formation of an addition product form switterion or from ozonide and PD accounting for the polymer-bound moiety.

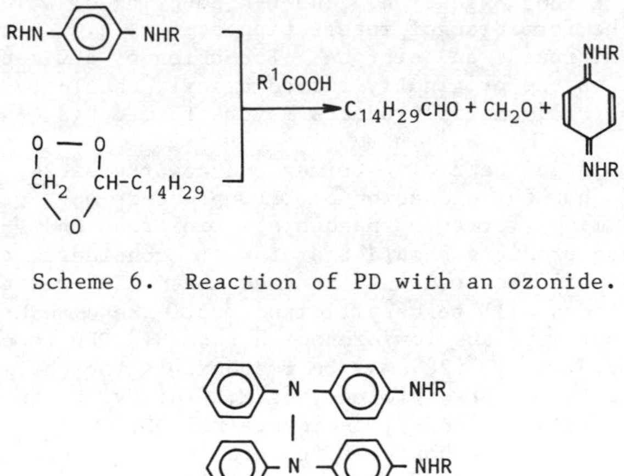

Scheme 6. Reaction of PD with an ozonide.

VI

Combinations of Individual Mechanisms. It is extremely difficult to obtain an undoubted proof of the superiority of one particular antiozonant stabilization mechanism and its unequivocal validity in complicated rubber vulcanizate systems exposed to a complex environmental attack. Using a product study concerned with BQDI chemistry, the possibility of mechanistic relations between the participation of PD in antiozonant, antioxidant and/or antiflex-crack processes during rubber weathering has been suggested: the corresponding BQDI are generally formed from PD during the stabilization of hydrocarbons against thermal oxidation as well as photo-oxidation (11, 12) and by the reaction with ozonides (39). More complicated BQDI of Bandrowski base type are formed in the PD ozonation (24, 40) and in the weak acid catalyzed decomposition of primarily formed BQDI (13). The latter have therefore to be considered as one of the clue compounds in the rubber stabilization against weathering. They are supposed to be more efficient scavengers of R· radicals than the corresponding PD derivatives (16) and contribute therefore to antiflex-cracking activity. Participation of a PD molecule in reactions wtih oxygen, ozone and rubber oxidation/ozonation products in the presence of other additives and/or acid impurities may account for the simultaneous and/or consecutive course of reactions resulting in a scavenger-protective film-chain relinking ternary combined mechanism. The predominance of any of these three particular mechanisms is substrate and environmental conditions dependent.

Rubber stabilization efficiency of amines results not only from an interaction of stabilizers with chemical deteriogens. The importance of physical factors (e.g. diffusion, migration, solubility, volatility or leaching of stabilizers), of physical or chemical consequences of the interaction of stabilizers with fillers (carbon black in particular) or catalytic impurities of various origin, and of environmental effects (acid rain in particular) is mostly underrated or even ignored in mechanistic discussions, although only a properly estimated influence of all factors and of their relative importance is an efficient approach targetting the explanation of rubber stabilization mechanisms. Unfortunately, many experimental data are still lacking and it is moreover uncertain if they will be available in the near future.

References

1. Pospisil, J., "Developments in Polymer Stabilization", Scott, C. (Ed.), Vol. 7, Chapter 1, Applied Science Publishers, Ltd., Barking, 1984.
2. Adamic, K., Bowman, D. F., Ingold, K. U., J. Amer. Oil Chemists' Soc. (1970), 47, 109.
3. Rozantsev, E. G., Sholle, V. D., Synthesis (1971), 190 and 401.
4. Forrester, A. R., Hay, J. M., Thomson, R. H., "Organic Chemistry of Stable Free Radicals", Academic Press, London, 1968.
5. Ingold, K. U., Advan. Chem. Ser. (1968), 75, 296.
6. Cholvad, V., Stasko, A., Tkac, A., Butchatchenko, A. L., Malik, M., Collect. Czech. Chem. Commun. (1981) 46, 823.
7. Bolsman, T. A. B. M., Blok, A. P., Frijns, J. H. G., Rec. Trav. Chim. Pays Bas (1978) 97, 310 and 313.
8. Nelsen, J. F., "Free Radicals", Kochi, J. K. (Ed.), Vol. 2, p. 257, J. Wiley & Sons, New York, 1973.
9. Katbab, A. A., Scott, G., Eur. Polym. J. (1981) 17, 559.

10. Lorenz, O., Parks, C. R., Rubber Chem. Technol. (1961) 34, 816.
11. Rotschova, J., Pospisil, J., Collect. Czech. Chem. Commun. (1982) 47, 2501.
12. Rotschova, J., Pospisil, J., J. Chromatogr. (1981) 211, 299.
13. Taimr, L., Pospisil, J., Angew. Makromol. Chem. (1980) 92, 53.
14. Taimr, L., Pospisil, J., Polym. Degradation Stab. (1984) 8, 23.
15. Cain, M. E., Gelling, I. R., Knight, G. T., Lewis, P. M., Rubber Ind. (1975) 9 (6), 216.
16. Mazaletskaya, L. I., Karpukhina, G. V., Maizus, Z. K., Izv. Akad. Nauk. USSR, otd. khim. (1981), p. 1988.
17. Pospisil, J., "Developments in Polymer Stabilization", Scott, G. (Ed.), Vol. 1, Chapter 1, Applied Science Publishers, Ltd., Barking, 1979.
18. Taimr, L., Pospisil, J., Polym. Degradation Stab. (1984) 8, 67.
19. Bailey, P. S., "Ozonation in Organic Chemistry", Vols. 1 and 2, Academic Press, New York, 1978 and 1982.
20. Murray, R. W., Kaplan, M. L., J. Amer. Chem. Soc. (1968) 90, 537.
21. Layer, R. W., Rubber Chem. Technol. (1966) 39, 1584.
22. Dibbo, A., Gummi, Asbest, Kunststoffe (1965) 18, 120.
23. Razumovskij, S. D., Podmasterov, V. V., Zaikov, G. E., Plaste Kaut. (1983) 30, 625.
24. Lattimer, R. P., Hooser, E. R., Diem, H. E., Layer, R. W., Rhee, C. K., Rubber Chem. Technol. (1980) 53, 1170.
25. Katbab, A. A., Scott, G., Polym. Degradation Stab. (1981) 3, 221.
26. Cox, N. L., Rubber Chem. Technol. (1959) 32, 364.
27. Ericsson, E. R., Berntsen, R. A., Hill, E. L., Kusy, P., Rubber Chem. Technol. (1959) 32, 1062.
29. Lorenz, O., Parks, C. R., Rubber Chem. Technol. (1963) 36, 194 and 201.
30. Braden, M., Gent, N. A., J. Appl. Polym. Sci. (1962) 6, 449.
31. Andries, J. C., Ross, D. B., Diem, H. E., Rubber Chem. Technol. (1975) 48, 41.
32. Andries, J. C., Rhee, C. K., Smith, R. W., Ross, D. B., Diem, H. E., Rubber Chem. Technol. (1979) 52, 823.
33. Lake, G. J., Rubber Chem. Technol. (1970) 43, 1230.
34. Braden, M., J. Appl. Polym. Sci. (1962) 6, S6.
35. Parfemov, V. M., Rakovski, S., Shopov, D., Popov, A. A., Zsikov, G. E., Izv. Khim. (1979) 11 (1), 180.
36. Braden, M., Gent, N. A., Rubber Chem. Technol. (1962) 35, 200.
37. Loan, N. D., Murray, R. W., Story, P. R., J. Inst. Rubber Ind. (1968) 2, 73.
38. Pobedimskij, D. G., Razumovskij, S. D., Izv. Akad. Nauk USSR, otd. khim. (1970), p. 602.
39. Taimr, L. Pospisil, J., Angew. Makromol. Chem. (1982) 102, 1.
40. Lattimer, R. P., Hooser, E. R., Layer, R. W., Rhee, C. K., Rubber Chem. Technol. (1983) 56, 431.
41. Andrews, E. H., Braden, M., J. Polym. Sci. (1961) 55, 787.

RECEIVED December 7, 1984

Polymer-Bound Antioxidants

GERALD SCOTT

Department of Chemistry, University of Aston in Birmingham, Birmingham B4 7ET, England

The reasons for the present interest in polymer-bound antioxidants is discussed, and some of the more important approaches to the chemical attachment of antioxidants and stabilizers to polymer molecules are briefly reviewed.

It is concluded that the modification of rubbers after manufacture with chemically reactive antioxidants offers the most promising procedure for producing concentrates of polymer-bound antioxidants that can be used as conventional additives.

Unexpected advantages of bound antioxidants have been observed due to the selective protection of the most oxidatively sensitive regions of the polymer in rubber-modified polymer blends.

Factors Determining the Effectiveness of Antioxidants. It was for many years a source of puzzlement to polymer chemists that the rating of antioxidants in polymers and in model hydrocarbons appeared to be very different. The widely used antioxidant BHT (I) is one of the most "efficient" antioxidants known for liquid hydrocarbons as measured by oxygen absorption but is virtually ineffective in rubbers

of plastics in an air oven heat-aging test, which is normally carried out with continual displacement of air over the surface of the sample. The bisphenol II is much more effective than BHT in a heat aging test in rubber but is not very effective in polypropylene in a similar test at somewhat higher temperature. A systematic study of the effect of structure variation in the homologous series (III) in

0097–6156/85/0280–0173$07.00/0

$$\text{III}$$

OH, tBu, tBu, CH_2CH_2COOR — structure III

which the normal alkyl chain, R, was varied from CH_3 to $C_{18}H_{37}$ threw
some light on this problem (1). Table 1 shows that in decalin all
the antioxidants have a very similar molar antioxidant activity* as
measured by oxygen absorption (D^c), but they are all considerably
less efficient than BHT at the same molar concentration.

In polypropylene in an air oven test at 140°C (PP^o), on the
other hand, the lower members of the series are completely ineffective, as is BHT. The highest member of the series, the commercial
antioxidant IRGANOX 1076 (III, R = $C_{18}H_{37}$) is highly effective under
these conditions.

Volatility and Migration Rate. A study of the half-lives ($T^{1/2}$) of
the antioxidants in the polymer (see Table 1) at the same temperature
suggests a reason for the lack of correlation between the two sets of
results. In an oxygen absorption test, volatilization cannot occur,
and the result is a true measure of the intrinsic activity of the
antioxidant molecule. In an air oven test, on the other hand,
physical loss of the antioxidant by migration and volatilization from
the surface must dominantly influence the test results. Billingham
and his coworkers (2) have shown that these two physical parameters
determine the rate of loss of antioxidants from polymers. Increase
in molecular mass generally decreases molecular mobility as well as
volatility, and which factor dominates depends on the thickness of
the sample (2).

Solubility. Table 1 also compared the behavior of the same series of
antioxidants by oxygen absorption in polypropylene film and in
decalin. In PP^c antioxidant activity is optimal at $C_{12}H_{25}$. The
commercial product, IRGANOX 1076, is less than half as effective.
The different order of activity in the polymer from that the model
substrate reflects the different solubilities of the additives in the
polymer. A correlation is therefore observed between PP^c and S,
the solubility of the additives in hexane at 25°C.

Solubility in the polymer has been shown to be of dominating
importance in the case of photoantioxidants (see Table 2) (3-5).

Antioxidant in Polymers Subjected to Aggressive Environments. The
methods used to evaluate the effectiveness of antioxidants described
above are accelerated tests and none of them can be assumed to
adequately represent the conditions to which most polymers will be
subjected in normal use. However, as polymers move into more and
more demanding engineering applications, they have to withstand increasingly aggressive environments (6). For example, whereas at one

---

* The higher members of the series are less effective on a weight
  basis.

Table 1. Antioxidant Effectiveness of the Hindered Phenol Series (III) at the Same Molar Concentration (2×10^{-4} M) at 140 °C

(III)	MOLAR MASS	$T^{\frac{1}{2}}$	S	INDUCTION PERIOD		
				D^C	PP^S	PP^O
R						
CH_3	292	0.28	32	25	95	2
C_6H_{13}	362	3.60	∞	23	312	2
$C_{12}H_{25}$	446	83.0	∞	20	420	2
$C_{18}H_{37}$	530	660.0	64	20	200	165
BHT(I)	220	0.1	100	150	140	2

$T^{\frac{1}{2}}$; Antioxidant half-life (h) in PP in N_2 stream at 140°C.

S ; Solubility in hexane at 25°C, g/100g.

D^C ; in decalin by oxygen absorption at 140°C, hr.

PP^S ; in PP film by oxygen absorption at 140°C, hr.

PP^O ; in PP film in a moving air stream at 140°C, by carbonyl measurement, hr

Table 2. Molar Antioxidant Effectiveness of the Zinc Dialkyl Dithiocarbamate Homologous Series as UV Stabilizers Polypropylene (3×10^{-4} mol/100 g)

$(R_2NCSS)_2Zn$	EMBRITTLEMENT TIME, hr	UV ABSORBANCE,* 285 nm
R		
CH_3	145	0.11
C_2H_5	170	0.15
C_4H_9	193	0.32
C_6H_{13}	330	0.63
C_8H_{17}	330	0.62

*UV Absorbance is directly proportional to the concentration of the additive in the polymer.

time it was not considered necessary to test rubbers at temperatures above 70°C, some rubbers now, when used in engine seals, gaskets and hoses, have to withstand service temperatures in excess of 100°C, often in the presence of lubricating oils, in critical components whose failure would not only be very costly but might also be highly dangerous. Tires represent another area of modern technology in which operating temperatures have steadily risen. It is not uncommon for temperatures in the shoulders of heavy duty truck tires to rise of over 100°C under high speed motorway conditions (7). Similar effects are also observed in the tires of fully laden wide-bodied aircraft even at relatively modest taxying speeds (8).

Both the above situations lead to rapid loss of antioxidants to the surrounding environment; in seals and hoses by simple solvent extraction and in tires by mechanical extrusion, with subsequent loss at the surface by water leaching (9).

Other areas where physical loss of additives from polymers is increasing in importance is in food contact applications and in medical uses of polymers. In Europe, legislation is steadily moving toward the situation where loss of any additives, irrespective of their toxicity, may be prohibitive above a certain level. It is, therefore, no longer an academic exercise to retain antioxidants and stabilizers in polymers merely to satisfy unrealistic accelerated tests. It is a practical necessity to improve the substantivity of antioxidants and stabilizers for a variety of quite different practical reasons.

The above arguments suggest that an ideal antioxidant, in addition to having a high intrinsic activity, should also satisfy the following physical criteria.

1. It should be completely soluble in the polymer at the concentration used.
2. It should be completely non-migrating from the polymer while still satisfying requirement I.

Published work suggests that there may be some mutual contradiction in the above requirements. For example, it is believed that antiozonants for rubbers function by migrating to the surface of the rubber where they react with ozone and are slowly converted to ozone impermeable films (9, 10). A second and more general problem is that when the size of an antioxidant molecule is increased in order to reduce volatility, it becomes increasingly insoluble in the host polymer. The ultimate example of this is a highly polymeric antioxidant; even minor differences in the repeat unit result in a two-phase system. Many polymeric antioxidants have been made, but little success has been achieved in developing highly effective antioxidants with sufficiently high molecular mass to be resistant to solvent leaching (11).

Polymer-Bound Antioxidants

In principle, an antioxidant which forms part of the host polymer should approach the ideal outlined above since it should

1. be molecularly dispersed in the polymer and hence completely soluble;

2.   not be lost to the surface; and
3.   be completely non-volatile.

   However, if migration is important to the function of an
antioxidant, then polymer-bound antioxidants should not be effective.
In the following sections, data will provide at least a partial
answer to the last point.

Approaches to Polymer-Bound Antioxidants.   Three basic methods of
producing bound antioxidants have been investigated.   These are

1.   copolymerization of vinyl antioxidants with vinyl monomers
     during polymer synthesis;
2.   reaction of conventional antioxidants with functional groups
     re-formed in the polymer; and
3.   direct reaction of antioxidants containing functional groups with
     normal polymers.

Polymer Modification by Copolymerization During Manufacture.   Since
the early 1960s, a variety of copolymerizable monomers have been
described (11), but only one successful product has been reported.
This results from research at Goodyear (12, 13) and a commercial grade
of nitrile rubber (Chemigum HR 665) is now produced containing a
small proportion (< 2%) of a copolymerized monomer, IV.

   There is no doubt that this grade of nitrile rubber is much more
resistant to hot lubricating oils and high temperatures than is
conventionally stabilized nitrile rubber (see Table 3).   The main
problem in this approach is the cost of producing a specialized
rubber in relatively small quantity.
   Some success has been reported in producing UV stable polymers
by copolymerizing vinyl UV stabilizers with other monomers (11) and
evidence from the literature suggests that many groups are
concentrating on this aspect (15).

Reactions of Antioxidants with Preformed Functional Groups.   The
reactive chlorine in epichlorohydrin polymers is a specific though
typical example of this approach (11).   A more general reaction is
the epoxidation of the double bonds in rubbers and subsequent
reaction of the epoxide group with an amine antioxidant (reaction 1)
(16).

The product formed in this reaction is a typical sec-alkylaminodiphenylamine (V) which is characteristic of the best rubber antidegradants. No commercial products based on this principle have so far been reported, possibly because of the cost of epoxidation, normally carried out in solution.

Reactions of Polymer-Reactive Antioxidants. Three different modifications of this approach have been reported.

1. Normal chemical reaction in solution or in latices (17, 18).
2. Reaction by post treatment of polymer artifacts (19-21).
3. Reaction during processing or fabrication procedures (18, 20-25).

Reactions of Reactive Antioxidants with Polymers by Normal Chemical Procedures. Grafting of vinyl antioxidants; e.g., VI, into rubbers has been used to produce modified rubber latices (26). Even simple

phenolic antioxidants such as I can be made to react with polymers to a limited extent in the presence of radical generators (27).
     A more significant development because quite high yields of adduct concentrates can be achieved is the Kharasch-Mayo thiol adduct process, reaction 2 (11, 17, 28).

A variety of antioxidants has been reacted with elastomers in the
latex in the presence of typical redox radical generators (27). The
mechanism of this process has been considered in detail elsewhere
(11, 17, 27, 28) and will not be discussed further here. The resulting products are much more age-resistant than conventional rubber
antioxidants under air oven aging conditions (see Figure 1), and in
the presence of aqueous detergents and dry cleaning solvents (27).
Figure 2 compares the loss of BHBM-B* with a commercial non-staining
antioxidant, WSP (II) in NR under leaching conditions. It is evident
that after removal of the non-bound material, no further reduction in
antioxidant occurs with either leachant. The same procedure has been
used to produce concentrates of thiol adducts in nitrile rubber (23).
In addition to BHBM (VII), the arylamine antidegradant MADA (VIII)
has also been grafted to NBR latices. Table 4 shows that the

$$\bigcirc\text{--NH--}\bigcirc\text{--NHCOCH}_2\text{SH} \qquad \text{VIII, MADA}$$

coagulum products contain substantial yields of adduct concentrates.
When these are diluted in unstabilized rubbers as normal additives,
they give interpenetrating networks when vulcanized in the conventional way (23). The latex concentrates can also be diluted to
normal antioxidant concentrations in unstabilized nitrile rubber
latex followed by normal coagulation and vulcanization (23). Some
typical results obtained by the latex concentrate (masterbatch) procedure are described in Table 5, before and after solvent extraction.
The copolymerized antioxidant (HR 665) and the thiol adduct
antioxidant adducts are much more effective than the best available
oligomeric antioxidant, Flectol H.
        Nitrile rubber is the preferred elastomer for use in seals and
gaskets in contact with hot hydrocarbon oils because of its resistance to swelling (6). However, conventional antioxidants are
readily removed by leaching during use and the polymer rapidly
becomes subject to oxidative deterioration. This leads to increase
in modulus and hardness and to loss of useful properties. This is an
obvious application for polymer-bound antioxidants. Figure 3 shows
the effect of a 20% latex concentrate of MADA-B when diluted to 2% in
NBR in a cyclical oil extraction (150°C)/oven aging (150°C) test
(27). It is evident that prior extraction has little effect on the
performance of MADA-B, which is very much superior in this test to
either the conventional antioxidant, Flectol H, or the copolymerized
antioxidant, Chemigum HR 665. The additive used as a coagulum
concentrate (LMC) was also more effective than the latex concentrate
(LML).
        Rubber-modified plastics (e.g, ABS), which are available in
latex form, have also been modified with thio antioxidants and UV
stabilizers. The 2-hydrobenzophennone (IX) reacts to give a 30%
concentrate of which 80% of the UV stabilizer used becomes chemically
attached to the polymer (24). The concentrates can be used either as

---

* The suffix -B indicates that the additive is substantially bound
  to the polymer.

Table 3. Comparison of Nitrile Rubbers Containing a Copolymerized
Antitoxidant (HR 665) and a Conventional Additive (Octylated
Diphenylamine, OD)

FORMULATION	TIME TO ABSORB 1% O_2 AT 100°C, HRS.	
	HR 665[*]	OD
UNVULCANISED (U)	676	250
UNVULCANISED (E)	620	10
VULCANISED (U)	290	185
VULCANISED (E)	415	16

[*] Contains copolymerized (◯)-NH-(◯)-NHCOC=CH$_2$ $\overset{CH_3}{|}$ (IV)

U, unextracted; E, extracted.

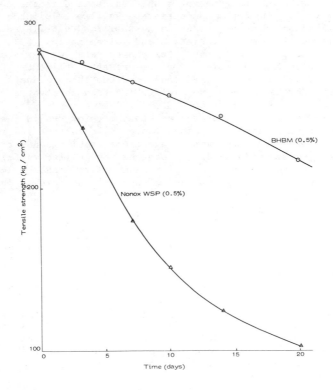

Figure 1. Comparison of bound antioxidant (BHBM) with a
conventional bisphenol in NR (air oven test at 100 °C).

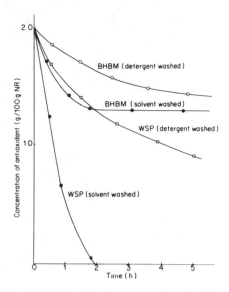

Figure 2. Residual antioxidant remaining during solvent (pet ether/toluene at 20 °C) and detergent (1% Tide at 100 °C) leaching of a typical bisphenol (WSP) and a polymer-bound phenol (BHBM).

Table 4. Effect of Antioxidant Concentration in NBR Lattices on the Yield of Adduct

ADDED ANTIOXIDANT CONCENTRATION y/100y NBR	PERCENTAGE BOUND (g/100g of antioxidant added)	
	BHBM	MADA
1	37	0
5	-	13
10	58	40
20	60	52
30	63	*
40	62	*
50	62	*

* Latices coagulated

Table 5.   Effectiveness of Bound Antioxidants from 20% Concentrate,
Reduced to 2 g/100 g in NBR After Vulcanization[a]

VULCANIZATE	TIME TO 1% OXYGEN ABSORPTION AT 150°C	
	UNEXTRACTED	EXTRACTED[b]
CONTROL	5	2
FLECTOL H[c]	38	6
CHEMIGUM HR 665[d]	45	41
BHBM-B	33	29
MADA-B	49	48

a High accelerator/low sulphur; b Hot extracted
   with MeOH (24 hr)
c Flectol H is an oligomeric antioxidant (ASTM standard)

d Commercial polymer containing a copolymerised antioxidant.

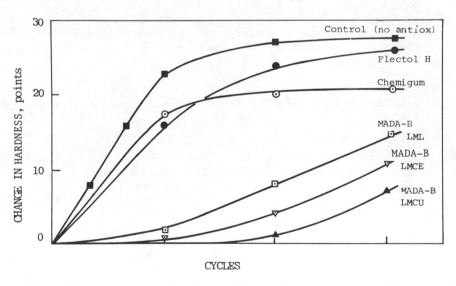

Figure 3.   Increase in hardness of nitrile containing rubber
antioxidants in a cyclical hot oil (150 °C)/hot air (150 °C)
test. LML, latex concentrate (20%) diluted to 2%; LMCE, latex
coagulum (20%), unextracted before incorporation into polymer;
LMCU, latex coagulum (20%), unextracted before incorporation into
polymer at 2%.

latex additives or as solid polymer additives during processing to give a superior antioxidant effect compared with the commercial UV stabilizer Cyasorb UV 531 (X) which contains the same functional group; see Table 6. It is evident that even after exhaustive extraction, EBHPT (IX) is considerably more effective than the conventional stabilizer which has not been extracted.

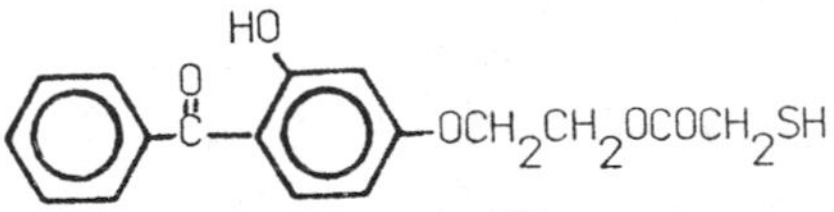

IX

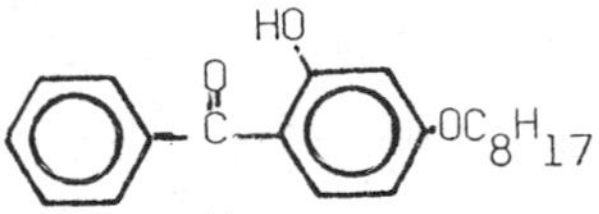

X

Post-Fabrication Treatment. UV light can be used to induce reaction of functional antioxidants with polymers. This can be achieved either by incorporating the reactive antioxidant during processing (20) or by subsequently surface treating the artifact (19). The former is more convenient, and Figure 4 shows that 68% of BHBM (VI) can be combined with polypropylene film within two hours of UV exposure. The reaction involved is a UV-initiated thiol addition process involving the tertiary hydrogens in PP (19). Table 7 compared the effectiveness of BHBM-B made by the above process in polypropylene and its lower molecular mass analogues (XI) added normally during processing (11).

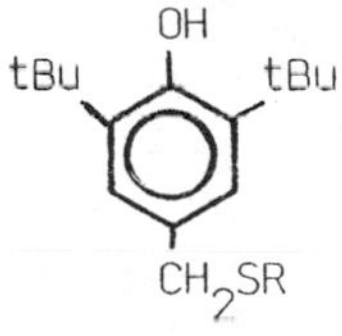

XI

This illustrates again the importance of non-volatility, and there is evidence (11) that BHBM may itself be partially converted to less volatile antioxidants during normal processing.

Reactions of Antioxidants with Polymers During Processing. One of the earliest polymer-bound antioxidants was obtained by reaction of nitroso antioxidants (ANO) with rubbers during vulcanization (29). The chemistry of this process is complex, but its discoverers proposed an "ene" reaction with the unsaturation in the polymer.

Table 6.   Comparison of EBHPT_B, Added as a 30% Concentrate to
ABS, with a Conventional UV Stabilizer at the Same Concentration
$(3.03 \times 10^{-3}$ mol/100 g)

UV STABILIZER	EMBRITTLEMENT TIME, hr
CONTROL, NO ADDITIVE	25
CYASORB UV 531 (U)	35
EBHPT-B (U)	90
EBHPT-B (E)	72

U, unextracted,   E, extracted

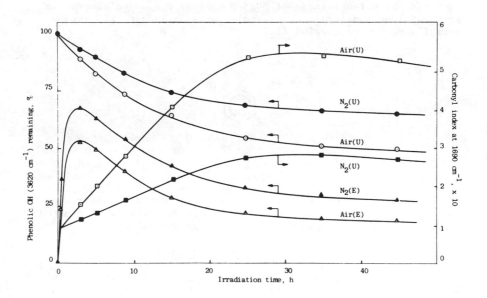

Figure 4.   Binding of BHBM to polypropylene in nitrogen and in air
by UV irradiation.   Phenolic OH measured by IR before (U) and
after (E) extraction.

$$
\begin{array}{c}
CH_3 \\
| \\
-CH_2C=CHCH_2-
\end{array}
\; + \; ANO \;\longrightarrow\;
\begin{array}{c}
CH_3 \\
| \\
-CH=CCHCH_2- \\
| \\
NOH \\
| \\
A
\end{array}
\; + \;
\begin{array}{c}
CH_3 \\
| \\
-CH=CC \\
\quad\; \| \\
N\rightarrow O \\
| \\
A
\end{array}
$$

$$
\begin{array}{c}
\downarrow \\
CH_3 \\
| \\
-CH=CCHCH_2- \\
| \\
NH \\
| \\
A
\end{array}
\tag{3}
$$

This process was not developed commercially, but closely related chemistry is used in a novel non-sulfur crosslinking agent for rubber (30, 31).

Another polymer reactive antioxidant which can be combined with rubber during vulcanization involves the 1,3 addition reaction of nitrones to the double bond in rubbers, reaction 4 (11).

$$
\begin{array}{c}
CH_3 \\
| \\
-CH_2C=CHCH_2-
\end{array}
\; + \;
\begin{array}{c}
O \\
\uparrow \\
R'N=CHR''
\end{array}
\;\longrightarrow\;
\begin{array}{c}
-CH_2 \quad CH_2- \\
\diagdown \quad \diagup \\
CH_3-C-CH \\
\diagup \quad \diagdown \\
O \qquad CHR'' \\
\diagdown \quad \diagup \\
N \\
| \\
R'
\end{array}
\tag{4}
$$

The antioxidant function may be in either R' or R''. Again, commercial development has not occurred due to an undesirable balance of other technological properties during vulcanization.

It has more recently been shown that there are advantages in separating the chemical reaction of antioxidants with solid polymers from the vulcanization process. It has long been recognized (32) that mechano-scission of polymer chains during processing or polymers gives rise to free radicals which can initiate radical chain reactions (32). Block copolymers can be synthesized in this way, reaction 5;

$$
\begin{array}{c}
CH_3 \quad\quad CH_3 \\
| \qquad\quad\; | \\
-C=CHCH_2CH_2C=CH-
\end{array}
\;\xrightarrow{\;Shear\;}\;
\begin{array}{c}
CH_3 \\
| \\
-C=CHCH_2^{\bullet}
\end{array}
\; + \;
\begin{array}{c}
CH_3 \\
| \\
{}^{\bullet}CH_2C=CH-
\end{array}
$$

$$
\qquad\qquad\qquad\qquad (R^{\bullet}) \qquad\qquad ({}^{\bullet}R') \tag{5}
$$

$$
R^{\bullet} \; + \;
\begin{array}{c}
CH_3 \\
| \\
CH_2=CCOOCH_3
\end{array}
\;\longrightarrow\;
\begin{array}{c}
CH_3 \quad\quad CH_3 \\
| \qquad\quad\; | \\
R(CH_2CH)_nCH_2CH^{\bullet}
\end{array}
$$

The same process can be used to graft vinyl antioxidants to polymer chains during processing of elastomer or to initiate the formation of polymer adducts with thiol antioxidants (21, 22, 24).

Table 8 shows that over 50% binding of MADA (VIII) can be achieved in both NR and SBR by the mechanochemical procedure (21) and that this increases during vulcanization to about 70%. Other sulfur-containing antioxidants, VII and IX, can be bound to similar levels in both polymers. Figure 5 shows the effect of extraction on MADA-B in NR in comparison with a conventional thermal antioxidant, WSP (II), at 100°C. Similar results for SBR are listed in Table 9 before and after solvent extraction. The hindered phenolic antioxidant, BHBM-B is in general somewhat less effective than MADA-B, but is nevertheless much less affected by extraction than are conventional additives.

BHBM-B and MADA-B concentrates, made by the mechanochemical procedure, are highly effective in preventing hardening of nitrile rubber in the cyclical hot oil/hot air test referred to earlier. Figure 6 shows that MADA-8 is again much superior to the oligomeric antioxidant Flectol H in retarding change in elongation at break (35). A summary of the effects of BHBM-B and MADA-B compared with commercial antioxidants is given in Table 10.

Rubber-modified plastics also react readily with thiol antioxidants mechanochemically to give high yields of bound antioxidants (18). An interesting feature of this method of preparation in ABS is the increase in effectiveness that occurs with increase in concentration (see Figure 7). Moreover, the effectiveness of the mechanochemically formed adducts is higher than the corresponding latex adducts.

Table 11 compares the effectiveness of a synergistic UV stabilizer (BHBM-B + EBHPT-B) with some commercial stabilizing systems for ABS added conventionally. The exceptional activity of the polymer-bound system is believed to be due to the fact that it is confined to the rubber phase of the polyblend (18), which is known to be more sensitive than the thermoplastic phase to the effects of both heat and light (36). This finding, if confirmed in other multiphase systems, could be of considerable importance for the stabilization of heterogeneous polymer blends.

Ethylene-propylene-diene terpolymers (EPDM) are being used increasingly as impact modifiers and solid phase dispersants (37) as well as being important elastomers in their own right. Although EPDM is relatively stable in comparison with the polydiene elastomers, the unsaturation in the polymer can be used to improve its thermal and UV stability even further (38). Table 12 shows that a substantial level of binding of both EBHPT and MADA can be achieved in EPDM by the mechanochemical procedure. The resulting antioxidant-modified EPDM, when added to normal EPDM, gave an impact-modifier for PP with considerably improved heat resistance (see Table 13) (39). MADA-B is also a reasonably effective photoantioxidant with similar activity to EBHPT-B (see Table 14). Before solvent extraction, both are less effective than a conventional UV stabilizer (UV 531), but they are more effective after extraction.

Although the major thermoplastics (PE, PP, PVC, and PS) do not formally contain unsaturation, they do undergo mechanomodification with reactive antioxidants (40 - 42). This process is particularly efficient in the case of PVC due to the unsaturation produced during processing (41 - 43). This is shown typically for BHBM in Figure 8 in the presence of a typical melt stabilizer, dibutyl tin maleate (DBTM, XII).

Table 7. Antioxidant Activity of Homologous Phenolic Sulphides (XI) in an Air Oven at 140 $^{\circ}$C (Concentration 6 x 10^{-4} mol/100 g)

R	INDUCTION PERIOD TO THERMAL OXIDATION, hr
H	21.0
C_2H_5	5.0
C_5H_{11}	6.0
C_8H_{17}	7.5
$C_{12}H_{25}$	11.0
$C_{18}H_{37}$	15.0
PP (BHBM-B)	81.0

Table 8. Mechanochemical Binding of MADA to Rubbers (10 g/100 g) at 70 $^{\circ}$C

RUBBER	EXTENT OF BINDING,% OF THEORETICAL	
	BEFORE VULCANISATION	AFTER VULCANISATION
NR	50	68
SBR	58	74

Table 9. Antioxidant Activity of 10% MADA Concentrate in SBR at 100 $^{\circ}$C (2 g/100 g)

PROCESSING AND POST-TREATMENT PROCEDURE	ACTUAL ANTIOXIDANT CONCENTRATION,g/100g	INDUCTION PERIOD,hr	TIME TO 1% O_2ABS,hr
U,V (U)	2.0	30	196
E,V (U)	1.16	29	143
U,V (E)	1.48	45	189
E,V (E)	1.16	42	107
CONTROL (U)	-	1	9.5
CONTROL (E)	-	0	7.0
WSP (U)	2.0	32	132
WSP (E)	-	4	13

U, unextracted before vulcanisation, E, extracted before vulcanisation
V, vulcanised, (U),vulcanisate unextracted, (E), vulcanisate extracted.

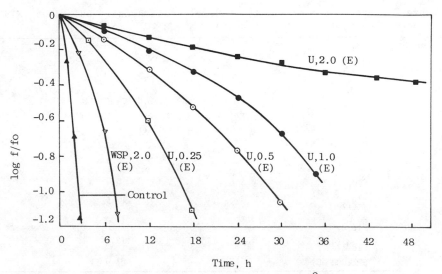

Figure 5.   Continuous stress relaxation at 100 °C of NR
vulcanizates containing a bound antioxidant (MADA-B).   U,
concentrate unextracted before dilution in rubber; E, extracted
after vulcanization.   Numbers on curves are concentrations, g/100 g.

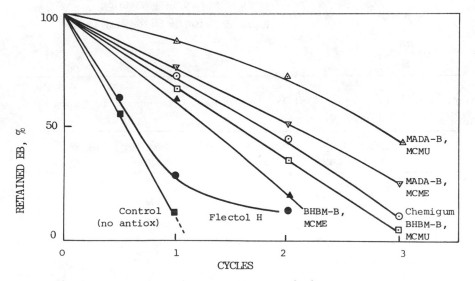

Figure 6.   Decay in elongation at break (E_B) of nitrile rubbers
containing antioxidants (2%) in a cyclical hot oil (150 °C)/
hot air (150 °C) test.   MCMU, mechanochemical concentrate,
unextracted before use; MCME, mechanochemical concentrate,
solvent extracted before use.

Table 10. Changes in Mechanical Properties of NBR Vulcanisates
Containing Bound Antioxidants in a Cyclical Oil/Air Oven Test at 150 ^{C}C.

ANTIOXIDANT	NUMBER OF CYCLES TO PROPERTY CHANGE			
	H_{10}	M_{50}	Ts_{50}	Eb_{50}
BHBM-B[*]	1	2	2	1
MADA-B[*]	3	3	3	3
FLECTOL H	1	1	1	1
CHEMIGUM HR 665	1	2	2	2
CONTROL (NO ANTIOX)	1	1	1	1

[*]Diluted from 20% masterbatch concentrate.

H_{10}, increase in hardness by 10 points.

M_{50}, 50% increase in modulus.

Ts_{50}, 50% decrease in tensile strength.

Eb_{50}, 50% decrease in elongation at break.

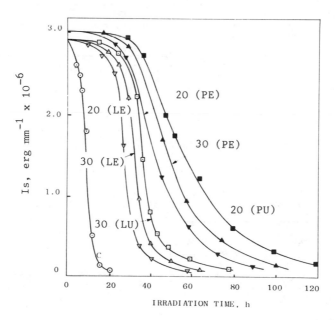

Figure 7. Loss of impact strength (I_s) of ABS during UV
exposure. L, latex concentrate (masterbatch); U, unextracted;
P, mechanochemical concentrate. Numbers on curves are the
concentrations of photoantioxidants in the concentrates.

Table 11.   Comparison of a Bound Synergistic UV Stabilizer with
Conventional Additives at 1 g/100 g ABS

ADDITIVE	INDUCTION PERIOD,hr	EMBRITTLEMENT TIME,hr
BHBM-B + EBHPT-B (U)	80	380
BHBM-B + EBHPT-B (E)	50	220
UV 531 (U)	10	40
BHT (U)	9	34
DLTP (U)	5	25
BHT + UV 531 (U)	25	65
BHT + UV 531 + DLTP (U)	25	85

Table 12.   Yields of Polymer-Bound Antioxidants in EPOM

ANTIOXIDANT CONCENTRATION REACTED IN POLYMER (g/100g)	EXTENT OF BINDING,% OF THEORETICAL	
	EBHPT[*]	MADA[+]
2	65	75
5	70	82
10	77	87

* Reacted at 100°C / 15min.   [+]Reacted at 150°C / 15 min.

Table 13.   Induction Periods (IP) and Time to 50% Loss of Impact
Strength (Is_{50}) for PP/EPDM (75/25) Blends in an Air Oven at 140 °C

ANTIOXIDANT	CONCENTRATION (10^4 mol/100g)	IP,hr	Is_{50},hr
NONE	–	0	5
MADA-B	3[*]	50	105
NiDBD[+]	3	50	93

[*] From 2% masterbatch concentrate, [+] Nickel dibutyl dithiocarbamate

Table 14. Effectiveness as UV Stabilizers of EBHP-B and MADA-B in PP/EPDM (75:25) Blends Before and After Solvent Extraction (from 10% Concentrates)

STABILIZER	CONCENTRATION g/100g	EMBRITTLEMENT TIME,hr	
		UNEXTRACTED	EXTRACTED
NONE	-	180	-
EBHPT-B	0.16	330	310
MADA-B	0.16	330	315
UV 531	0.20	400	200

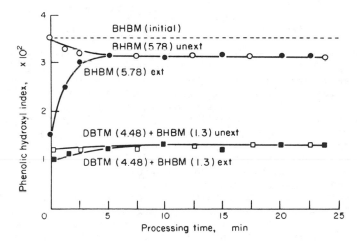

Figure 8. Retention of BHBM in PVC during processing at 170 °C in the presence and absence of dibutyl tin maleate (DBTM). Numbers on the curves are additive concentrations (10^3 mol/100 g).

$$Bu_2Sn \underset{O-C}{\overset{O-C}{\diagdown}} \overset{\overset{O}{\diagup}}{\underset{\diagdown O}{\underset{CH}{\overset{CH}{\|}}}} \qquad XII\ (DBTM)$$

The unsaturation and hydroperoxides formed initially in the polymer during the early stages of processing both disappear as a result of the peroxide catalyzed thiol addition to the double bond (41, 42). The resulting BHBM-B is a very powerful thermal antioxidant in an air oven test at 140°C (41), see Figure 9. EBHPT can be similarly bound to PVC to give an effective UV stabilizer (42).

Thiol antioxidants can also be reacted with polyethylene and polypropylene (40). Figure 10 illustrates the superiority of MADA-B in PP compared with conventional thermal antioxidants when subjected to continuous hot water leaching. It is also an effective processing stabilizer and photoantioxidant. Table 15 shows that the latter effect is resistant to solvent extraction. The loss of activity after extraction corresponds to the amount of unbound MADA present.

Theoretical Implications of Polymer-Bound Antioxidants

The study of bound antioxidants in polymers has provided fundamental information about the way antioxidants function, particularly in heterogeneous systems. It has been accepted almost without question that in order to scavenge alkylperoxyl radicals in the polymer, a chain-breaking antioxidant has to migrate through the matrix. If this is correct, then it should follow that bound antioxidants must be less effective than low molecular mass analogues. Moreover, the effectiveness of the bound system should decrease as the concentration of the bound concentrate increases. The evidence does not support such a conclusion. Figure 6 shows that in a fully extracted polymer, antioxidant effectiveness actually increases with the concentration of the adduct in the concentrate. Similar conclusions have been reached for rubbers (34), and it must be assumed that the ability to migrate through the polymer matrix is not an essential requirement of an antioxidant. However, it must be recognized that in rubbery polymers, an antioxidant attached to a polymer chain does not have the localized mobility associated with the segment of the polymer chain to which it is attached. This may well be sufficient to protect a micro-volume of the polymer.

The requirements for an antiozonant are believed to be more stringent. Since ozone attack is essentially a surface phenomenon, it is assumed (44-46) that an antioxidant can only be effective in the surface of the rubber. However, it has been found (47) that extraction makes very little difference to the antiozonant activity of MADA-B in rubber (see Table 16). The mechanism of antiozonant action may therefore require some revision.

Perhaps the most important conclusion to emerge from the study of bound antioxidants is that it is much more cost-effective to attach antioxidants to those domains in the polymer that are most susceptible to oxidation. In the case of rubber-modified polyblends,

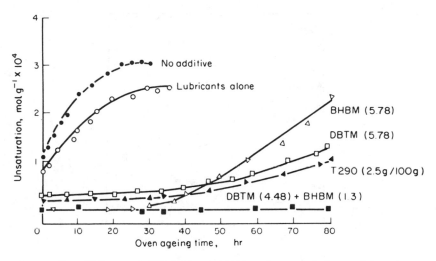

Figure 9. Effect of BHBM-B and DBTM, alone and in combination on the formation of unsaturation in PVC during oven aging at 140 °C. All formulations contain lubricants. Numbers on curves are concentrations (10^3 mol/100 g). T290 is a stabilizer containing a tin maleate and a conventional hindered phenol.

Table 15. Effect of Solvent Extraction on the Photoantioxidant Activity of Bound Antioxidants

ANTIOXIDANT	EMBRITTLEMENT TIME, hr/ % REDUCTION[*] (ET)		
	3.0 / %	4.5 / %	6.0 / %
MADA–B (U)	220 ⎱ 23	280 ⎱ 31	360 ⎱ 22
MADA–B (E)	190 ⎰	220 ⎰	300
UV 531 (U)	340 ⎱ 96	–	–
UV 531 (E)	100 ⎰	–	–

[*] % Reduction on extraction = $\dfrac{ET(U) - 2ET(control) - ET(E)}{ET(U) - ET\ (control)}$

(U), unextracted, (E), extracted

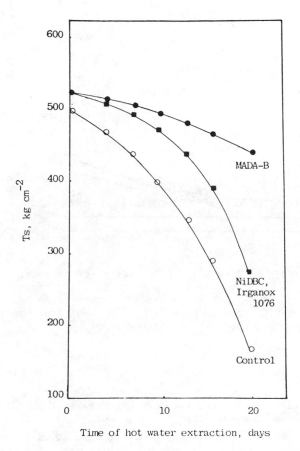

Figure 10. Comparison of a bound antioxidant, MADA-B, and conventional heat stabilizers in the stabilization of polypropylene in an air/hot water extraction test.

Table 16. Antioxidant and Antiozonant Activity of MADA-B

FORMULATION	TIME TO 1% O_2 ABS.,(70°C) hr	TIME TO GRADE 5 CRACKING IN OZONE (20% EXTENSION),hr
NO ANTIOXIDANT	9	5
MADA-B (U)	60	22
MADA-B (E)	51	22

(U), unextracted, (E), extracted

this is normally the rubbery domain (48). Attachment through the double bonds of the rubber leads, therefore, to molecular dispersion of the protective agent in high concentration exactly where it is most required.

Literature Cited

1. S. Al-Malaika and G. Scott, Degradation and Stabilization of Polyolefins, Ed. N. S. Allen, App. Sci. Pub., Ltd., London, 1983, p. 247.
2. N. C. Billingham and P. D. Calvert, Developments in Polymer Stabilization-3, Ed. G. Scott, App. Sci. Pub., Ltd., London, 1980, p. 139.
3. K. B. Chakraborty, G. Scott and W. R. Poyner, Plast. Rubb. Proc. and App., 3, 59 (1983).
4. K. B. Chakraborty, G. Scott and W. R. Poyner, Europ. Polym. J., in press.
5. S. Al-Malaika, K. B. Chakraborty, G. Scott and S. B. Tao, Polym. Deg. and Stab., in press.
6. D. K. Thomas, Developments in Polymers Stabilization-1, Ed. G. Scott, App. Sci. Pub., Ltd., London, 1979, p. 137.
7. R. M. Russell, Brit. Polym. J., 1, 53 (1969).
8. R. N. Dodge and S. K. Clark, Soc. Automotive Eng., Tech. Paper No. 821392, Oct., 1982.
9. A. S. Kuzminsky, Developments in Polymer Stabilization-4, Ed. G. Scott, App. Sci. Pub., Ltd., London, 1981, p. 71.
10. S. D. Razumovskii and G. E. Zaikov, Developments in Polymer Stabilization-6, Ed. G. Scott, App. Sci. Pub., Ltd., London, 1983, p. 239.
11. G. Scott, Developments in Polymer Stabilization-4, Ed. G. Scott, App. Sci. Pub., Ltd., London, 1981, p. 181.
12. G. D. Meyer, R. W. Kavchok and J. F. Naples, Rubb. Chem. Tech., 46, 106 (1973).
13. J. W. Horvath, D. C. Grimm and J. A. Stevick, Rubb. Chem. Tech., 48, 337 (1975).
14. J. W. Horvath, J. R. Burdon, J. R. Meyer and J. A. Stevick, Paper presented to the ACS Meeting (Chicago, Aug., 1973).
15. Anon., Chem. Eng. News, 38 (April 7, 1980).
16. V. P. Kenpicher, L. K. Yajubishi and C. N. Malglysh, Rubb. Chem. Tech., 43, 1228 (1970).
17. K. W. S. Kularatne and C. Scott, Europ. Polym. J., 15, 827 (1979).
18. M. Ghaemy and G. Scott, Polym. Deg. and Stab., 3, 405 (1980-81).
19. B. W. Evans and G. Scott, Europ. Polym. J., 10, 453 (1974).
20. G. Scott., Brit. Pat. 1,503,501 (1975; U.S. Pat. 4,213,892 (1980); U.S. Pat. 4,354,007 (1982).
21. G. Scott and M. F. Yusoff, Polym. Deg. and Stab., 3, 53 (1980).
22. G. Scott and S. M. Tavakoli, Polym. Deg. and Stab., 4, 343 (1982).
23. O. Ajiboye and G. Scott, Polym. Deg. and Stab., 4, 415 (1982).
24. G. Scott and E. Setoudeh, Polym. Deg. and Stab., 5, 81 (1983).
25. P. J. Burchill and D. M. Pemberton, J. Polym. Sci., Symp. 55, 185 (1976).
26. G. Scott, Plast. and Rubb. Proc., 41, (June, 1977).

27. K. W. S. Kularatne and G. Scott, Europ. Polym. J., $\underline{14}$, 835
     (1978).
28. O. Ajiboye and G. Scott, Polym. Deg. and Stab., $\underline{4}$, 397 (1982).
29. M. E. Cain, G. T. Knight, P. M. Lewis and B. Saville, Rubb. J.,
     $\underline{50}$, 204 (1968).
30. Anon., Rubber Developments, $\underline{28}$, 90 (1975).
31. P. M. Lewis, Developments in Polymer Stabilization-7, Ed.
     G. Scott, App. Sci. Pub., Ltd., London, 1984, Chap. 3.
32. M. Pike and W. F. Watson, J. Polym. Sci., $\underline{9}$, 229 (1952).
33. G. Scott and S. M. Tavakoli, Polym. Deg. and Stab., $\underline{4}$, 279
     (1982).
34. G. Scott and S. M. Tavakoli, Polym. Deg. and Stab., $\underline{4}$, 267
     (1982).
35. O. Ajiboye and G. Scott, Polym. Deg. and Stab., $\underline{4}$, 415 (1982).
36. M. Ghaemy and G. Scott, Polym. Deg. and Stab., $\underline{3}$, 233 (1981).
37. G. Scott, Polym. Plast. Tech. and Eng., in press.
38. G. Scott and E. Setoudeh, Polym. Deg. and Stab., $\underline{5}$, 8 (1839).
39. G. Scott and E. Setoudeh, Polym. Deg. and Stab., $\underline{5}$, 11 (1839).
40. G. Scott and E. Setoudeh, Polym. Deg. and Stab., $\underline{5}$, 1 (1839).
41. B. B. Cooray and G. Scott, Europ. Polym. J., $\underline{16}$, 1145 (1980).
42. B. B. Cooray and G. Scott, Europ. Polym. J., $\underline{17}$, 379 (1981).
43. B. B. Cooray and G. Scott, Europ. Polym. J., $\underline{17}$, 385 (1981).
44. B. S. Biggs, Rubb. Chem. Tech., $\underline{31}$, 1015 (1959).
45. W. L. Cox, Rubb. Chem. Tech., $\underline{32}$, 364 (1959).
46. G. Scott, Atmospheric Oxidation and Antioxidants, Elsevier,
     London, New York, 1965, p. 499 et seq.
47. A. A. Katbab and G. Scott, Polym. Deg. and Stab., $\underline{3}$, 221 (1981).
48. G. Scott, Developments in Polymer Stabilization-1, Ed. G. Scott,
     App. Sci. Pub., Ltd., London, 1979, p. 309.

RECEIVED December 7, 1984

Polymerizable, Polymeric, and Polymer-Bound (Ultraviolet) Stabilizers

OTTO VOGL, ANN CHRISTINE ALBERTSSON[1], and ZVONIMIR JANOVIC[2]

Polytechnic Institute of New York, Brooklyn, NY 11201

A number of polymerizable ultraviolet stabilizers have been synthesized and their homo- and copolymerization studied. The initial work consisted of the synthesis of vinyl derivatives of methyl salicylate (three isomers), 2,4-dihydroxybenzophenone, and ethyl α-cyano-β-phenylcinnamate. More recently, vinyl derivatives (two isomers) and one isopropenyl derivative of 2(2-hydroxyphenyl)2H-benzotriazole have been prepared. A number of benzotriazoles with more than one benzotriazole ring in the molecule, or compounds with both benzo(or aceto)phenone and 2(2-hydroxyphenyl)2H-benzotriazole groups in one molecule, have also been synthesized. Acryloyl and methacryloyl derivatives of benzotriazole-substituted polyphenols have been prepared and homo- and copolymerized.

Other polymerizable ultraviolet stabilizers and antioxidants have also been synthesized and incorporated into polymeric structures.

Many polymers undergo thermal and thermal oxidative degradation during fabrication processing and require stabilization. Over longer periods of time and at ambient temperature, polymers also deteriorate in the solid state (aging, weathering) through autooxidation and photooxidation. Stabilization against these two types of deterioration is also necessary. In outdoor applications, where the materials

NOTE: This paper was written in cooperation with Eberhard Borsig, Amitava Gupta, Shanjun Li, Zohar Nir, Witold Pradellok, Fu Xi, and Shohei Yoshida.

[1] Current address: Department of Polymer Technology, Royal Institute of Technology, Stockholm, Sweden.

[2] Current address: INA-Research Institute, Zagreb, Yugoslavia.

0097–6156/85/0280–0197$06.00/0

are exposed to ultraviolet solar radiation above 290 nm, the energy of this radiation is primarily absorbed by functional groups, particularly carbonyl groups (impurities in polymers:carbonyl groups are impurities in polyolefins, but major components in acrylics, for example); their excited states can lead to the initiation of photochemical reactions and, ultimately, to polymer degradation. Fortunately, the quantum yields of many photochemical reactions are low, but over a period of years, degradation occurs. Some of this degradation is aided by moisture; the total exposure to the environment is called weathering. The deterioration of plastic materials: oxidation, chain cleavage, crosslinking, and the elimination of small molecules, leads to the loss of the desirable properties and to the failure of the polymer.

Plastics are commonly protected against such deterioration (1-4) by the addition of antioxidants against thermal oxidation, and of ultraviolet stabilizers which can absorb the damaging radiation and re-emit it in a harmless form or function by trapping radicals or decomposing hydroperoxides.

The stabilizers must be effective over a long period of time. It is important that the stabilizers do not volatilize, be leached out, or can otherwise be removed from the plastic material. It is also important that the stabilizer be distributed in the polymeric matrix where it is most needed, particularly on the surface of the materials. These conditions require that the stabilizers be compatible, particularly within the amorphous fraction of the polymer. In semicrystalline materials, the stabilizers are substantially excluded from the crystalline part and the spherulites of such polymers.

Low molecular weight molecules readily diffuse through the amorphous portion of the polymer; even higher molecular weight materials often move through the amorphous matrix. In polyolefins, the solubility of stabilizers is relatively low because most stabilizers, both antioxidants and ultraviolet stabilizers, are aromatic compounds of relatively high polarity which are not very compatible with the paraffinlike hydrocarbon polymers.

Significant amounts of stabilizers are often lost from polymeric materials because of volatilization during fabrication or because of exudation, leaching, or solvent extraction during end use. This problem is especially severe with semicrystalline polymers which have relatively small amorphous fractions and with articles having a high surface-to-volume ratio, such as films or fibers.

Mobility and compatibility of chemical compounds in a polymer matrix are functions of molecular weight, molecular size, and relative polarity of the stabilizers (antioxidants, ultraviolet stabilizers). To increase compatibility and decrease the loss of stabilizers, high molecular weight products having several stabilizing groups in the same molecule have been synthesized. Molecules with stabilizer characteristics have been used as capping agents for oligomers or low molecular weight polymers, or have been attached to compatibilizing groups such as linear alkyl groups of 12-15 carbon atoms for stabilizers to be used in polyolefins.

All this work has led to stabilizers that are less volatile, more compatible, and are consequently less readily lost during fabrication and exposure to the environment. The myth that stabilizers must have relatively high mobility to be effective has, however, persisted, and only recently have indications become compelling that high mobility of low molecular weight compounds is not essential for the effectiveness of stabilizers in polymeric materials. In fact, incorporation of polymerizable stabilizers into polymer chains has been found to make them effective, especially in long-term exposure. Sometimes a flexible, rather short spacer group between the polymer backbone chain and the active stabilizer moiety is necessary; this general approach has become the most up-to-date target of investigations.

Recently, epoxy and olefin monomers with spacer groups between the functional group and the polymerizable groups have been synthesized and their polymerization established; functional epoxides and olefins polymerize well when the functional ester group is separated from the polymerizable group by a spacer group of at least three methylene groups (5,6), onto which the polymer stabilizer groups can now be attached.

Polymerizable and Bound Ultraviolet Stabilizers

When ultraviolet radiation is absorbed by a macromolecule which is excited to a higher energy state, the excited compound or the macromolecule usually returns to the electronic ground state because the energy is dissipated by photophysical pathways by: (a) radiationless conversion to the ground state and release of vibrational energy (heat), (b) emission of the energy of longer wavelengths (fluorescence or phosphorescence), (c) transfer of the energy to another molecule, as in the case of radical trappers or hydroperoxide decomposers.

Typical ultraviolet absorbers are salicylate esters, 4-aminobenzoate esters, α-cyano-β-phenylcinnamate esters, 2-hydroxybenzophenones, 2(2-hydroxyphenyl)2H-benzotriazoles (Figure 1), and more recently diaminodiphenyloxamides. Some pigments and various types of carbon blacks are also known as effective ultraviolet stabilizers. Another mechanism by which the stabilization of the polymers can be accomplished is by quenching the excited state of the molecules. Very few compounds are known where the energy of the excited state is transferred to another molecule; nickel chelates are the best-known examples.

Review of Approaches. Over the last two decades, observations have been made suggesting that polymers containing ultraviolet stabilizing groups directly attached to the polymer chain may be very effective in elastomeric or amorphous (glassy) polymers. Some experiments have shown that such stabilizers incorporated by grafting did not diminish the activity of the stabilizers.

Most attempts in the past to prepare higher molecular weight ultraviolet stabilizers have used the reaction of a preformed polymer with an ultraviolet-absorbing small molecule, e.g., of polymeric methacrylic acid salts with the 2-bromoethyl ether of the 4-hydroxyl group of 2,4-dihydroxybenzophenone (2).

Probably the first polymerizable ultraviolet absorbers for general use were acryloyl or methacryloyl derivatives of the 4-hydroxy group of 2,4-dihydroxybenzophenone (7). Acrylic and methacrylic esters have also been prepared from 2(2-hydroxyphenyl)2H-benzotriazoles with a phenolic hydroxyl group in the carbocyclic ring of a benzotriazole group (8).

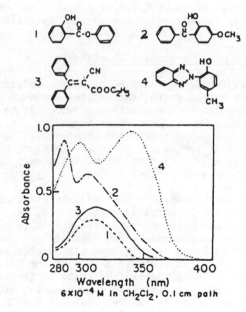

Figure 1.   Typical ultraviolet absorbers.

Allyl groups have also been introduced in the 3 position of 2,4-dihydroxybenzophenone. Nucleophilic displacement reaction of the chloride in 4-chloromethylstyrene by phenolates of 2,4-dihydroxy-benzophenone (9) or 2(2-hydroxy-5-methylphenyl)4'-hydroxybenzotriazole also provided a method of producing styrene-type polymerizable ultraviolet stabilizers (9,10).

Another polymerizable group was introduced by allowing the non-hindered phenolic hydroxyl group in 4 position of 2,4-dihydroxybenzo-phenone to react with polyepichlorohydrin. Allyloxy, vinyloxy, vinyl-sulfonyl, and acryloamido groups have also been utilized in making polymerizable light stabilizers (10).

When stabilizing groups, including ultraviolet-absorbing groups, are introduced into polymers by polymer reactions, they are accompanied by side reactions causing color formation. By using an azo initiator (which contains an ultraviolet-absorbing group, 2-hydroxy-benzophenone) as polymerization initiator, it was possible to introduce a stabilizer at the end of the polymer chain (1).

Ultraviolet-absorbing groups were also modified in such a way that they could be incorporated into condensation polymers, particularly into polyesters, polyamides, and polyurethanes by using tri- and tetra-substituted 2-hydroxybenzophenones with polymerizable hydroxyl and carbonyl groups in the 4 and 4' positions, leaving thus the hydrogen-bonding hydroxyl group in the orthoposition to the carbonyl group.

Very recently it was found that substituted diaminophenyloxa-mides are effective ultraviolet stabilizers. We had now prepared regular copolyoxamides with 4,4'-diaminodiphenyloxamide and 3,3'-di-aminodiphenyloxamide with various diacid chlorides (11).

Polymerizable Ultraviolet Stabilizers — Miscellaneous Types. In our research on polymerizable ultraviolet stabilizers, we have decided to prepare styrene-type monomers in which the vinyl (or isopropenyl) group is directly attached to the phenyl group of the stabilizer, which might be polymerized similarly to styrene. These monomers can indeed be polymerized and copolymerized successfully with styrene, acrylic and methacrylic acid derivatives with azobisisobutyronitrile (AIBN) as the radical initiator (12-14).

Syntheses of styrene derivatives are normally carried out by two general methods (12): (a) Dehydration of a 1-hydroxyethyl group attached to the benzene ring, which in turn is easily obtained by Friedel-Crafts acylation followed by sodium borohydride reduction. The dehydration technique requires that the reaction be carried out in a good vacuum and the product formed be sufficiently volatile to be removed immediately from the reaction, otherwise polymerization (or oligomerization) occurs which reduces the yield of the monomer. (b) Bromination of the ethyl group with N-bromosuccinimide followed by dehydrobromination of the 1-bromoethyl derivative, conveniently carried out with aliphatic tertiary amines in aprotic solvents, such as acetonitrile or dimethylacetamide (Structure 1).

For the preparation of methyl vinylsalicylates (3-, 4-, 5-iso-mers) both synthetic routes were used. Methyl 5-vinylsalicylate was synthesized in about 50% overall yield from methylsalicylate by the first route (dehydration) (15), the synthesis of methyl 3-vinylsali-cylate (16) and methyl 4-vinylsalicylate (17) starting from 2-ethyl-

Structure 1.

phenol or from 3-ethylphenol was accomplished by the second route
(bromination and dehydrobromination) (Structure 2).

Structure 2.

The synthesis of 2,4-dihydroxy-4'-vinylbenzophenone was achieved
from 4-ethylbenzoic acid and resorcinol. The last steps consisted of
bromination of the ethyl group (free phenolic groups protected by
acetylation) to the 1-bromoethyl compound and dehydrobromination (18,
19).

Ethyl 4-vinyl-α-cyano-β-phenylcinnamate (20,21) was synthesized
in a sequence of five steps from 4-ethylbenzoic acid. 4-Ethylbenzo-
phenone was condensed with cyanoacetate by the Knoevenagel reaction;
again, bromination and dehydrobromination were the last two steps to
ethyl 4-vinyl-α-cyano-β-phenylcinnamate.

These ultraviolet absorbers (Structure 2) have been homopolymerized with AIBN as the initiator and copolymerized with styrene,
methyl methacrylate, and to a limited extent with n-butyl acrylate.
The amount of incorporation of the functional monomer depended on the
type of comonomer. A more extensive study of the copolymerization of
methyl 5-vinylsalicylate was carried out with a substantial number of

styrene, acrylic, or methacrylic acid derivatives (22). The type of derivative determined very much the copolymerizability of the comonomer pairs. For methyl 5-vinylsalicylate (15 mol %) and methyl methacrylate, the copolymerization even at high conversion seems to give a comonomer incorporation similar to that of the comonomer feed.

Polymerizable 2(2-Hydroxyphenyl)2H-Benzotriazole Ultraviolet Stabilizers. Synthesis. Some of the most important ultraviolet stabilizers belong to the class of 2(2-hydroxyphenyl)2H-benzotriazoles. Ever since the early work on benzotriazoles by Elbs (23,24), a number of 2(2-hydroxyphenyl)2H-benzotriazoles [2H-benzotriazole-2-yl)2-hydroxybenzene] have been prepared (25). Even some polymerizable compounds of this class had been synthesized. The benzotriazole stabilizing groups had in the past been attached to a polymerizable group via a phenolic hydroxyl group placed in the benzotriazole ring (26).

We have studied the synthesis of 2(2-hydroxyvinylphenyl)2H-benzotriazoles (Structure 3) and synthesized two of the isomers with the vinyl group in 5 and 4' position and the 5- isopropenyl derivative (Figure 2).

2-(2-Hydroxyphenyl) Benzotriazoles

Structure 3.

For the synthesis of 2H5V (27,28), o-nitroaniline was diazotized, the diazonium solution coupled with 4-ethylphenol, and the azo compound reduced, which also caused ring closure; this method is the traditional synthesis of 2(2-hydroxyphenyl)2H-benzotriazole derivatives. In order to prepare for the bromination of 2(2-hydroxy-5-ethylphenyl)2H-benzotriazole (2H5E), the compound was O-acetylated with acetic anhydride in pyridine and the acetyl compound, 2H5E, brominated with N-bromosuccinimide to the 1-bromoethyl derivative (2A5B). Dehydrobromination of 2A5B in dimethylacetamide with tri-n-butylamine was followed by hydrolysis to 2H5V. It was important that 2H5V be handled in the presence of a free radical polymerization inhibitor, preferably picric acid, otherwise 2H5V polymerized. Under less careful conditions of isolation, 2H5V was often contaminated with oligomeric and polymeric products of 2H5V.

A second synthesis for the preparation of 2H5V (29) starts by condensation of o-nitrobenzenediazonium chloride with 4-hydroxyacetophenone. The azo compound was reduced with zinc and sodium hydroxide without reduction of the carbonyl group of the acetyl group. After acetylation, 2(2-acetoxy-5-acetylphenyl)2H-benzotriazole (2A5A) was reduced with sodium borohydride and the secondary alcohol was dehydrated with potassium hydrogen sulfate; after hydrolysis, 2H5V was

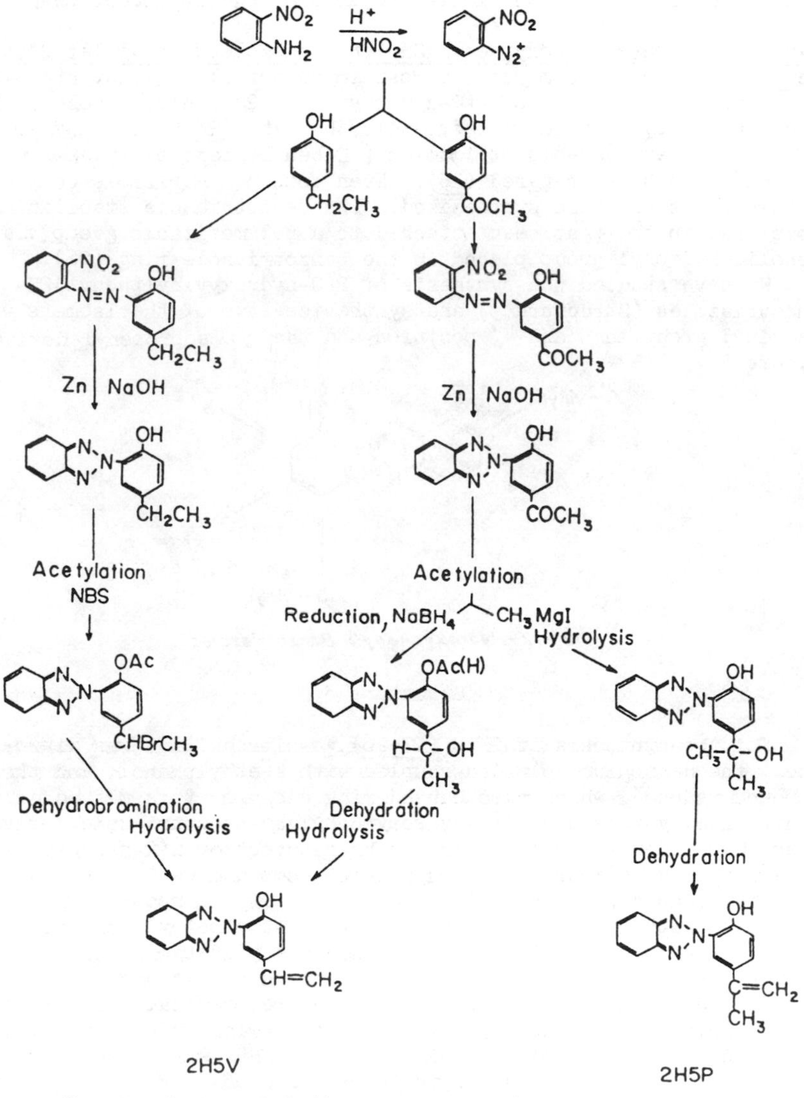

Figure 2. Synthesis of 2-(2-hydroxyphenyl)-2H-benzotriazoles with polymerizable vinyl or isopropenyl groups.

obtained. One of the intermediates of this synthesis, 2A5A, was also a useful intermediate for the preparation of 2(2-hydroxy-5-isopropenyl phenyl)2H-benzotriazole (2H5P). 2A5A was allowed to react with methyl Grignard reagent to give the tertiary alcohol. Dehydration of this compound occurred very readily even by vacuum sublimation; hydrolysis gave 2H5P.

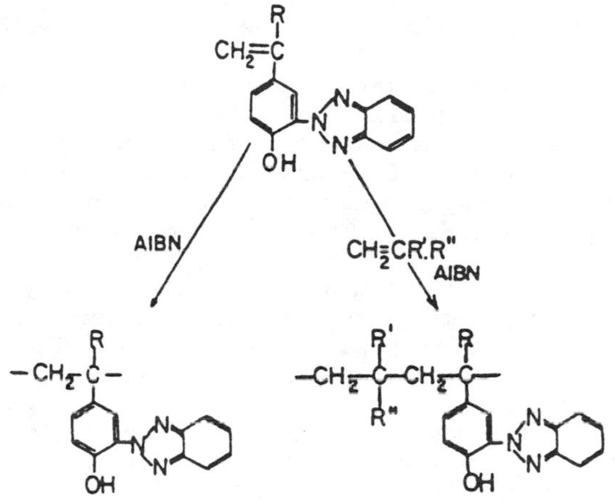

R : H   (4)

R : H , CH$_3$

R': C$_6$H$_5$, COOCH$_3$

R": H , CH$_3$

Structure 4.

Attempts to prepare 2(2-hydroxy-3-vinylphenyl)2H-benzotriazole or 2(2-hydroxy-4-vinylphenyl)2H-benzotriazole failed. When 3-ethylphenol or 4-ethylphenol were allowed to condense with o-nitrobenzenediazonium salts, the condensation did not occur in the 2 position but rather in the 4 position. As a consequence, after ring closure, bromination and dehydrobromination, 2(4-hydroxy-2-vinylphenyl)2H-benzotriazole (4H2V) and 2(4-hydroxy-3-vinylphenyl)2H-benzotriazole (4H3V) were obtained (30) instead of the desired 2H4V and 2H3V.

The synthesis of 2(2-hydroxy-5-methylphenyl)2H-4-vinyl-benzotriazole was accomplished (31) by a sequence of reactions similar to those which gave 2H5V. The starting material for this synthesis was, however, not o-nitroaniline but 4-ethyl-o-nitroaniline. After diazotiation, the diazonium compounds were allowed to react with p-cresol; the condensation product gave 2(2-hydroxy-5-methylphenyl)2H-4-ethyl-benzotriazole after reductive cyclization. This compound was acetylated, brominated, dehydrobrominated, and hydrolyzed to 2(2-hydroxy-5-methylphenyl)2H-4-vinyl-benzotriazole.

Polymerization, Copolymerization, and Grafting. In this work we have prepared vinyl and isopropenyl derivatives of 2(2-hydroxyphenyl)2H-benzotriazoles: 2H5V, 2H5P, and 2(2-hydroxy-5-methylphenyl)2H-4-vinyl-benzotriazole; 2H5V and 2(2-hydroxy-5-methylphenyl)2H-4-vinyl-benzotriazole were homopolymerized and copolymerized with styrene,

methyl methacrylate, and in some cases n-butyl acrylate (Structure 4).
Even when the polymerizations were carried out to high conversion,
copolymers were obtained whose copolymer compositions were similar to
that of the initial comonomer feed. The polymerizations had to be
carried out with complete exclusion of oxygen, otherwise the phenolic
hydroxyl group became an inhibitor of the polymerization since oxygen
radicals are readily terminated by phenols.

2H5P, an α-methylstyrene derivative, seems to have a low ceiling
temperature and consequently did not homopolymerize but underwent copolymerization with styrene, methyl methacrylate, and n-butyl acrylate. Based on the homopolymerization attempts, it appears that 2H5P
is present as isolated monomer units in these copolymers. The copolymerization parameters of 2H5V and 2H5P with styrene, methyl methacrylate, and n-butyl acrylate have also been determined. The
results are shown in Figure 3. The copolymerization experiments were
done to 5% conversions.

2H5V was used as grafting monomer for atactic polypropylene,
ethylene/propylene copolymer, ethylene/vinyl acetate copolymer,
poly(methyl acrylate) and poly(methyl methacrylate) (32). In all
cases, grafting was achieved in chlorobenzene with ditertiarybutyl
peroxide as the initiator. Up to now, we have not succeeded in
grafting 2H5P under the same conditions onto these polymers.

2H5V and methyl-5-vinylsalicylate were also grafted onto cis-
1,4-poly(butadiene) and 1,2-poly(butadiene). Solution grafting has
been accomplished when the grafting reaction was carried out in low
polymer concentration. Grafting efficiency for the base polymers as
well as for the monomer has been established (33,34).

2H5V and 2H5P have also been used as monomers for incorporation
into polymerizing mixtures of unsaturated polyesters and styrene.
Experiments were carried out with oligomeric polyesters made from
maleic anhydride, phthalic anhydride, 1,3-dihydroxypropane (70 wt. %)
and styrene (30 wt. %) (35). Under the normal reaction conditions
(benzoyl peroxide as the initiator), the two polymerizable benzotriazole derivatives were essentially quantitatively incorporated when
the feed ratio was between 0.5 and 5 mol %.

Initial accelerated tests have shown copolymers of 2H5V and 2H5P
to be effective as ultraviolet stabilizers, particularly for acrylic
polymers, and to protect the polymers for more than 20 years' normal
outdoor exposure to the environment.

<u>Multichromophoric 2(2-Hydroxyphenyl)2H-Benzotriazole Ultraviolet</u>
<u>Stabilizers</u>. Up to now, we have discussed primarily the preparation
of polymerizable 2(2-hydroxyphenyl)2H-benzotriazoles. In our most
recent work, we became also interested in developing entirely new
ultraviolet-absorbing systems based on 2(2-hydroxyphenyl)2H-benzo-
triazoles; the main emphasis was on the preparation of multibenzo-
triazolized phenols or polyphenols. Of particular interest was the
preparation of 2(2-hydroxyphenyl)2H-benzotriazoles with more than one
phenolic hydroxyl group in the benzene ring. By condensing o-nitrobenzenediazonium salts with resorcinol or phloroglucinol under carefully controlled conditions, monobenzotriazolized products were obtained; the nonhindered phenolic groups in the 4 position were now
available for reaction with acryloyl chloride or methacryloyl chloride (8).

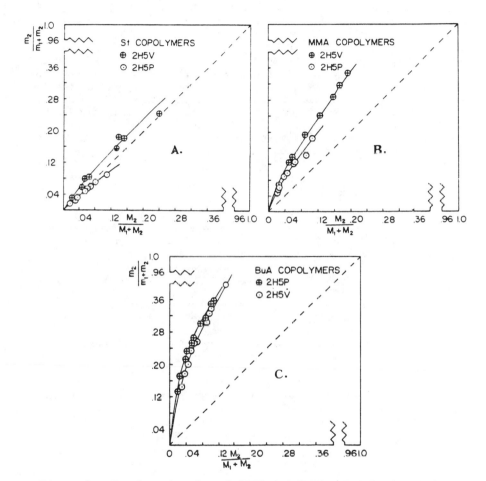

Figure 3. Copolymerization of 2H5V and 2H5P. Initiator: AIBN. Temperature: 60 °C. Conversion: ~5%. (A) Styrene. (B) Methyl methacrylate. (C) n–Butyl acrylate.

We have established that normal condensation of o-nitrobenzenediazonium chloride with resorcinol in strongly basic media gave di-
benzotriazolization (after reductive coupling of the azo-compounds).
Benzotriazolization occurred in the 2 and 4 positions of resorcinol
(36). With phloroglucinol, the reaction (also in basic solution)
leads to di- and if done with excess diazonium chloride even to tri-
benzotriazolization. It is obvious that these compounds are of
interest because of their very broad ultraviolet absorption spectra
and their extremely high extinction coefficients, while at the same
time having a sharp cutoff at 370 nm.

Ultraviolet-absorbing compounds have also been prepared having
the 2(2-hydroxyphenyl)2H-benzotriazole and the 2-hydroxybenzophenone
(or acetophenone) chromophor in the same molecule, by benzotriazolization of 2,4-dihydroxybenzophenone or 2,4-dihydroxyacetophenone. In
either case, the dibenzotriazolized products were obtained: 2[2,4-
hydroxy-5-acetyl(benzoyl)phenyl]1,3-2H-dibenzotriazole. These compounds are very powerful UV absorbers over a broad range from 250 to
370 nm at very high extinction coefficient and with a sharp cutoff
(37) (see also 38,39).

Polymerizable Antioxidants

In the last few years, it has become fully appreciated that polymeric
antioxidants are effective in retarding the thermal and autooxidation. Such polymer-bound stabilizers are similar in efficiency to
the low molecular weight stabilizer incorporated into the polymer by
blending. The polymer-bound stabilizer should have a flexible spacer
between the point of attachment to the polymer and the functional
group of the phenolic antioxidants.

Antioxidants based on 2,6-ditertiarybutyl-4-vinylphenol or 2,6-
ditertiarybutyl-4-isopropenylphenol are the only monomeric stabilizers that have been synthesized and studied. We have developed
efficient synthetic methods for the preparation of such compounds and
have polymerized them with styrene or methyl methacrylate in solution
or in bulk with AIBN as the initiator. More importantly, we have
developed a good emulsion polymerization of 2,6-ditertiarybutyl-4-
vinylphenol and 2,6-ditertiarybutyl-4-isopropenylphenol with butadiene or isoprene. The copolymers of good molecular weights had
comonomer contents between 6 mol % and 20 mol % of the vinyl or isopropenyl monomer. The polymers were effective at a 0.1 weight percent level in retarding autooxidation of polybutadiene and polyiso-
prene.

Catalytic hydrogenation with cobalt/aluminum catalyst gave polyethylene copolymers (from the hydrogenation of butadiene copolymers)
or ethylene/propylene copolymers (from isoprene copolymers) containing 2,6-ditertiarybutyl-4-vinylphenol or 2,6-ditertiarybutyl-4-iso-
propenylphenol in the polymer. These polymers have been used as
polymeric antioxidants and are effective in retarding autooxidation
of polyolefins (40).

We have shown that polymeric stabilizers, polymeric ultraviolet
stabilizers, and polymeric antioxidants are effective in retarding
photooxidation and thermal oxidation in polymers. Polymeric stabilizers are not volatile, not extractable or leachable, and do not provide the undesirable characteristics of toxicity, potential carcino-

genicity and allergenicity, and other potential side effects of low molecular weight compounds.

Acknowledgments

We appreciate the contributions of A. Gupta, W. Dickstein, W. Bassett Jr., P. Grosso, and C. P. Lillya. We are also indebted to Mrs. E. Cary and L. S. Corley for their assistance in preparing this manuscript.

Literature Cited

1. Bailey, D.; Vogl, O. J. Macromol. Sci., Reviews 1976, C14(2), 267.
2. Bailey, D.; Tirrell, D.; Vogl, O. J. Macromol. Sci., Chem. 1978, A12, 661.
3. Tirrell, D.; Bailey, D.; Vogl, O. In "Polymeric Drugs"; Donaruma, L. G.; Vogl, O., Eds.; Academic Press: New York, 1978; p. 77.
4. Hawkins, W. L. "Polymer Stabilization"; Wiley: New York, 1972.
5. Vogl, O.; Muggee, J.; Bansleben, D. Polymer J. (Japan) 1980, 12, 677.
6. Vogl, O.; Loeffler, P.; Bansleben, D.; Muggee, J. In "Coordination Polymerization"; Price, E. C.; Vandenberg, E. J., Eds.; POLYMERIC SCIENCE AND TECHNOLOGY, Plenum Press; New York and London, 1983, Vol. 19, p. 95.
7. Tocker, S. Makromol. Chem. 1967, 101, 23.
8. Mannens, M. G.; Hove, J. J.; Aarschot, W. J.; Priem, J. J. U.S. Patent 3 813 255, 1974; Germ. Patent 2 128 005, 1971; Chem. Abstr. 1972, 76, 119908m.
9. Balaban, L.; Borkovec, J.; Rysary, D. Czech. Patent 108 792, 1963; Chem. Abstr. 1964, 61, 3267n.
10. Belousa, J.; Janousek, Z.; Knoflickova, H. Chem. Zvesti 1974, 28, 673.
11. Stevenson, D.; Beeber, A.; Gaudiana, R.; Vogl, O. J. Macromol. Sci., Chem. 1977, A11, 779.
12. Vogl, O. Pure and Applied Chem. 1979, 51, 2409.
13. Vogl, O.; Yoshida, S. "Preprints. Plen. Lect., Ann. Meeting, Soc. Pol. Sci."; Kyoto, 1980, 29(4), 648.
14. Vogl, O.; Yoshida, S. Rev. Roum. de Chim. 1980, 7, 1123.
15. Bailey, D.; Tirrell, D.; Vogl, O. J. Polym. Sci., Polym. Chem. Ed. 1976, 14, 2725.
16. Iwasaki, N.; Tirrell, D.; Vogl, O. J. Polym. Sci., Polym. Chem. Ed. 1980, 18, 2755.
17. Tirrell, D.; Vogl, O. Makromol. Chem. 1980, 181, 2097.
18. Tirrell, D.; Bailey, D.; Pinazzi, C.; Vogl, O. Macromolecules 1978, 11, 213.
19. Tirrell, D.; Bailey, D.; Vogl, O. Polymer Preprints, ACS Division of Polymer Chemistry 1977, 18(1), 542.
20. Sumida, Y.; Yoshida, S.; Vogl, O. Polymer Preprints, ACS Division of Polymer Chemistry 1980, 21(1), 201.
21. Sumida, Y.; Vogl, O. Polymer J. (Japan) 1981, 13, 521.
22. Yoshida, S., unpublished results.

23. Elbs, K.; Keiper, W. J. Prakt. Chem. [2] 1905, 67, 580.
24. Elbs, K. J. Prakt. Chem. 1924, 108, 209.
25. Hardy, W. B. In "Developments in Polymer Photochemistry";
    Allen, N. S., Ed.; Applied Science Publishers Ltd.: London,
    1981; Chap. 8.
26. Milionis, J. P.; Hardy, W. B.; Baitinger, W. F.   U.S. Patent
    3 159 646, 1964.
27. Yoshida, S.; Vogl, O. Polymer Preprints, ACS Division of
    Polymer Chemistry 1980, 21(1), 203.
28. Yoshida, S.; Vogl, O. Makromol. Chem. 1982, 183, 259.
29. Nir, Z.; Vogl, O.; Gupta, A. J. Polym. Sci., Polym. Chem. Ed.
    1982, 20, 2737.
30. Yoshida, S.; Lillya, C. P.; Vogl, O. Monatsh. Chem. 1982, 113,
    603.
31. Yoshida, S.; Lillya, C. P.; Vogl, O. J. Polym. Chem., Polym.
    Chem. Ed. 1982, 20, 2215.
32. Pradellok, W.; Gupta, A.; Vogl, O. J. Polym. Chem., Polym.
    Chem. Ed. 1981, 19, 3307.
33. Kitayama, M.; Vogl, O. J. Macromol. Sci., Chem. 1983, 19(3),
    875.
34. Kitayama, M.; Vogl, O. Polymer J. (Japan) 1982, 14, 537.
35. Borsig, E.; Ranby, B.; Vogl, O., unpublished results.
36. Li, S. J.; Gupta, A.; Vogl, O. Monatsh. Chem. 1983, 114, 93.
37. Li, S. J.; Gupta, A.; Vogl, O. J. Polym. Sci., Polym. Chem.
    Ed., in press.
38. Eastman Kodak Co. U.S. Patent 4 256 626, 1981.
39. Eastman Kodak Co. U.S. Patent 4 271 307, 1981.
40. Grosso, P., Ph.D. Thesis, University of Massachusetts, Amherst,
    1983.

RECEIVED December 17, 1984

Computer Modeling Studies of Polymer Photooxidation and Stabilization

A. C. SOMERSALL and J. E. GUILLET

Department of Chemistry, University of Toronto, Toronto, Canada M5S 1A1

A computer model has been developed which can generate real-istic concentration versus time profiles of the chemical species formed during photooxidation of hydrocarbon polymers using as input data a set of elementary reactions with corresponding rate constants and initial conditions. Simulation of different mechanisms for stabilization of clear, amorphous linear poly-ethylene as a prototype suggests that the optimum stabilizer would be a molecularly dispersed additive in very low concen-tration which can trap peroxy radicals and also decompose hydroperoxides.

The oxidative deterioration of most commercial polymers when exposed to sunlight has restricted their use in outdoor applications. A novel approach to the problem of predicting 20-year performance for such materials in solar photovoltaic devices has been developed in our labo-ratories. The process of photooxidation has been described by a qual-itative model, in terms of elementary reactions with corresponding rates. A numerical integration procedure on the computer provides the predicted values of all species concentration terms over time, with-out any further assumptions. In principle, once the model has been verified with experimental data from accelerated and/or outdoor ex-posures of appropriate materials, we can have some confidence in the necessary numerical extrapolation of the solutions to very extended time periods. Moreover, manipulation of this computer model affords a novel and relatively simple means of testing common theories related to photooxidation and stabilization. The computations are derived from a chosen input block based on the literature where data are available and on experience gained from other studies of polymer photochemical reactions. Despite the problems associated with a somewhat arbitrary choice of rate constants for certain reactions, it is hoped that the study can unravel some of the complexity of the process, resolve some of the contentious issues and point the way for further experimentation.

The Computer Model

A complex chemical mechanism can be expressed as a system of NR elementary (fundamental) chemical reactions comprising NS different

0097–6156/85/0280–0211$07.00/0

chemical species, $\{Y_i\}$, $i = 1$, NS. The resulting mechanistic scheme takes the concise form given by

$$\sum_{i=1}^{NS} v_{ij} Y_i = 0 \quad ; \quad j = 1, \, NR \tag{1}$$

in which the stoichiometric coefficients, v_{ij}, are positive for product and negative for reactant species. Furthermore, each reaction j ($j=1$, NR) is characterized by a rate constant k_j. From the molecularity of the elementary reactions comprising the reaction scheme it is then possible to write explicit expressions for the concentration-time derivatives of all chemical species, Y_i, as shown by Equation 2. In this

$$\frac{d[Y_i]}{dt} = \sum_{j=1}^{NR} \left\{ v_{ij} k_j \prod_{l=1}^{NS} \left([Y_l]^{-v_{lj}} \right) \right\} ; \quad i = 1, \, NS \tag{2}$$

equation the product Π includes only those l-values for reactant species (v_{lj} negative). The time behavior of a derived reaction scheme can be obtained by integration of the resulting system of ordinary differential equations (ODE), given a specified set of initial conditions. This yields explicit concentration-time data for all species.

Generally speaking, the system of ODE is nonlinear, necessitating numerical solution. Furthermore, the rates of the individual reactions in the usual kinetic scheme commonly vary by many orders of magnitude, giving rise to the so-called stiff system of ODE. Attempts to employ classical integration algorithms require limiting the integration step size so as to keep pace with the fastest transients in the system. This has the undesirable consequence that extremely small step sizes are taken and exorbitant amounts of computing time are required to complete the problem. Clearly, for the usual chemical kinetics problem, integration algorithms especially designed to tackle a stiff system of ODE must be employed. For this work, the original Gear routine (1) and its various modifications (2) have been employed in the simulation package. Essentially, the algorithms are multi-step predictor-corrector methods utilizing the backward differencing formulation as applied by Gear with automatic error-controlled, order-step size selection.

To validate our numerical procedure, the data base given for the cesium flare system (which is becoming a standard in the literature) was used, and the curves generated were identical to those of Edelson (3) for the same system. The excellent agreement between predicted and actual rate curves showed that the program itself (irrespective of the data base) performs in a satisfactory manner.

A similar computational modelling approach has been shown to be useful, for example, in studying the mechanism of low-temperature oxidation of alkanes (4), pyrolysis of alkanes (5-7), other gas-phase reactions (8), the formation of photochemical smog (9,10), and peroxide decomposition (11), among others. It is not uncommon to begin with all possible species and by permutation and combination derive a complete set of reactions, and then eliminate a subset by chemical intuition or by sensitivity analysis. Polymer photooxidation with over 40 different species involved is too complex a process for such an approach. Our model is built up sequentially from the basic chemistry of the system and we presume that it approaches the real process in its essential elements.

The Reaction Scheme

As a starting point for this computational approach to the photooxidative process in polymeric materials, we have examined the simplest prototype: neat, amorphous, linear polyethylene above its glass transition temperature. In practice, polyethylene is partially crystalline, and contains truly linear olefins, vinylidene groups, ketones and hydroperoxides in addition to the short side chains. Much insight has already been gained into the photooxidation process by conventional experimentation on such polymers (12,13), yet several important questions still remain. Several good reviews have appeared recently (14-16).

A mechanistic scheme has been devised which includes most, if not all of the apparent non-trivial elementary processes. The 56 relevant reactions and their corresponding rates are listed in Table I and are shown schematically in Scheme I.

Initiation. Three types of initiation have been considered in the model: the photolytic cleavage of ketones and/or hydroperoxide and some general fortuitous generation of alkyl radicals (such as thermal C–H homolysis, ultrasonics, mechanical treatment, ionizing radiation, etc.). More often the Norrish type I cleavage of ketones has been used as standard, assuming initial ketone concentrations of 10^{-3} M ketone. The Norrish type II process is very important for chain scission but no participation is presumed in the initiation step for the suspected biradical intermediate. Although "pure" saturated polyolefins should not be expected to absorb beyond 2000 Å, the absorption in the far UV tail in the atmospheric sunlight spectrum around 3000 Å has commonly been attributed to the low concentrations of ketone and/or hydroperoxide groups introduced in the commercial polymers during processing. No consideration has been given to other proposed initiation processes such as absorption by dyes, pigments, catalyst residues, oxygen adducts (18), charge-transfer or polynuclear aromatics absorbed from polluted urban atmospheres (19). The possibility of energy transfer from excited ketones leading to photosensitized decomposition of hydroperoxide (20) has also been included. The extinction coefficient for the hydroperoxide is very low but the quantum yield of cleavage is very near to unity. The quantum yield for ketones is low for carbonyl incorporated in the backbone of the polymer chain, but is much higher when at a chain end or branch (21). The radicals generated are presumed to have free access to oxygen, and no special cage effect in the bulk polymer has been assumed. All of the rates for the initiation reactions follow directly from the absorption spectra and known photochemical quantum yields. The standard light intensity employed refers to the average solar intensity which is one-third of the typical peak value of $0.15 \text{ Em}^{-2}\text{h}^{-1}$ incident on a 1 mm film.

Table I. Data Set: Photooxidation Reaction Scheme and
Activation Parameters

Reaction matrix $\underline{a}$	A	E kcal/mol	Remarks
1. Ketone $\xrightarrow{h\nu}$ KET*	0.70×10^{-9}	0	--
2. KET* $\longrightarrow$ SMRO$_2$ + SMRCO	0.59×10^{9}	4.8	Ref. 36
3. SMRCO $\longrightarrow$ SMRO$_2$ + CO	0.80×10^{17}	15	Ref. 37
4. KET* $\longrightarrow$ Alkene + SMKetone	0.56×10^{8}	2.0	Ref. 36
5. SMKetone $\xrightarrow{h\nu}$ SMKET*	0.70×10^{-9}	0	--
6. SMKET* $\longrightarrow$ SMRO$_2$ + CH$_3$CO	0.32×10^{13}	8.5	Ref. 38
7. SMKET* $\longrightarrow$ Alkene + Acetone	0.56×10^{9}	2.0	Ref. 36
8. ROOH $\xrightarrow{h\nu}$ RO + OH	0.13×10^{9}	0	--
9. RO$_2$ + RH $\longrightarrow$ ROOH + RO$_2$	0.10×10^{10}	17.0	Ref. 39
10. SMRO$_2$ + RH $\longrightarrow$ SMROOH + RO$_2$	0.10×10^{10}	17.0	Ref. 39
11. SMROOH $\xrightarrow{h\nu}$ SMRO + OH	0.13×10^{-9}	0	--
12. SMRO + RH $\longrightarrow$ SMROH + RO$_2$	0.16×10^{10}	6.2	Ref. 39
13. RO + RH $\longrightarrow$ ROH + RO$_2$	0.16×10^{10}	6.2	Ref. 39

Table I. Continued

---

14. $RO \longrightarrow SMRO_2 + Aldehyde$

$$0.32 \times 10^{16} \qquad 17.4 \qquad Ref. 39$$

15. $KET^* + ROOH \longrightarrow Ketone + RO + OH$

$$0.25 \times 10^{10} \qquad 11.6 \qquad Ref. 40$$

16. $SMKET^* + ROOH \longrightarrow SMKetone + RO + OH$

$$0.25 \times 10^{10} \qquad 11.6 \qquad Ref. 40$$

17. $SMRCO + O_2 \longrightarrow SMRCOOO$

$$0.80 \times 10^{-4} \qquad 9.6 \qquad Ref. 41$$

18. $SMRCO + RH \longrightarrow RO_2 + Aldehyde$

$$0.10 \times 10^{10} \qquad 7.3 \qquad Ref. 42$$

19. $SMRCOOO + RH \longrightarrow SMRCOOOH + RO_2$

$$0.10 \times 10^{10} \qquad 17.0 \qquad cf. 36, 37$$

20. $SMRCOOOH \xrightarrow{h\nu} SMRCOO + OH$

$$0.13 \times 10^{-9} \qquad 0 \qquad --$$

21. $SMRCOO \longrightarrow SMRO_2 + CO_2$

$$0.10 \times 10^{15} \qquad 6.6 \qquad Ref. 43$$

22. $SMRCOO + RH \longrightarrow Acid + RO_2$

$$0.10 \times 10^{10} \qquad 17.0 \qquad cf. 36, 37$$

23. $OH + RH \longrightarrow RO_2 + Water$

$$0.10 \times 10^{10} \qquad 0.5 \qquad Ref. 44$$

24. $CH_3CO + RH \longrightarrow RO_2 + CH_3CHO$

$$0.10 \times 10^{10} \qquad 7.3 \qquad Ref. 42$$

25. $CH_3CO + O_2 > CH_3COOO$

$$0.89 \times 10^{10} \qquad 9.6 \qquad Ref. 41$$

26. $CH_3COOO + RH \longrightarrow CH_3COOOH + RO_2$

$$0.10 \times 10^{10} \qquad 17.0 \qquad cf. 36, 37$$

27. $CH_3COOOH \xrightarrow{h\nu} CH_3COO + OH$

$$0.13 \times 10^{-9} \qquad 0 \qquad --$$

28. $CH_3COO + RH \longrightarrow CH_3COOH + RO_2$

$$0.10 \times 10^{15} \qquad 6.6 \qquad Ref. 43$$

Continued on next page

Table I. Continued

---

29. KET* —> Ketone

$$0.10 \times 10^9 \qquad 0 \qquad --$$

30. SMKET* —> SMKetone

$$0.10 \times 10^9 \qquad 0 \qquad --$$

31. KET* + O_2 —> Ketone + SO_2

$$0.89 \times 10^{14} \qquad 9.6 \qquad Ref. 41$$

32. SMKET* + O_2 —> SMKetone + SO_2

$$0.89 \times 10^{14} \qquad 9.6 \qquad Ref. 41$$

33. RO_2 + RO_2 —> ROH + Ketone + SO_2

$$0.25 \times 10^{10} \qquad 11.6 \qquad Ref. 39$$

34. RO_2 + ROH —> ROOH + Ketone + HOO

$$0.10 \times 10^{10} \qquad 15.3 \qquad Ref. 39$$

35. HOO + RH —> HOOH + RO_2

$$0.32 \times 10^9 \qquad 15.0 \qquad Ref 45$$

36. HOO + RO_2 —> ROOH + SO_2

$$0.32 \times 10^9 \qquad 2.1 \qquad Ref. 38$$

37. RO_2 + Ketone —> ROOH + Peroxy CO

$$0.13 \times 10^5 \qquad 8.9 \qquad Ref. 46$$

38. Peroxy CO + RH —> PER OOH + RO_2

$$0.10 \times 10^{10} \qquad 17.0 \qquad cf. 36, 37$$

39. PER OOH $\xrightarrow{h\nu}$ PER O + OH

$$0.13 \times 10^{-9} \qquad 0 \qquad --$$

40. PER O + RO_2 —> DIKetone + ROOH

$$0.25 \times 10^{10} \qquad 11.6 \qquad Ref. 40$$

41. RO_2 + ROOH —> ROOH + Ketone + OH

$$0.25 \times 10^8 \qquad 11.6 \qquad Ref. 39$$

42. RO_2 + SMROH —> ROOH + Aldehyde + HOO

$$0.10 \times 10^{10} \qquad 15.3 \qquad Ref. 39$$

43. RO_2 + Aldehyde —> ROOH + SMRCO

$$0.25 \times 10^{10} \qquad 11.6 \qquad Ref. 40$$

Table I.   Continued

---

44. $RO_2 + RO_2 \longrightarrow ROOR + SO_2$

0.38×10^{12}	16.0	Ref. 39

45. $SO_2 \longrightarrow O_2$

0.63×10^5	0	--

46. SO_2 + Alkene $\longrightarrow$ ROOH

0.20×10^{14}	10.0	Ref. 47

47. RO_2 + Alkene $\longrightarrow$ Branch

0.16×10^9	11.6	Ref. 39

48. $SMRO_2$ + Alkene $\longrightarrow$ ROOH

0.16×10^9	11.6	Ref. 39

49. RO_2 ⁞ QH $\longrightarrow$ ROOH + Q

0.16×10^8	5.2	Ref. 39

50.   KET* + Q1 $\longrightarrow$ Ketone + Heat

0.80×10^{13}	9.5	Ref. 48

51.   ROOH + QD $\longrightarrow$ PRODS

0.80×10^{13}	9.5	Ref. 48

52.   ROOH $\overset{\Delta}{\longrightarrow} RO\cdot + OH\cdot$

0.63×10^{15}	35	Ref. 49

53.   SMROOH $\overset{\Delta}{\longrightarrow}$ SMRO + OH

0.63×10^{15}	35	cf. 52

54.   SMRCOOOH $\overset{\Delta}{\longrightarrow}$ SMRCOO + OH

0.63×10^{15}	35	cf. 52

55. $CH_3COOOH \overset{\Delta}{\longrightarrow} CH_3COO + OH$

0.63×10^{15}	35	cf. 52

56.   PER OOH $\overset{\Delta}{\longrightarrow}$ PER O + OH

0.63×10^{15}	35	cf. 52

---

[a] SMProduct = product from chain cleavage, $SO_2 - {}^1O_2$.

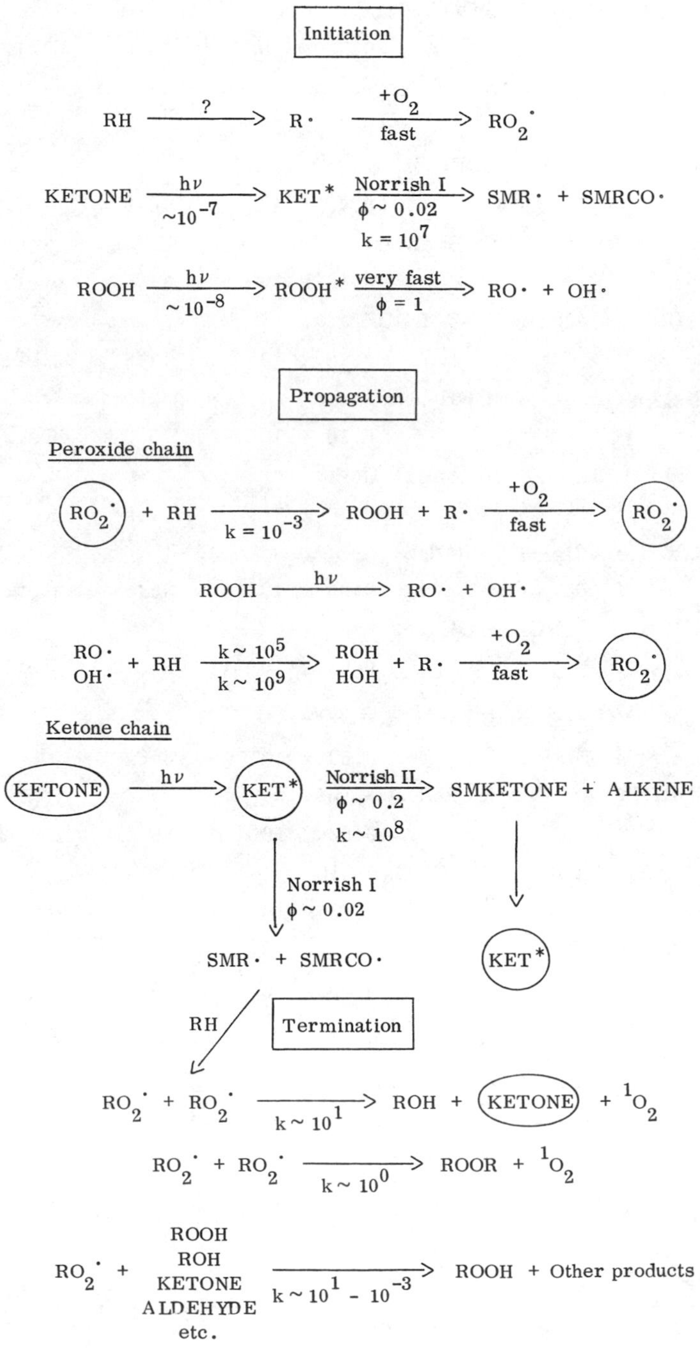

$$\boxed{\text{Initiation}}$$

$$RH \xrightarrow{\quad ? \quad} R\cdot \xrightarrow[\text{fast}]{+O_2} RO_2\cdot$$

$$KETONE \xrightarrow[\sim 10^{-7}]{h\nu} KET^* \xrightarrow[\substack{\phi \sim 0.02 \\ k = 10^7}]{\text{Norrish I}} SMR\cdot + SMRCO\cdot$$

$$ROOH \xrightarrow[\sim 10^{-8}]{h\nu} ROOH^* \xrightarrow[\phi = 1]{\text{very fast}} RO\cdot + OH\cdot$$

$$\boxed{\text{Propagation}}$$

Peroxide chain

$$\bigcirc\!\!\!RO_2\cdot + RH \xrightarrow[k = 10^{-3}]{} ROOH + R\cdot \xrightarrow[\text{fast}]{+O_2} \bigcirc\!\!\!RO_2\cdot$$

$$ROOH \xrightarrow{h\nu} RO\cdot + OH\cdot$$

$$\begin{array}{c} RO\cdot \\ OH\cdot \end{array} + RH \xrightarrow[k \sim 10^9]{k \sim 10^5} \begin{array}{c} ROH \\ HOH \end{array} + R\cdot \xrightarrow[\text{fast}]{+O_2} \bigcirc\!\!\!RO_2\cdot$$

Ketone chain

$$\bigcirc\!\!\!KETONE \xrightarrow{h\nu} \bigcirc\!\!\!KET^* \xrightarrow[\substack{\phi \sim 0.2 \\ k \sim 10^8}]{\text{Norrish II}} SMKETONE + ALKENE$$

Norrish I
$\phi \sim 0.02$

$$SMR\cdot + SMRCO\cdot \qquad \bigcirc\!\!\!KET^*$$

RH

$$\boxed{\text{Termination}}$$

$$RO_2\cdot + RO_2\cdot \xrightarrow[k \sim 10^1]{} ROH + \bigcirc\!\!\!KETONE + {}^1O_2$$

$$RO_2\cdot + RO_2\cdot \xrightarrow[k \sim 10^0]{} ROOR + {}^1O_2$$

$$RO_2\cdot + \begin{array}{c} ROOH \\ ROH \\ KETONE \\ ALDEHYDE \\ \text{etc.} \end{array} \xrightarrow[k \sim 10^1 - 10^{-3}]{} ROOH + \text{Other products}$$

Scheme I. Polyethylene photooxidaton scheme.

Propagation. The key radical in the propagation sequence is the peroxy radical formed by the addition of oxygen to alkyl radicals. Molecular oxygen $O_2(^3\Sigma_g^-)$ found in nature is paramagnetic, with two paral-
lel-spin electrons, giving it the characteristics of a biradical. As a
consequence the oxygen molecule combines very rapidly with free
radicals to form the peroxy radicals (22). The solubility of oxygen in
amorphous polyethylene is low ($10^{-3} M$) but practically constant due to
the steady diffusion of oxygen from the atmosphere to replenish what
is consumed during oxidation over long periods of time. Thus we have
assumed that all carbon-centered radicals will inevitably add oxygen to
form peroxy radicals so we have incorporated this step directly in those
reactions involving alkyl radical products, to simplify the computation
(23).

The key propagation step is the abstracton of a hydrogen atom by
the peroxy radical to generate a new peroxy radical and a hydroperoxide group. Alkoxy radicals and hydroxy radicals formed by cleavage
of the hydroperoxide will also abstract a hydrogen atom to produce
further peroxy radicals. These reactions are well studied in solution
(24) and the same rates have been assumed in our model. Other radicals formed in Norrish type II chain scission and subsequent reactions
can also abstract hydrogen atoms to generate peroxy radicals. Rearrangement of alkoxy radicals in the β-scission process leads to aldehydes, whereas thermal cleavage of intermediates leads to small molecule fragments and products such as carbon dioxide and formaldehyde.
Acids are formed in the model via the addition of oxygen to acyl radicals, followed by hydrogen atom abstraction, then cleavage of the resulting peroxy acid and a further H-atom abstraction. Esters would
develop from the condensation of acid with the alcohols formed by H-
atom abstraction of alkoxy radicals.

In linear amorphous polyethylene, all the C–H bonds are presumed
to be secondary and on the backbone of the polymer chains, with neglect of the few chain ends. The rates of hydrogen abstraction by
various radical species are taken from similar processes in solution,
but subsequent processes are modified to allow for the higher internal
viscosity of the medium.

Termination. Just as peroxy radicals are key to the propagation sequence, so the bimolecular recombination of these radicals is the major
termination process in the unstabilized polymer. The existence of an
intermediate tetroxide has been established in solution (25). Several
factors influence the competitive pathways of subsequent decomposition
to form alcohols, ketone and singlet oxygen or to form alkoxy radicals
which can couple before separation from the reaction center to form a
peroxide. This latter process is a route to crosslinking in the case of
polymeric peroxy radicals. The effect of steric control, viscosity and
temperature have been studied in solution. However, in the solid phase
the rates of bimolecular processes which require the mutual diffusion
of the reactant groups will be limited by the diffusion process. As a
standard, we have assumed a value close to that determined from oxygen absorption (26) and by ESR spectra (27) for oxidized polypropylene films.

For those bimolecular reactions that do not require mutual diffusion such as H-atom abstraction, in which one reactant is a virtual solvent, the rates are similar to solution values. The peroxy radical may react with many other species and in some cases the rate constants are even lower than the diffusion rate and hence the value is not limited in that way.

Modelling Photooxidation

Photooxidation of amorphous polyethylene. Figure 1 shows the general behavior predicted for exposure to different intensities of UV light. Assuming initiation by 10^{-3} M ketone groups and constant ambient oxygen pressure, the model shows that the C—H bonds become oxidized slowly at first and then more rapidly later on. A 5% C—H bond (1% monomer units) oxidation level has been used to assign a point of failure which is within the range one would anticipate for marked change in mechanical properties. Under typical conditions, the time to failure (τ_f) of unstabilized polyethylene would be three to four months in temperate climates and shorter in regions of high solar intensity. Product formation and other observations are consistent with the authors' experimental knowledge of polyethylene weathering (28). Table II summarizes the concentration of all chemical species in the model at the chosen time of failure. The major products are ketones, hydroperoxides, alcohols and water. The only anomaly is the relatively high concentration of hydroperoxide which the authors were unable to eliminate in their model predictions and which may point to the inadequacy of experimental methods for monitoring —OOH groups in oxidized films. The time to failure was plotted as a function of the intensity of light to find the relationship that $\tau_{\frac{1}{2}} \propto I^{-\frac{1}{2}}$, as one would expect for photochemical initiation producing two radical species.

Also, one finds that the behavior is almost unaffected by both the initiator type and concentration (Figure 2). This is not surprising for an autocatalytic process since the result is dominated by relative rates of propagation and termination.

Two interesting conclusions can be drawn. First, there has been much controversy in the literature about the relative importance of the various possible mechanisms for initiation (ketone groups, hydroperoxide, catalyst residues, singlet oxygen), but the controversy is practically irrelevant if initiation does not much influence the course, rate or extent of photooxidation. Second, in terms of stabilization, minute amounts of photoinitiation will lead to practical failure so that the purification of polymers to minimize the initiating residue is not a practical alternative for stabilization. Both points underline the value of this approach to the understanding of the complex photooxidation process.

Figure 3 shows the dependence of the time to failure on the rate of propagation, i.e., the rate of hydrogen abstraction from the polymer backbone. The log-log plot is linear with a negative slope less than unity. The rate of abstraction of ^{t}C—H as in branched polyethylene and ethylene copolymers would only be a factor of two to three higher than for ^{s}C—H in linear polyethylene. The time to failure in these

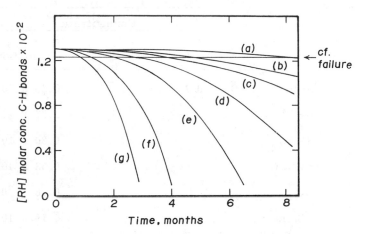

Figure 1. Photooxidation of unstabilized PE: (a) hν/10; (b) hν /2; (c) hν/3; (d) hν average daylight; (e) hν x 5; (f) hν x 10; (g) hν x 10².

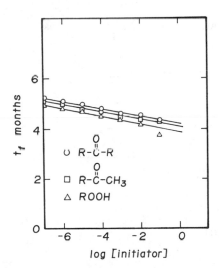

Figure 2. Photooxidation of unstabilized PE: time to failure as a function of initiator type and concentration.

Table II.   Concentration of All Species, Initially and
at Failure (5% C–H Oxidation)

	Species	Concentration	
		Initial	Final (5% C–H oxidation)
1.	Ketone	0.1×10^{-2}	0.39×10^{1}
2.	KET*	--	0.26×10^{-16}
3.	$SMRO_2$	--	0.20×10^{-9}
4.	SMRCO	--	0.53×10^{-17}
5.	CO	--	0.16×10^{-4}
6.	Alkene	---	0.19×10^{-4}
7.	SMKetone	--	0.17×10^{-3}
8.	SMKET*	--	0.95×10^{-21}
9.	CH_3CO	--	0.33×10^{-20}
10.	Acetone	--	0.51×10^{-7}
11.	ROOH	--	0.24×10^{1}
12.	RO	--	0.61×10^{-16}
13.	OH	--	0.20×10^{-16}
14.	RO_2	--	0.51×10^{-5}
15.	RH	0.13×10^{3}	0.123×10^{3}
16.	SMROOH	--	0.33×10^{-4}
17.	SMRO	--	0.70×10^{-21}
18.	SMROH	--	0.11×10^{-7}
19.	ROH	--	0.26×10^{-2}
20.	Aldehyde	--	0.70×10^{-7}
21.	SMRCOOO	--	0.84×10^{-12}
22.	SMRCOOOH	--	0.14×10^{-6}
23.	SMRCOO	--	0.12×10^{-25}
24.	CO_2	--	0.49×10^{-10}
25.	Acid	--	0.16×10^{-20}

Table II. Continued

26.	Water	--	0.42×10^{1}
27.	CH_3CHO	--	0.53×10^{-8}
28.	CH_3COOO	--	0.58×10^{-15}
29.	CH_3COOOH	--	0.79×10^{-10}
30.	CH_3COO	--	0.53×10^{-31}
31.	CH_3COOH	--	0.24×10^{-13}
32.	$^{1}O_2$ (SO_2)	--	0.55×10^{-14}
33.	HOO	--	0.20×10^{-11}
34.	HOOH	--	0.38×10^{-5}
35.	Peroxy CO	--	0.17×10^{-5}
36.	PER OOH	--	0.23×10^{0}
37.	PER O	--	0.66×10^{-6}
38.	DIKetone	--	0.69×10^{-4}
39.	ROOR	--	0.12×10^{-3}
40.	Branch	--	0.15×10^{-3}

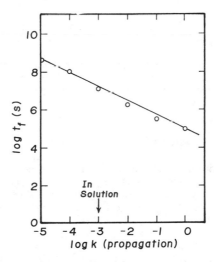

Figure 3. Photooxidation of unstabilized PE: time to failure as a function of propagation rate constant.

$$RO_2{}^{\cdot} + RH \xrightarrow{k_p} ROOH + R^{\cdot} \xrightarrow{O_2} RO_2{}^{\cdot}$$

polymers should therefore not be very different from polyethylene, all other factors being equal. But for some other polymers, other factors are not equal. For example, the rate for polypropylene is enhanced by the presence of sequences of $^{t}C{-}H$ allowing extensive intramolecular hydrogen migration, hydroperoxide sequences, more stable tertiary peroxy radicals, and subsequent reactions.

Figure 4 shows the dependence of the time to failure on the rate of bimolecular radical termination which is directly related to diffusion of the polymeric segments in the solid matrix. Again, the log-log plot is linear with a _positive_ slope and this time, again less than unity. Increased diffusion shortens the kinetic chain length and increases the time to failure.

Effect of diffusion on chemical degradation. In principle, diffusion in a polymer matrix will affect the rates of all of the bimolecular reactions to some degree. Four categories of such reactions have been treated.

1. *Small molecule-small molecule reactions.* These reactions are almost non-existent in the mechanism except for the addition of oxygen to small radical fragments such as the acyl radical from type I cleavage of methyl ketones. The relatively high solubility of oxygen and the mobility of these small molecules in a medium that is essentially equivalent to a very viscous liquid in local regions, minimizes any special "polymer effect". The other stable small molecule product is water which may have effects on electrical properties but does not participate in the photooxidation sequence per se.

2. *Small molecule-polymer "moiety" reactions.* Groups formed on the backbone of polymer chains (such as ROO˙ peroxy, RCO keto, RO˙ alkoxy, etc.) do react with "small molecule" species, including oxygen and some radical fragments. The effect of the polymer medium apparently is to reduce the bimolecular collision (reaction) rates by a factor of about 10^{-2} relative to fluid solution rates, due to the reduced mobility of the chains.

3. *Small molecule-polymer "solvent" reactions.* Many radical reactions involve H-atom abstraction from the hydrocarbon backbone of the "solvent" polymer matrix. It is likely that all species will have C-H bonds in the immediate vicinity so that there would be no real sensitivity to diffusion and rates would then be similar to fluid solution rates.

4. *Polymer "moiety"-polymer "moiety" reactions.* Chemical groups formed on the backbone of polymer chains such as peroxy radicals, hydroperoxides, ketones, etc. can undergo bimolecular reactions such as disproportionation, energy transfer, etc. The effect of the polymeric medium apparently is to reduce the bimolecular rates by a factor of about 10^{-4} relative to fluid solution rates, due to the reduced mobility of both reactants.

What has been shown in Figure 3 is that increasing the rates of all of the bimolecular reactions, proportionately and together, causes a net increase in the induction time (time to failure). This result probably reflects the importance of the bimolecular termination of polymeric peroxy radicals which can then compete more effectively with the hydrogen abstraction propagation step, when diffusion is increased.

The Effect of Temperature

For outdoor applications, the variation of temperature will have important effects on the useful lifetimes of hydrocarbon polymers. To accommodate this, the rate constants were expressed in terms of the Arrhenius parameters, A (the pre-exponential factor) and E (the activation energy). This naturally multiplied the mathematical complexity in the program with now more than 100 different parameters. Many of the species in the complex process undergo a number of competitive pathways, the relative importance of each being often sensitive to small changes in the calculated rate constant values. In addition, the different magnitudes of the activation energy caused various changes in the relative importance of the different key processes of propagation and termination with changes in operating temperature. No attempt has been made as yet to model thermal oxidation at elevated temperature, but even at moderate outdoor temperatures (100°F) the decomposition of hydroperoxides does become significant. Five such decomposition reactions (numbers 52 to 56) were then included in the basis set (Table I).

Figure 5 shows the variation of time to failure (5% oxidation) with temperature. The decrease in lifetime with no stabilizer is more or less as expected, ranging from a few months in hot tropical weather, 310 K (100°F), to almost two years in cool weather, 280 K (45°F). An attempt at a typical Arrhenius plot (Figure 6) shows an "apparent net activation energy" of 10-16 kcal/mol near atmospheric temperatures (280–310 K). Experimental values of 16-35 kcal/mol for the dependence of the induction period in polyethylene oxidation have been reported by Wilson (29) and Blum et al. (30) at temperatures above 380 K. For thick films the observed value is as low as 10 kcal/mol (31).

Modelling Stabilization

To examine the different stabilization mechanisms reviewed by Carlsson et al. (30) we have included some appropriate reactions in the model (Scheme II).

UV shielding. Effective stabilization of colorless films by this mechanism would require an additive which absorbs only in the ultraviolet (to be colorless), high concentrations (0.5-2% for adequate coverage of 10-20 μm films) and chromophores of high extinction coefficient ($>10^5$ M^{-1} cm^{-1}). A simple calculation under such typical conditions shows that the incident UV light available to be absorbed by the initiating species is reduced to about 1-5% of the actual intensity. From the variation of time to failure (5% oxidation) with the light intensity, the net effect of such a stabilizer is to increase the useful lifetime by a factor of five to ten (Figure 7). In addition, the front surface layer remains essentially unshielded so that surface oxidation becomes the limiting factor.

In the case of reflective or opaque pigment particles, the screening effect can be made very effective, especially when very finely divided

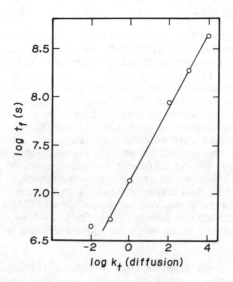

Figure 4.   Photooxidation of unstabilized PE: effect of bi(macro)-
molecular diffusion (radical termination) on time to failure, e.g.,

$$RO_2^{\cdot} + RO_2^{\cdot} \xrightarrow{k_t} \text{Products}$$

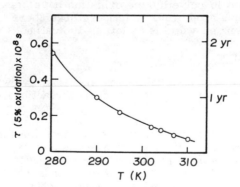

Figure 5.   Time to failure (5% oxidation) as a function of temperature.

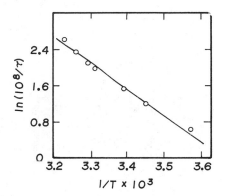

Figure 6. Arrhenius plot of the rate of oxidation (k vs. 1/T).

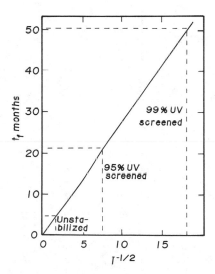

Figure 7. Stabilization of PE with a UV screen.

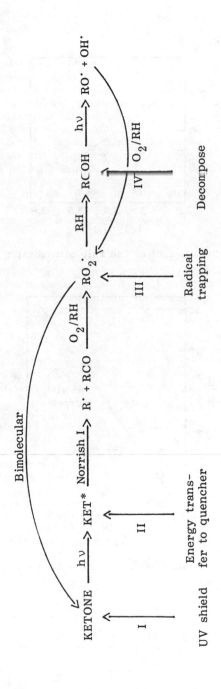

Scheme II. Stabilization mechanism.

stabilizer is dispersed in a good solid solution or coating on thick specimens. The success of carbon blacks in polyethylenes is now well accepted, but the mechanism of action may not be exclusively that of a UV shield (33).

Energy transfer. To model this mechanism of stabilization, a reaction (number 50, Table I) was included to allow for quenching of the excited ketone by an additive (Q1) with a rate constant comparable to the upper limit for diffusion of a small molecule in a polymer matrix. Figure 8 shows that up to 1 M concentration (about 8 wt-%) of quencher had minimal effect on the time to failure (5% oxidation). This assumes completely random distribution of both the excited ketones and the stabilizer as in the calculation of Heller and Blattman (34). Such a bimolecular process is too slow to compete with the fast unimolecular reactions of the excited ketone, and thus stabilization by such transfer is predicted to be ineffective in polyethylene. Allowance must be made, however, for special cases in which the excitation energy can effectively migrate (e.g., in some aromatic polymers), in which case the bimolecular process may become competitive with the other chemical processes from the excited states.

Radical trapping. To allow for stabilizaton by this mechanism, another reaction (number 49) was included to allow easy abstraction of a hydrogen atom from an additive (QH) by a peroxy radical to form a hydroperoxide and a harmless adduct. With the same value of the rate constant as for energy transfer and for concentrations as low as 10^{-7} M, the photooxidation process was efficiently slowed. Figure 9 shows the linear dependence of the time to failure (5% oxidation) as the concentration of QH is altered. Note that the trap is consumed in the process and the apparent induction time is associated with its removal. The stabilization is less effective for higher intensity (and probably higher temperature) because the faster photo (or thermal) decomposition of ROOH continues the degradation process.

The model suggests that under similar conditions, peroxy radical trapping is a far more effective mechanism of stabilization than energy transfer. This could relate directly to the lifetime of the key species being removed. The natural decay of peroxy radicals in the dark was monitored by removing all photochemical steps from the reaction sequence. The decay is bimolecular and the time required to decrease the concentration of peroxy radicals to one-third of its original value was monitored as a function of the latter. The results in Figure 10 show a linear inverse dependence with lifetimes of up to several hours, thereby allowing sufficient time for diffusion of the stabilizing scavenger in the bimolecular process. This is not inconsistent with some recent experimental results on the ESR decay studies of peroxy radicals in polymers reported by Gupta and coworkers (35).

The model further suggests that for effective stabilization to 20 years a molecular concentration of 10^{-5} M would be adequate. This corresponds to 4×10^{-4} wt-%, assuming a molecular weight of 400 for this type of stabilizer. In fact, the model predicts that for each 10^{-6} wt-% of stabilizer used, the time to failure (5% oxidation) increases systematically by two years.

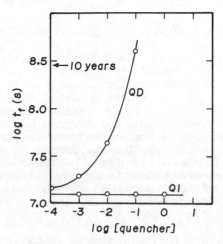

Figure 8. Stabilization of PE: ROOH + QD $\xrightarrow{10^6}$ Products;

KET* + Q1 $\xrightarrow{10^6}$ Ketone + Heat

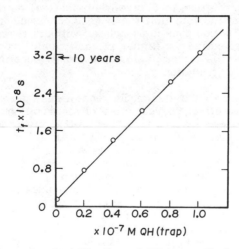

Figure 9. Stabilization of PE by radical trapping
(very low molecular concentration).

$$RO_2{}^{\cdot} + QH \xrightarrow{10^6} ROOH + Q$$

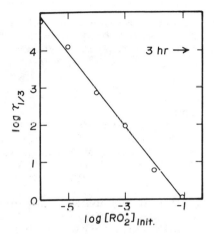

Figure 10. Lifetime of peroxy radicals in the dark ($\tau_{1/3}$ in s).

Hydroperoxide decomposition. Yet another reaction (number 48) was added to the basic mechanism scheme to allow for the stabilization of amorphous linear polyethylene by the harmless removal of the hydroperoxide by an additive (QD). Reactions of the products (alcohol, $^1O_2$) were not considered further. Also included in Figure 8 is the effect of this type of stabilizer on the time to failure (5% oxidation) as a function of its concentration. The mechanism is effective as low as 10^{-4} M (about 10^{-3} wt-%) and the increasing effectiveness at higher concentrations could reflect the autocatalytic participation of the hydroperoxide which normally decomposes to produce two destructive radicals per molecule.

The model suggests that optimum photoprotection for clear polyethylene films would be obtained with a stabilizer which effectively scavenges peroxy radicals and also decomposes hydroperoxide. This agrees with the conclusions put forward earlier by Carlsson et al. (32).

Conclusions

In principle, the computational approach to the kinetics of the complex photooxidation process can give meaningful insight into the effects of outdoor weathering of hydrocarbon polymers. For clear amorphous linear polyethylene, the model suggests that the optimum stabilizer would be a molecularly dispersed additive in very low concentration which could trap peroxy radicals. An additive which decomposes hydroperoxides would also be effective but would require higher concentrations. The useful lifetime of unstabilized polyethylene is predicted to vary from a few months in hot weather (100°F) to almost two years in cool weather (45°F), which correlates well with experimental results and general experience.

Acknowledgments

This work was performed as part of a research contract for the Jet Propulsion Laboratory, California Institute of Technology, sponsored by the U. S. Department of Energy through an agreement with the National Aeronautics and Space Administration. The computer program was adapted for use here by Dr. J. W. Gordon, now at Canadian General Electric Company, Toronto. The authors wish to acknowledge the invaluable advice of Dr. K. U. Ingold, National Research Council of Canada, Ottawa, Dr. R. R. Hiatt, Brock University, St. Catharines, Ontario, Dr. J. R. MacCallum, St. Andrews University, Scotland, and Dr. A. Gupta, Jet Propulsion Laboratory, Pasadena, California.

Literature Cited

1. (a) Gear, C. W. ACM 1971, 14, 176; (b) Hindmarsh, A. C. "GEAR Ordinary Differential Equation Solver," Lawrence Livermore Laboratories, Rep. UCID-30001-R3, 1974.
2. (a) Stabler, R. N.; Chescick, J. HAVCHM, Haverford College, 1977; (b) Carver, M. B.; Hanley, D. V.; Chaplin, K. R. MAKSIMA-CHEMIST, Atomic Energy of Canada CRN2, 1979; (c) Whitten, G. Z. CHEMK, S. A. I., San Rafael, Calif., 1979.
3. Edelson, D. J. Chem. Ed. 1975, 52, 642.

4. Allara, D. L.; Edelson, D.; Irwin, K. C. Int. J. Chem. Kinet. 1972, 4, 345.
5. Allara, D. L.; Edelson, D. Int. J. Chem. Kinet. 1975, 7, 479.
6. Olson, D. B.; Tanzawa, T.; Gardiner, W. C. Int. J. Chem. Kinet. 1979, 11, 23.
7. Sundaram, K. M.; Froment, G. F. Ind. Eng. Chem. Fundam. 1978. 17, 174.
8. Ebert, K. H.; Ederer, H. J.; Isbarn, G. Angew. Chem., Int. Ed. Engl. 1980, 19, 333.
9. Farrow, L.; Edelson, D. Int. J. Chem. Kinet. 1974, 6, 787.
10. Carter, W. P. L.; Lloyd, A. C.; Sprung, J. L; Pitts, J. N. Int. J. Chem. Kinet. 1979, 11, 45.
11. Hiatt, R.; Nair, V. G. K. Can. J. Chem. 1979, 57, 450.
12. Adams, J. H. J. Polym. Sci., Part A 1970, 8, 1279.
13. Heacock, J. F.; Mallory, F. B.; Gay, F. P. J. Polym. Sci., Part A 1968, 6, 2921.
14. Garton, A.; Carlsson, D. J.; Wiles, D. M. In "Developments in Polymer Photochemistry"' Allen, N. S., Ed.; Applied Science, Barking, 1980, vol. 1, pp. 93-123.
15. Shlyapintokh, V. Ya. In "Developments in Polymer Photochemistry"; Allen, N. S., Ed.; Applied Science, Barking, 1980; vol. 2, p. 215.
16. Hardy, W. B. In "Developments in Polymer Photochemistry"; Allen, N. S., Ed.; Applied Science, Barking, 1980; vol. 3, p. 187.
17. Cicchetti, O.; Gratini, F. Eur. Polym. J. 1972, 8, 561.
18. Stenberg, V. I.; Sneeringer, P. V.; Niu, C.; Kulevsky, N. Photochem. Photobiol. 1972, 16, 81.
19. Aspler, J.; Carlsson, D. J.; Wiles, D. M. ADVANCES IN CHEMISTRY SERIES no. 169, American Chemical Society: Washington, D. C., 1976, p. 5.
20. Ng, H. C.; Guillet, J. E. Macromolecules 1978, 11, 937.
21. Sitek, F.; Guillet, J. E.; Heskins, M. J. Polym. Sci., Polym. Symp. 1976, 57, 343.
22. Walling, C. "Free Radicals in Solution"; Wiley: New York, 1957.
23. In earlier computations we allowed for reactions of allyl radicals without oxygen and found extremely low radical concentrations and little or no photooxidation after very many iterative steps.
24. Mill, T.; Hendry, D. G. Comp. Chem. Kinet. 1980, 16, 1.
25. Howard, J. A.; Ingold, K. U. J. Am. Chem. Soc. 1968, 90, 1056.
26. Mayo, F. R. Macromolecules 1978, 11, 942.
27. Garton, A.; Carlsson, D. J.; Wiles, D. M. Macromolecules 1979, 12, 1071.
28. Guillet, J. E. Pure Appl. Chem. 1980, 52, 285.
29. Wilson, J. E. Ind. Eng. Chem. 1955, 47, 2201.
30. Blum, G. W.; Shelton, J. R.; Winn, H. Ind. Eng. Chem. 1951, 43, 464.
31. Jellinek, H. H. G.; Lipovac, S. N. Macromolecules 1970, 3, 231, 237.
32. Carlsson, D. J.; Garton, A.; Wiles, D. M. In "Developments in Polymer Stabilisation"; Scott, G., Ed.; Applied Science: Barking, 1979; vol. 1, pp. 219-259.
33. Heskins, M.; Guillet, J. E. Macromolecules 1968, 1, 97.
34. Heller, J. H.; Blattman, H. R. Pure Appl. Chem. 1973, 36, 141.
35. Gupta, A., personal communication.

36. Hartley, G.; Guillet, J. E. Macromolecules 1968, 1, 165.
37. Watkins, K. W.; Thompson, W. W. Int. J. Chem. Kinet. 1973, 5, 791.
38. Golemba, F.; Guillet, J. E. Macromolecules 1972, 5, 63.
39. Mill, T.; Hendry, D. G. Comp. Chem. Kinet. 1980, 1, 16.
40. Mayo, F. R. Macromolecules 1978, 11, 942.
41. Rabek, J. F. Comp. Chem. Kinet. 1978, 14, 463.
42. Hendry, D. G.; Mill, T.; Piszkiewics, L.; Howard, J. A.; Eigenmann, H. K. J. Phys. Chem., Ref. Data 1974, 3, 937.
43. Braun, W.; Rajbenbach, L.; Eirich, F. R. J. Phys. Chem. 1962, 66, 1591.
44. Wilson, W. F. J. Phys. Chem., Ref. Data 1972, 1, 570.
45. Ingold, K. U., personal communication.
46. Denisov, E. G. Comp. Chem. Kinet. 1980, 16, 169.
47. Hoch, E. Tetrahedron 1968, 24, 6295.
48. Braun, J.-M.; Guillet, J. E. J. Polym. Sci., Polym. Lett. Ed. 1976, 14, 257.
49. Brandrup, J.; Immergut, E. H., Ed. "Polymer Handbook"; Wiley: New York, 1975; 2nd ed., pp. II-12 ff.

RECEIVED December 7, 1984

Investigation of Thermal Oxidation and Stabilization of High-Density Polyethylene

PAUL-LI HORNG and PETER P. KLEMCHUK

Additives Department, Plastics and Additives Division, CIBA-GEIGY Corporation, Ardsley, NY 10502

A commercial HDPE, unstabilized or stabilized with a hindered phenolic antioxidant, was subjected to thermal oxidation at 140°, 100° and 40°C by oxygen uptake. The oxidative induction times of the unstabilized samples were found to fit into a linear apparent Arrhenius relationship. The calculated activation energy for thermooxidative degradation of the HDPE agrees with literature data. Ultimate elongation, carbonyl formation and molecular weight distribution were found to change little before the induction time was reached. The degree of chain breaking, calculated from molecular weight data, shows an average of about one scission per molecule caused the polymer to lose its elongation property totally. Stabilization provided by a phenolic antioxidant demonstrated a relatively long induction time; e.g., 4700 versus 35 hours at 100°C. Within the induction time, chain scissioning and elongation were nearly unaffected. After the induction time, chain scissioning became uninhibited and was manifested by loss of elongation. Mechanisms of chain scissioning and stabilization are discussed.

Polyolefins are sensitive to heat- and light-induced oxidative degradation. Studies in the past on thermal oxidative stability of high density polyethylene (HDPE) have generated information on how HDPE is oxidized under thermal stress (1-4). Alkyl and peroxy radicals, hydroperoxides, beta-scission after hydroperoxide decomposition to carbonyl and an alkyl radical end group are recognized as the major elements in the general oxidation pathway. Stabilization through interruption of the degradation cycle has resulted in the development of effective stabilizer systems for the many uses of this polymer.
    Recently we have been studying both the molecular weight changes and the physical property changes in HDPE as a function of oxidation. Unstabilized and stabilized HDPE were evaluated by oxygen uptake at

140°, 100° and 40°C. This paper presents the results of our findings
which provide some insights into the relationships among degree of
oxidation, molecular weight changes, physical property retention, and
stabilization of the polymer.

Experimental

The polymer used was Alathon 5496, DuPont HDPE. The stabilizer used
was tetrakis[methylene-(3,5-di-tert-butyl-4-hydroxyhydrocinnamate)]
methane, referred to as AO1. Powdered samples of HDPE, stabilized
and unstabilized, were prepared and subjected to oxidation in a
closed system with oxygen. Oxygen uptake was monitored periodically
at given temperatures. The induction period was picked from the
curve where the onset of autocatalytic oxidation took place.
    Elongation data of 5-mil films were generated on an Instron
Tensile Tester according to ASTM D882 at a pulling rate of 50mm/min.
Molecular weights of polymer samples were determined by high temperature gel permeation chromatography calibrated with polyethylene standards, except the results from 40°C were done using the Q factor from
polystyrene standards. Chain scission was calculated as
$[\overline{Mn}(\text{unoxidized})/\overline{Mn}(\text{oxidized})]-1$. Carbonyl absorbances were determined
by Infrared Spectroscopy at 1710 cm^{-1}.

Results and Discussion

Oxidation Curves at 140°C, 100°C and 40°C. Samples of HDPE, unstabilized or stabilized with 0.1% of AO1, were oxidized in a closed
system with oxygen. The oxidation curves at 140°, 100° and 40°C are
shown in Figures 1-3. These data indicate the effectiveness of AO1
in preventing oxygen consumption at both high and low temperatures.
Once the induction period was passed at 140°C and 100°C, the oxygen
consumption rates were virtually the same for the unstabilized and
stabilized samples. The unstabilized HDPE consumed oxygen at significant rates, even at 40°C, with the induction period lasting about
two years. The need to stabilize polymers for use at all temperatures is evident from these data.

Correlation of Induction Times at 140°, 100° and 40°C. The induction
times obtained for the unstabilized HDPE at three temperatures are
tabulated in Table I and also plotted against reciprocal absolute
temperature in an apparent Arrhenius relationship in Figure 4.

Table I. Oxidation of HDPE by Oxygen Uptake.
          Influence of Temperature on Oxidative Induction Time.

Stabilizer	Induction Time at Temperature, Hours		
	140°C	100°C	40°C
None	1	35	15,960*
0.1% AO1	120	4,700	39,480+**

*95 weeks; **235+ weeks (no indication of degradation)

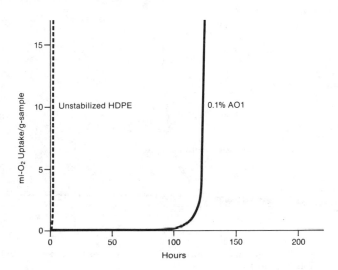

Figure 1.   Oxygen Uptake of Stabilized and Unstabilized HDPE at 140°C and 1 Atmosphere

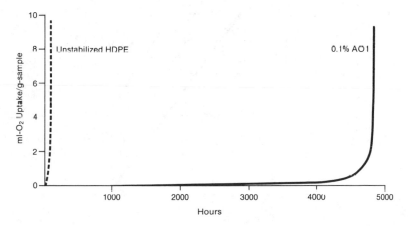

Figure 2.   Oxygen Uptake of Stabilized and Unstabilized HDPE at 100°C and 1 Atmosphere

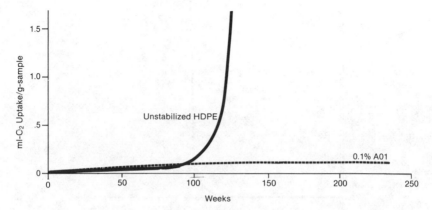

Figure 3. Oxygen Uptake of Stabilized and Unstabilized HDPE at
40°C and 1 Atmosphere

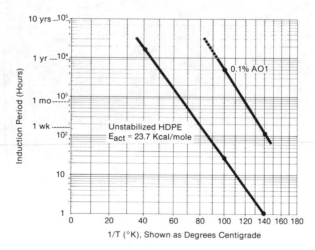

Figure 4. Arrhenius Plot of Stabilized and Unstabilized HDPE
During Oxidation, Oxygen Uptake at 1 Atmosphere and Three
Temperatures

It is interesting to note that they fall into a linear relationship
with a calculated activation energy of 23.7 Kcal/mole. Literature
data showed values of 21–26 Kcal/mole. (5,6) This is one of the few
instances that oxidative induction times from 140°C to near ambient
temperature have been available. Notably, the temperature range includes the melting point of this HDPE polymer so the range from 140°C
to 40°C includes a phase change. In this instance, we did not find a
discrepancy in plotting above-melting experimental results along with
those obtained at much lower temperatures. Our finding on the linearity of the Arrhenius relationship suggests that extrapolation over
a moderate temperature range is warranted at least for unstabilized
HDPE, provided several data points are available.

Property Correlation During Oxidation at 100°C and 40°C. A series of
unstabilized and 0.1% AO1-stabilized HDPE samples were oxidized at
100°C and removed periodically for evaluation. A monotonic change in
the retention of elongation in relation to oxygen uptake was found as
in Figure 5. Carbonyl formation during oxidation was also found to
correlate with oxygen uptake in a linear relationship. 50% retention
of elongation was found to correlate with 0.25 carbonyl absorbance of
5-mil film and zero elongation was found to correspond to 0.5 carbonyl absorbance. Oxidation to 6.5 ml-O_2/g-HDPE or 2.1 mmole O_2/
mmole HDPE destroyed elongation completely for both unstabilized and
stabilized HDPE. However, the time involved for such a catastrophic
change was dramatically prolonged from the unstabilized to AO1-
stabilized HDPE; 75 vs. 4700 hours (Figure 6).
   At 40°C, oxidation took place at a relatively low, but measurable rate. Samples stabilized with 0.1% AO1 showed no loss in
elongation retention in over four years as compared to the
unstabilized sample which showed a catastrophic decrease in elongation after the 95-week induction period (Table II).

Table II.   Change in Elongation of HDPE Oxidized at 40°C

Sample	Time (wks)	Oxygen Uptake (ml/g)	Elongation (%)
Unstabilized	0	0	750
	105	0.25	200
	115	0.53	165
	120	1.0	55
Stabilized with	0	0	720
0.1% AO1	105	0.10	690
	150	0.12	800
	235	0.1	650

The molecular weight distribution change of the unstabilized HDPE was
also monitored and is plotted in Figure 7. At 105 weeks, oxygen
consumption of the unstabilized HDPE was about 0.25 ml/g and the
sample showed a slight reduction in the high molecular weight fraction. Continuing oxidation showed evidence of more chain scissioning
and lowering of average molecular weights. Again, AO1-stabilized

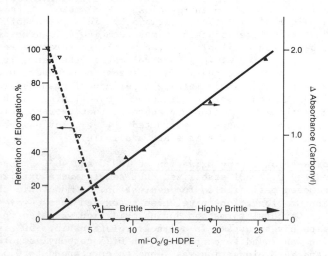

Figure 5.   Correlation of Percent Retention of Elongation and
Carbonyl Absorbance to the Degree of Oxidation, Oxygen Uptake of
Unstabilized HDPE at 100°C and 1 Atmosphere

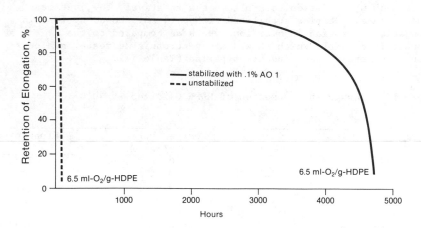

Figure 6.   Comparison of Elongation Retention of Stabilized and
Unstabilized HDPE, Oxygen Uptake at 100°C and 1 Atmosphere

HDPE after four years at this temperature exhibited no change in both elongation retention and molecular weight retention.

Chain Scissioning and its Effect on Molecular Weight. Table III shows the molecular weight data of the unstabilized HDPE when oxidized at 100°C to different oxygen uptake levels.

Table III. Molecular Weight Data and Chain Scission of Unstabilized HDPE Oxidized at 100°C

Oxygen Uptake (ml-O_2/g-HDPE)	$\overline{Mw}$	$\overline{Mn}$	$\overline{Mw}$/$\overline{Mn}$	S
0	151,000	8,270	18.1	0
0.12	137,000	6,450	21.1	0.3
2.3	29,100	5,610	5.2	0.5
3.8	15,900	4,620	3.4	0.8
9.2	13,800	3,700	3.7	1.2
11.1	10,400	3,200	3.2	1.6
19.1	8,460	2,670	3.2	2.1
25.7	7,080	2,390	3.0	2.5

S = Average Scissions per Molecule = $[\overline{Mn}(0)/\overline{Mn}(t)]-1$

Weight average molecular weight ($\overline{Mw}$) was found to be more pronouncedly affected at the early stage of oxidation, while number average molecular weight ($\overline{Mn}$) was affected to a lesser degree. Based on the available data, the greatest change in molecular weight took place between 0.12 and 2.3 ml O_2/g HDPE oxygen uptake. In that interval, $\overline{Mw}$ was reduced 80% but $\overline{Mn}$ only about 15%. Closer examination of the molecular weight distribution (MWD) curves (Figure 8) indicates this was the interval where the loss of the high molecular weight fraction was greatest and so was the formation of lower molecular weight species. Statistically speaking, longer chains have greater probability for oxidative attack and chain rupture than do shorter chains. The mathematical moment of Mw reflects a heavier contribution from higher molecular weight species than $\overline{Mn}$. Therefore, the statistically greater scissioning of longer chains had a greater impact on changes in $\overline{Mw}$ than on $\overline{Mn}$. Figure 8 and Table I further indicate when samples were oxidized from 0.12 ml O_2/g HDPE (35 hours) to 2.3 ml O_2/g HDPE oxygen consumption (40 hours), they had passed through the induction period and entered into the autocatalytic region of oxidation.

An examination of the sequential change of the MWD curves of both unstabilized and stabilized samples (Figures 8 and 9) shows oxidation up to about 0.1 ml-O_2/g-HDPE had little effect on average molecular weight as compared to the original. However, it took about 4300 hours for the stabilized sample to reach that point as opposed to merely 35 hours for the unstabilized sample. When oxidation continued past the induction period, a significant change in molecular weight became evident as a result of chain scissioning. Oxidation thereafter was autocatalytic, and the molecular weight reduction was catastrophic within a short period of time.

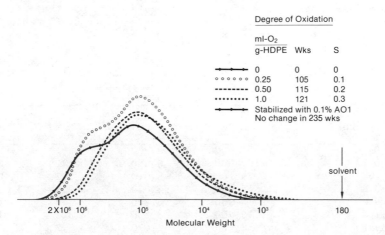

Figure 7. Molecular Weight Distribution of Oxidized and
Unoxidized HDPE, Oxygen Uptake of Unstabilized HDPE at 40°C and
1 Atmosphere

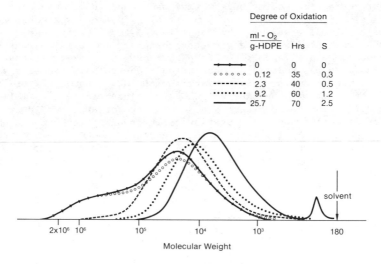

Figure 8. Molecular Weight Distribution of Oxidized and
Unoxidized HDPE, Oxygen Uptake of Unstabilized HDPE at 100°C and
1 Atmosphere

Average chain scissions per molecule as calculated from Mn changes are plotted against oxygen uptake in Figure 10. This graph demonstrates by linear regression that the chain scission rate of HDPE was similar for both the stabilized and unstabilized samples when compared by degree of oxidation in terms of oxygen uptake. Again, the stabilized samples showed a long induction period during which chain scissioning was quite insignificant. After the induction period, chain scissioning continues with increasing oxygen consumption in a linear fashion. The embrittlement point, zero elongation, was found to coincide with only 1 and 1.2 average scissions per molecule for stabilized and unstabilized HDPE, respectively.

Chain Scissioning and its Effect on Elongation. Since a linear relationship was found between oxygen uptake and chain scission rate (Figure 10) and also between oxygen uptake and retention of elongation (Figure 5), it is obvious that the chain scission rate and retention of elongation can enter into another first order relationship. Chain scissioning effected not only molecular weight, but also mechanical properties, such as elongation: after the induction period, elongation decreased rapidly with both time and chain scissioning (Figure 6).

Oxidation of semicrystalline polymers is generally considered to occur within the amorphous region which can be treated as a boundary phase of the neighboring crystalline regions. Peterlin's model (7) of tensile deformation explained the contribution of tie molecules in the amorphous region to the necking elongation of a semi-crystalline polymer. Since oxidation takes place mostly in amorphous regions, tie molecules which connect crystallites through amorphous regions may be scissioned in the oxidation process resulting in a decrease of elongation and other physical properties. At later stages of oxidation when many chains in the amorphous phase and also at the crystalline boundary are ruptured, samples exhibit brittleness upon external stress.

Mechanism of Chain Scissioning and Stabilization. A review of the average number of chain scissions as a function of oxygen consumption shows the number of oxygen molecules consumed per chain scission increased with increasing oxygen consumption. Beyond the induction period, the mmoles of oxygen per chain scission increased rapidly with time; the data are summarized in Table IV.

Table IV.   Calculated Oxygen Molecules Consumed per Chain Scission
During Oxidation of Unstabilized HDPE at 100°C

| Time | Oxygen Uptake | | | Oxygen Molecules Consumed |
(hours)	A	B	S	per Chain Scission
0	0	0	0	–
35	0.12	0.04	0.3	0.2
40	2.3	0.7	0.5	1.6
50	3.8	1.2	0.8	1.6
60	9.2	3.1	1.2	2.5
66	19.1	6.3	2.1	3.0
70	25.7	8.5	2.5	3.5

A = ml O_2/g HDPE; B = mmole O_2/mmole HDPE.

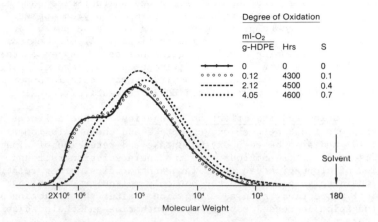

Figure 9.   Molecular Weight Distribution of Oxidized and Unoxidized HDPE, Oxygen Uptake of 0.1% AO1-Stabilized HDPE at 100°C and 1 Atmosphere

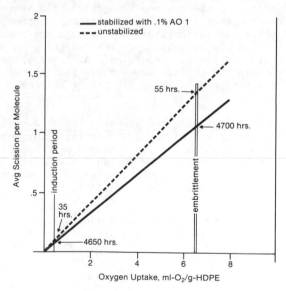

Figure 10.   Comparison of Chain Scission Rates in Stabilized and Unstabilized HDPE, Oxidation by Oxygen Uptake at 100°C and 1 Atmosphere

This result is in agreement with those reported by M. Iring, et al (2) for polypropylene and polyethylene. The oxygen consumption data indicate the oxidation of polyethylene consists of a complex group of reactions beyond the induction period with no single, simple relationship between the number of oxygen molecules consumed and the number of chain scissions.

Using the data of Chien (8), we calculated the half-life of polyethylene hydroperoxide to be 6.4 hours at 100°C. Thus, since our shortest induction period at 100°C was over 35 hours, it is reasonable to postulate the scissioning of polyethylene results from unimolecular homolytic dissociation of polyethylene hydroperoxide, a major oxidation product:

$$ROOH \longrightarrow RO^{\bullet} + HO^{\bullet}$$
$$RO^{\bullet} \longrightarrow R'CHO + R''^{\bullet}$$

Chain scissioning of polyethylene via the alkoxy radical would yield a terminal aldehyde and shorter-chain polyethylene free radical. Both products will react with oxygen. The aldehyde should be oxidized readily to the peracid via a chain reaction:

$$R'CHO + RO_2^{\bullet} \longrightarrow R'CO^{\bullet} \xrightarrow{O_2} R'CO\underset{OO^{\bullet}}{} \xrightarrow{RH} R'CO\underset{OOH}{} + R^{\bullet}$$

The polyethylene radical would be expected to react readily with oxygen and contribute to oxidative propagation as another peroxy radical:

$$R''^{\bullet} + O_2 \longrightarrow R''O_2^{\bullet} \xrightarrow{RH} R''O_2H + R^{\bullet}$$

Since this resultant hydroperoxide is a terminal hydroperoxide, its decomposition will not result in chain scissioning.

Assuming each reaction to be 100% efficient, i.e., hydroperoxide decomposition to alkoxyl radical; alkoxyl radical chain scission to aldehyde and shorter polyethylene radical; aldehyde oxidation to peracid; and oxygen reaction with the primary polyethylene radical to yield a peroxy radical; the maximum number of oxygen molecules associated with each chain scission would be three. Not suprisingly, the oxidation process is too complex to have shown a single relationship between oxygen molecules consumed and the number of chain scissions. At the time of embrittlement, about 2 molecules of oxygen had been consumed per chain scission and increased beyond this point. The increase in the number of oxygen molecules per chain scission beyond the induction period is not surprising since in the ultimate oxidation of polyethylene to carbon dioxide and water, 1.5 molecules of oxygen would be required for each methylene group. This means a polymer with an initial number average molecular weight of 8,270 would require 886 molecules of oxygen for complete conversion to carbon dioxide and water.

The addition of the phenolic antioxidant AO1 to the polymer demonstrated a long induction period (Table I) during which chain scissioning and physical property changes were negligible. The equally complex stabilization chemistry (9) for phenolic antioxidants can be summarized by the trapping of peroxy and alkoxy radicals and also by the decomposition of hydroperoxide with phenols and its

transformation products. Figure 7 at 40°C oxidation and Figure 9 at
100°C oxidation showed the MWD was almost unchanged during the
induction period of the stabilized polymer. The oxidative chain
reaction leading to chain scissioning as discussed above, is
interrupted at the expense of AO1 which is nearly consumed at the
end of the induction period. Beyond the induction period, oxidation
continues uninhibitedly similar to the unstabilized undergoing
oxidation at the beginning.

Conclusions

1. During the induction period, the stabilized and unstabilized
   samples underwent little chain scissioning and loss of elongation. When the induction period was passed, both stabilized and
   unstabilized samples exhibited considerable chain scissioning and
   loss of elongation during this uninhibited period of oxidation.
2. The time difference in the induction periods for AO1-stabilized
   and unstabilized HDPE was phenomenal; 120 versus ~1 hour at 140°C
   and 4700 versus 35 hours at 100°C, respectively.
3. At 40°C the oxidation rate was slow but measurable for unstabilized HDPE. The stabilization provided by AO1 for more than four
   years was evident from the oxygen uptake, elongation and
   molecular weight data.
4. The Arrhenius plot of induction period vs. temperatures ranging
   from 140°C to 40°C for unstabilized HDPE suggests extrapolation
   is permissable provided several data points are available.

Literature Cited

1. Chan, M., Proc. Int. Wire Cable Symp. 23rd, 1974, pp. 34-41.
2. Iring, M.; Laszlo-Hedvig, S.; Barabas, K.; Kelen, T.; Tudos, F.,
   Eur. Polym. J., 1978, 14, 439-442.
3. Holmstrom A., in "Durability of Macromolecular Materials"; Eby,
   R. K., Ed.; ACS Symp. Ser. No. 95, ACS: Washington, D.C., 1979;
   Chap. 4.
4. Holmstrom, A.; Sorvik, E. M., J. Polym. Sci. Polym. Chem. Ed.,
   1978, 16, 2555-86.
5. Hawkins, W. L.; Matreyek, W.; Winslow, E. H., J. Polym Sci.,
   1959, 41, 1-11.
6. Grieveson, B. M.; Howard, R. N.; Wright, B., in "Thermal
   Degradation of Polymers"; Society of Chemical Industry:
   London, 1961; pp. 413-20.
7. Peterlin, A., Int. J. Polym. Mater., 1980, 8, 285-301.
8. Chien, J. C. W., J. Polym. Sci., 1968, 6, 375-379.
9. Pospisil, J. in "Development in Polymer Stabilization - I";
   Scott, G. Ed.; Applied Science; London, 1979; Chap. 1.

RECEIVED October 26, 1984

Bis- and Trisphosphites Having Dioxaphosphepin and Dioxaphosphocin Rings as Polyolefin-Processing Stabilizers

JOHN D. SPIVACK, STEPHEN D. PASTOR, AMBELAL PATEL, and
LEANDER P. STEINHUEBEL

CIBA-GEIGY Corporation, Ardsley, NY 10502

In a previous paper, the effectiveness of certain
hindered mono phosphites and phosphonites as processing stabilizers for polyolefins(1) was discussed.
The present paper describes several new classes of
hindered phenyl bis- and tris-phosphites having dibenzo[d,f][1,3,2]dioxaphosphepin and dibenzo[d,g]-
[1,3,2]dioxaphosphocin rings. The latter compounds
exhibit superior effectiveness as processing stabilizers together with greater resistance to discoloration and hydrolysis. The di- and tri-alkanolamine
esters are of particular interest because of their
even greater resistance to hydrolysis at 50°C for
extended time periods previously achievable only in
the case of certain di-hindered phenyl phosphonites.

Processing stabilizers are a special class of antioxidants used
to inhibit polymer degradation during the processing steps subsequent to polymerization such as extrusion, injection molding,
spinning, etc. These steps are carried out at relatively high
temperatures (220-320°C) in the presence of some entrapped oxygen
and the resulting radical species.
    Attempts to counteract degradation by the use of 2,6-di-tert-
butyl-4-methylphenol (BHT) and various organophosphorus compounds
such as tris-nonylphenyl phosphite (1), 3,9-dioctadecyloxy-2,4,8,10-
tetraoxa-3,9-diphospha[5.5]-spiroundecane (2) and the 3,9 bis(2,4-
di-tert-butylphenyl) analog (3) have only been partially successful. BHT is volatile at high temperatures and contributes to discoloration during processing. 1, 2, and 3 can undergo hydrolysis and loss of processing stabilizer effectiveness if stored im-
properly. Products of hydrolysis can lead to problems with extrusion and spinning as well as contamination of extrudates with
hydrolysis products.

0097-6156/85/0280-0247$06.00/0

$$\left[C_9H_{19}\!-\!\!\bigcirc\!\!-O \right]_3\!\!P \qquad\qquad RO-P\begin{matrix} O-CH_2 \\ \diagdown \\ O-CH_2 \end{matrix}\begin{matrix} CH_2-O \\ C \\ CH_2-O \end{matrix}P-OR$$

<u>2a</u>

$$\underline{2}\quad R = n\text{-}C_{18}H_{27}\text{-}\ ;\quad \underline{3}\quad R = \text{---}\!\!\bigcirc\!\!\text{---}$$

A generally accepted mechanism for the oxidation of polyolefins, RH, in the presence of antioxidants of the hydrogen-atom donor type, AH, involves initiation, propagation and termination steps (2). In the initiation step, the polyolefin, RH, is converted to the polymeric radical, R·, while the polymeric peroxy radical, ROO·, is formed in the propagation step by rapid reaction with oxygen. In the absence of an efficient antioxidant, the peroxy radical is converted to the hydroperoxide, ROOH, by hydrogen-atom abstraction from the polymer chain giving rise to another polymeric radical, R· , and peroxy radical ROO·. Propagation via chain transfer is also promoted by homolysis of ROOH to RO· and ·OH in the absence or depletion of the antioxidant AH. Chain transfer will take place to an insignificant degree if A· is a stable free radical such as is provided by a hindered phenol as is shown in Figure 1. A simplified version of the reaction sequence in the presence of hindered phenols (AH) and a tertiary phosphorus (III) compound is shown in Figure 1.

It is thus apparent that selected antioxidants, AH, in the present case hindered phenolics, and hydroperoxide decomposers, such as PR''_3, act synergistically to inhibit radical initiated polymer-chain oxidations.

The two classes of hindered substituted bis- and tris- phosphites having either (a) dibenzo[d,f][1,3,2]dioxaphosphepin or (b) dibenzo[d,g][1,2,3]dioxaphosphocin rings were selected for study as processing stabilizers because their bicyclic structures promised stabilizers of increased thermal and hydrolytic stability compared to the acyclic hindered phenylphosphonites previously studied (1).

The specific compounds submitted for comparative studies are shown in Figure 2.

Experimental Section

Evaluation of Experimental Organophosphorus Esters in Polyolefins. The compounds were tested in the following polyolefins as listed below, the results being shown in the Tables in parenthesis:
     Polypropylene (Table I, compounds 4 to 9 inclusive). High Moleculer Weight High Density Polyethylene.

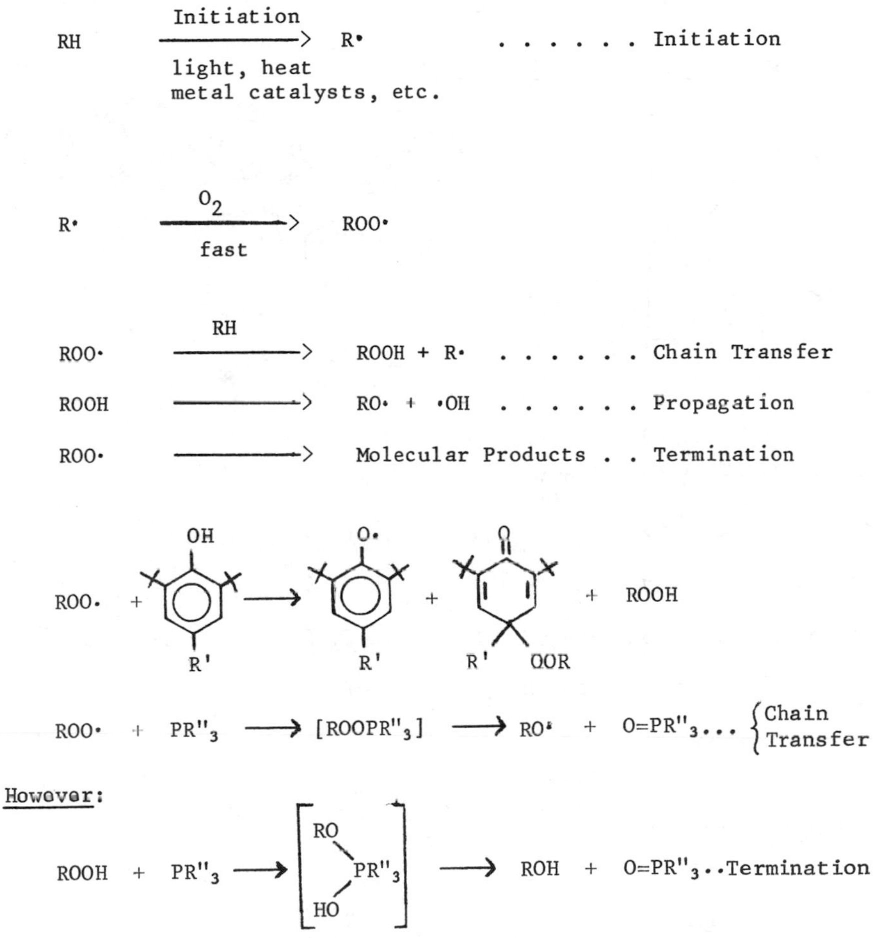

Figure 1 - Inhibited Oxidation of Polyolefins

Experimental Processing Stabilizers

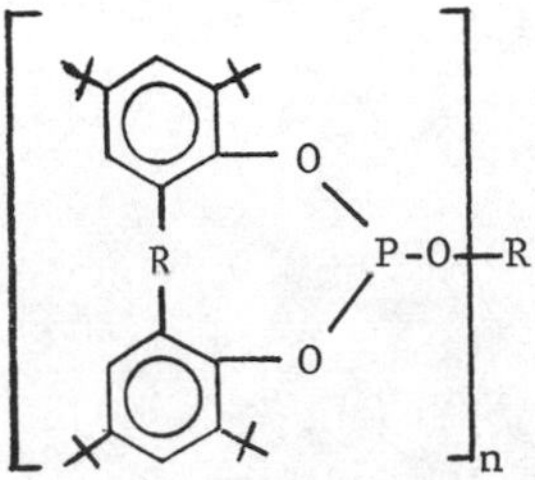

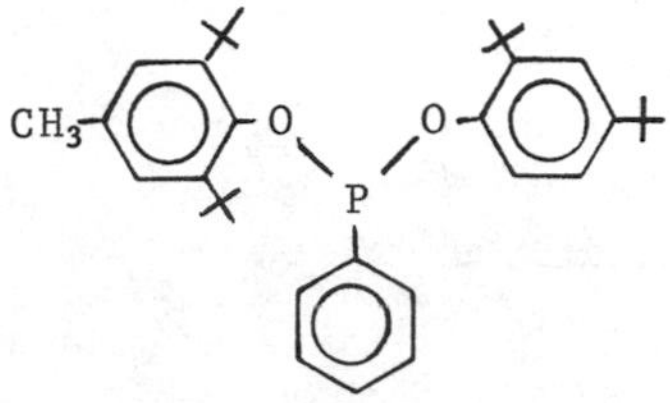

4 (Ref. 1)

No.	R	n	R'	
5	Direct Bond	2	$(CH_2)_6$	(Ref. 4)
6	Direct Bond	2	$-CH_2CH_2)_2NCH_3$	(Ref. 5)
7	CH_2	2	$-CH_2CH_2)_2N$	(Ref. 5)
8	Direct Bond	3	$-CH_2CH_2)_3N$	(Ref. 5)
9	CH_2	3	$-CH_2CH_2)_3N$	(Ref. 5)

Antioxidants

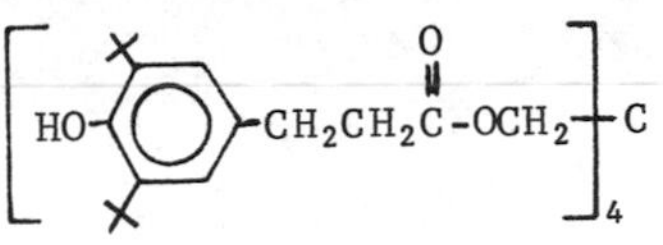

AO-1 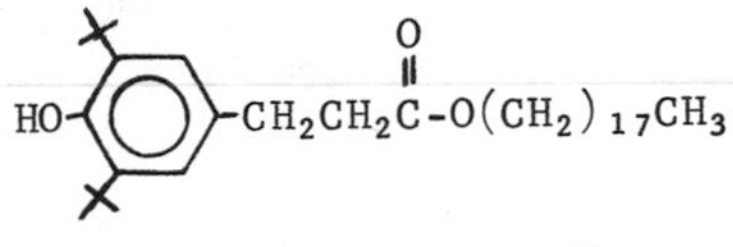   AO-2

Figure 2 - Processing Stabilizers and Antioxidants
used in Polyolefin Testing

(Tables II, compounds 5 to 9 inclusive.) Linear-Low Density Polyethylene (Table III, compounds 6 and 8) Ethylene Propylene Diene Rubber (EPDM) (Table IV, compounds 6 and 8).

In addition, all compounds were tested neat for hydrolytic stability at room temperature and at 50°C both at a relative humidity of 80%.

Test methods are described in some detail as shown in the following studies.

Processing Stability of Polypropylene

Sufficient AO-1 and processing stabilizers were dissolved in methylene chloride to give a final concentration in polypropylene of 0.1% and 0.05%, respectively. The methylene chloride solution was added to the Profax 6501 (about 1000 g) containing 0.1% calcium stearate and the mixture was mixed with a Hobart Kitchen Aid mixer at room temperature until all the solvent had evaporated. The blend was either pelletized on a Modern Plastics Machinery 1" extruder at 500°F or 550°F for 1, 2, 3, 4 and 5 extrusions. Samples of each formulation were collected after the first, third and fifth extrusions. The extrusion temperature profile was kept as follows:

Screen Pack   20-60-100-60-20

	For 500°F	For 550°F
Zone 1	450°F	500°F
Zone 2	475°F	525°F
Zone 3	500°F	550°F
Die 1	500°F	550°F
Die 2	500°F	550°F

The yellowness index (YI) was determined using the standard ASTM procedure (XL-10). Melt flow rates were also determined for the selected extrusion runs. A Tinius-Olson melt indexer was operated at standard ASTM conditions L(230°C - total weight 2160 grams).

Processing Stability for High Density Polyethylene (HDPE). The specific polymer used in the test described below was a high molecular weight HDPE made by BASF Lupolen 5260 Z. The following test runs were carried out by the test laboratories of CIBA-GEIGY Corporation Basle, Switzerland.

Procedure: Thirty-eight grams of Lupolene 5260 Z containing 0.05% AO-1 and 0.05% processing stabilizer were mixed in a Brabender Plasticorder 50 operating at a temperature of 220°C, the rotor rotating at a speed of 50 r.p.m. in an air atmosphere.

Two properties were recorded as a result of processing each sample in the Brabender Plasticoder: (1) time in minutes to reach a rapid increase in torque and (2) the color of each test sample as determined by yellowness index (ASTM D1925-70).

Resistance to Discoloration of Linear-LDPE During Long-Term Heat Aging at 90°C

Procedure: Sufficient AO-1 and processing stabilizer to give a final concentration of 0.02% of each compound were dissolved in methylene chloride. The mixture was solvent blended into Dow linear-LDPE (Dowlex 2552) and extruded at 100 RPM and a temperature of 425°F. The extrudate was then pelletized and injection moulded into 2" x 2" x 0.125" plaques. The plaques were oven aged over a period of six weeks and the Lba color was determined on a Gardner XL-type Tristimulus colorimeter. The lower the number obtained, the less the discoloration. More details on the Lba system is given in Publications No. 007 and No. 010, Gardner Laboratory Inc., Bethesda, Maryland 20014.

Prevention of Discoloration of Ethylene Propylene Diene Rubber

Procedure. Additive-free EPDM was dissolved in cyclohexane to give a cement containing 10% solids by weight. The additive formulations under investigation were dissolved in cyclohexane and were added to the cement. The stabilized cement was steam coagulated to remove the cyclohexane. The resultant stabilized rubber crumb was dried overnight at 40 °C both under nitrogen and in vacuo (about 100 mm Hg). To simulate field storage conditions, 10 grams of the stabilized test samples were put in 50 mL beakers and they were placed in pint Mason jars on the top of a layer of glass beads sitting in 10 mL of distilled water. The jars were sealed and were placed in an oven thermostatically controlled at 90°C, thus providing a controlled environment for each test sample of 90°C and 100% relative humidity. The test samples were observed for Gardner color changes for a period of 20 days.

Hydrolytic Stability Test

The hydrolytic stability test was designed to determine the storage stability of organophosphorus processing stabilizers at two temperature/humidity conditions:
    (a) room temperature (25 $\pm$ 1°C) at 80% relative humdity.
    (b) 50°C at 80% relative humidity.
Eighty percent relative humidity was achieved in each case by providing a saturated ammonium chloride solution in each of the test chambers.

Procedure at Room Temperature/80% relative humidty. A 200 mg sample of processing stabilizer was weighed into a Petri dish and it was placed in a desiccator having about 600 mL of staturated ammonium chloride in the bottom. The water uptake was measured at regular intervals until about 1% weight gain had been recorded. The disappearance of starting material was measured by both thin layer chromotography (TLC) and infrared spectroscopy and the test continued until complete disappearance of starting materials. Phosphite or phosphonite groups hydrolyse to give products having prominent absorption peaks at about 3600 cm^{-1} due to a hydroxyl group and another absorption peak at 2400 cm^{-1} due to

a >P-OH group. Thus, the IR absorption bands serve to identify
the type of reactions responsible for the disappearance of the
processing stabilizer tested. The infrared spectrum of the sample was obtained on 1% solutions of the sample in methylene chloride if the sample was completely soluble. If this was not the
case, the IR spectrum was obtained on a KBr pellet. Time in
days to complete hydrolysis can be readily determined by TLC.

Hydrolysis Tests at 50°C/80% Relative Humidity. One gram of
sample was weighed into an open vial measuring 1" diam. x 2 1/4"
high and placed on a bed of glass beads just immersed in saturated KCl solution to a level of 1 inch in a 1 pint Mason jar.
The Mason jar was sealed with a screw cap and placed in an oven
thermostatically controlled at 50°C. The increase in weight of
this vial was monitored and the hydrolysis of the test sample
measured as described above.

Discussion and Results

Comparative Effectiveness of Experimental Processing
Stabilizers in Polyolefins

Polypropylene. The six organophosphorus compounds (Figure 1,
0.05 weight %) were each formulated together with AO-1 (0.10%)
in Hercules general purpose Profax 6501 polypropylene and were
subjected to multiple extrusion in two series, one at 500°F
(260°C) and another at 550°F (288°C). Comparison of the test
compounds in the presence of AO-1 were made at each temperature
by measuring the melt flow rate and color development after the
first, third and fifth extrusions (6,7). The results after the
fifth extrusion are shown in Table I.

TABLE I

PROCESSING STABILITY OF GENERAL PURPOSE POLYPROPYLENE

Temp. °F.	500 MFR[3] g/10 min.	550 MFR[3] g/10 min.	500 Color[3] Y.I.	550 Color[3] Y.I.
Series	I	II	I	II
Compound				
4	2.8	4.1	10.8	12.3
5	3.7	5.3	10.9	10.3
6	3.4	5.8	9.8	10.8
7	3.7	6.4	10.6	10.0
8	3.6	5.0	10.1	9.6
9	3.6	6.0	10.6	10.1
1	4.8	6.7	12.0	11.8
2	6.4	13.2	7.4	7.7
Base Resin A[1]	20.6	78.5	8.1	8.6
Base Resin B[2]	6.6	15.6	13.6	12.5

Notes:
(1) Profax 6501 + 0.1% Ca Stearate.
(2) Base Resin A + 0.1% AO-1.
(3) Melt flow rate (MFR) after 5th extrusion.

Conclusions

All the processing stabilizers, both experimental (compounds 4
to 9) as well as commercial (1 and 2), provided improved melt
stability over base resin A at both 500°F (260°C) and 550°F
(288°C). Experimental compounds 4 and 9 and commercial compounds 1
showed improved melt stability over base resin B at 500°F while
commercial compound 2 provided no improvement at this same temperature. At 550°F experimental compounds 4 to 9 and commercial
compound 1 showed marked improvement over base resin B, while
compound 2 provided only comparable performance and failed to
inhibit the rapid rise in melt flow.
   Despite its deficiency as a melt stabilizer, commerical compound 2 provided the best color stability being equivalent in this
regard to base resin A containing no AO-1. Experimental compound
8 is the next best providing marked color improvement over base
resin B.
   The phenylphosphonite derivative, 4, provided the best melt
stabilization at 500°F and 550°F, but provided no improvement in
color over that obtained with base resin B.
   Compound 8 is comparable in melt stability to 4 both at 500°F
and 550°F providing additionally an improvement of color of about
3 YI units over base resin B.

Polyethylene

High Density Polyethylene (HDPE). The preferred method of evaluating HDPE was found to be the time to cross link at 220°C using
the Brabender apparatus, as is made evident by the rapid increase
in torque. Determining melt viscosity by multiple extrusion
showed only small differences compared to the base polymer.
   The results are presented in Table IV.

TABLE II
PROCESSING STABILIZERS for HMW-HDPE PROCESSING STABILITY[1]

Additive	Lupolene 5260Z (BASF) Processing Stability - 220°C Brabender (Minutes to Crosslink)	
	0.05%[2]	0.1%[2]
7	8.5	13.5
6	9	15
5	9	16
9	9	15
8	11	16.5

(1) Data provided by CIBA-GEIGY Corporation, Basle, Switzerland.
(2) +0.05% AO-1

The data of Table II demonstrates that in HMW HDPE compound 8 in
combination with AO-1 is the most effective combination. 5, 6
and 9 are about equally effective when tested similarly with
AO-1.

Linear-Low Density Polyethylene (LLDPE) and Ethylene -
Propylene Diene (EPDM) Stabilization. Prevention of discoloration
of LLDPE and EPDM are prime objectives in addition to thermal
stabilization. While thermal stabilization is readily achieved
by the use of phenolic antioxidants, prevention of discoloration
requires a co-additive such as a phosphite.

TABLE III
PREVENTION OF DISCOLORATION OF
L-LDPE at 90°C

0.02% AO-1 + 0.02% Co-Additive	Lba Initial	Lba After 6 Weeks
AO-1 alone	0.60	4.50
2	-2.70	0.70
8	-2.80	0.60
6	-3.70	0.00

Note: (a)   Dow grade 2552
       (b)   Lba Color.

It is apparent from the data of Table III that while all
three compounds 2, 6, and 8, are effective in preventing discoloration initially, the most effective in counteracting the
color contributed by AO-1 is compound 6.
    Similarly, compounds 2 and 8 are most proficient in pre-
venting yellowing of EPDM plaques containing AO-2 kept at 90°C
at 100% R.H. for 20 days (Table IV).

TABLE IV
PREVENTION OF DISCOLORATION OF EPDM
AT 90°C and 100% R.H.

Antioxidant	phr	Gardner Color Initial	13 Days	20 Days
Blank	-	3	4	4
AO-2	0.05	3	5	5
AO-2 + 1	0.05 + 0.05	3	5	5
AO-2 + 2	0.05 + 0.05	3	3	4(1)
AO-2 + 6	0.05 + 0.05	3	5	6
AO-2 + 8	0.05 + 0.05	3	4	4

Note: (1) very yellow

Hydrolytic Stability of Experimental Processing Stabilizers

Compounds 4 and 5, two of the experimental processing stabilizers
listed in Figure 2, remained essentially unchanged when exposed
neat to an air atmosphere of 80% relative humidity at room temperature for 50 days. This is in contrast to some of the commercial
compounds such as 1, 2, and 3 which hydrolyze rapidly under the
same conditions (Table V). Hydrolysis tests at 50°C/80% relative
humidity show that compounds 4 to 9 inclusive are much more resistant to hydrolysis than 3 (Table VI). Of particular interest in
this respect is the dramatic increase in hydrolytic stability of
compounds 6 and 9 all of which have tertiary amino functions in
their phosphite structures capable of preferentially neutralizing

protons from acidic moieties. The protonation of trivalent phos-
phorous esters, which is a precursor to hydrolysis, is avoided.
Such protonation has been implicated in the cleavage of trialkyl
phosphites (8), monocyclic phosphite esters (9) as well as phosphonite esters (10).

TABLE V
HYDROLYTIC STABILITY AT ROOM TEMPERATURE (ca 25°C)
– 80% RELATIVE HUMIDITY

	HYDROLYSIS %	TIME DAYS
$\frac{2}{3}$	100	<5
	100	5
$\frac{1}{4}$	100	<4
	0	50
$\frac{5}{}$	0	50

TABLE VI
HYDROLYTIC STABILITY – 50°C/80% RELATIVE HUMIDITY
CRYSTALLINE SOLID

Additive	Time to Complete Hydrolysis (Days)
$\frac{3}{}$	2
$\frac{4}{}$	17
$\frac{5}{}$	22(1)
$\frac{6}{}$	>41(1)
$\frac{7}{}$	>41
$\frac{8}{}$	>62
$\frac{9}{}$	>62

Note: (1) Slight incipient hydrolysis

General Conclusions

It is apparent that of the compounds studied herein the hindered
substituted bis- and tris-phosphites having dibenzo[d,f][1,3,2]
dioxaphosphepin rings, are the most effective compounds in the
specific polyolefin substrates tested. Of particular interest
in this regard is the triethanolamine tris-phosphite compound 8.

Acknowledgments

The assistance of Messrs. Thomas M. Chucta and Peter W. Stewart
is gratefully acknowledged for their help in providing some of
the evaluation data presented herein. The authors wish to thank
Victoria Rivera and Teresa Schaeffer for preparation of the
manuscript.

Literature Cited

1. Previous Paper, Spivack, J. D.; Patel, A.; and Steinhuebel, L. P."Phosphorus Chemistry" - ACS Symposium 171 1981, pp. 351-354.
2. For Example, Chapter 12, Howard, J. A., in "Free Radicals Vol. II", Kochi, Jay K.; Ed. John Wiley, N.Y. 1973
3. Spivack, J. D., U.S. 4,233,207, 1980; Chem. Abstr., 94, 6670y.
4. (a) Spivack, J. D., U.S. 4,288,391, 1981; Chem. Abstr., 96, 7586m.
   (b) U.S. 4,351,759, 1982; Chem. Abstr. 97, 199111t.
   (c) Odorisio, P. A.; Pastor, S. D.; Spivack, J.D.; Bini, D.; Rodebaugh, R. K. Phosphorus and Sulfur, 1984, 19, 285
5. (a) Spivack, J. D.; Dexter, M.; Pastor, S. D., U.S. 4,318,845 1982; Chem. Abstr., 96, 182292
   (b) Odorisio, P. A.; Pastor, S. D.; Spivack, J.D., Phosphorus and Sulfur 1984, 19, 1.
6. ASTM METHOD 1238 Condition L
7. ASTM METHOD D1925-70T
8. Hudson, H. R.; Roberts, J.C., J. Chem. Soc. Perkin II, 1974, 1575-1580.
9. Weiss, R.; Vande Griend, L. J.; Verkade, J.G. J. Org. Chem. 1979, 44, 1860-1863.
10. Hudson, H.R.; Kow, A.; and Roberts, J.C. Phosphorus and Sulfur 1984, 19, 375-378.

RECEIVED January 31, 1985

Formation of Anomalous Structures in Poly(vinyl chloride) and Their Influence on the Thermal Stability

Effect of Polymerization Temperature and Pressure

THOMAS HJERTBERG and ERLING M. SORVIK

The Polymer Group, Department of Polymer Technology, Chalmers University of Technology, S–412 96 Goteborg, Sweden

Vinyl chloride was polymerized at subsaturation conditions in the temperature range of 45–80°C. The ratio between the monomer pressure (P) and the saturation pressure (P_o) varied from 0.53 to 0.97. The rate of dehydrochlorination decreased with decreasing values of P/P_o and showed a minimum at 55°C. To explain this behavior, the content of structures with thermally labile chlorine was determined: internal double bonds by ozonolysis and the branching structure by ^{13}C–NMR after reduction with tributyltin hydride. The content of tertiary chlorine was determined as the total content of ethyl, butyl and long chain branches. The anomalous structures are formed by different inter- and intramolecular transfer reactions and generally, their content increased with decreasing monomer concentration and increasing temperature in accordance with the proposed mechanisms. Instead, the content of internal double bonds decreased with increasing temperature. This is suggested to be due to a decreased tendency of chlorine atoms to selectively attack methylene groups at increased temperatures. If the content of labile chlorine is calculated as the sum of tertiary and internal allylic chlorine, a good relation is obtained between rate of dehydrochlorination and labile chlorine. Extrapolation to zero content of labile chlorine supports the existence of random dehydrochlorination. In all samples, the content of tertiary chlorine is considerably higher than the content of internal allylic chlorine. It is obvious that tertiary chlorine contributes most to the instability of PVC, while the contribution from internal allylic chlorine is of the same order as that of random dehydrochlorination.

The low thermal stability of poly(vinyl chloride) (PVC) has been ascribed to the pressure of low amounts of anomalous structures with

NOTE: This chapter is part 4 in a series.

0097–6156/85/0280–0259$07.50/0
© 1985 American Chemical Society

more labile carbon–chlorine bonds than those in the nominal structure
(1-3). There has been considerable controversy concerning the exact
nature and concentration of these anomalous structures in the polymer. However, the advancements achieved with different analytical
techniques during the last years have led to a substantial increase
in our knowledge of different anomalous structures in PVC (see e.g.
refs. 4-9).

In our work we have used the polymerization of vinyl chloride at
pressures (P) below the saturation value (P_o) as a way to produce
polymers, subsaturation PVC (U-PVC), with increased amounts of defects.
This system is also a model for the later stages in a conventional
batch polymerization of vinyl chloride, i.e. after the pressure drop.
With decreasing relative monomer pressure, P/P_o, the thermal
stability of PVC deteriorates strongly (6-8, 10). In a series of
investigations we have determined different structures in several
U-PVC samples and, as a reference, in a series of fractions of a
commercial suspension PVC (S-PVC) (6-8, 11).

Using ^{1}H-NMR we could show that 1 and 2 are the most frequent
end groups in ordinary PVC (6). By ^{13}C-NMR measurements on
reductively dehalogenated PVC, Bovey et al. (4,5) had earlier shown
that 3 is the dominating short chain structure on PVC.

~ CH$_2$-CH=CH-CH$_2$ ~CH$_2$-CH-CH$_2$ ~ CH-CH-CH$_2$~
            | | | | |
            Cl   Cl Cl   Cl CH$_2$Cl

         1   2   3

These findings as well as other investigations (see e.g. refs.
9, 12-14) have resulted in a completely new view of the mechanism for
chain transfer to monomer: an occasional head-to-head addition is
followed by 1,2-Cl-migration and subsequent elimination of a chlorine
atom results in structure 1 whereafter 2 is formed by addition of the
chlorine atom to a monomer molecule. If propagation occurs before
the elimination of chlorine, a chloromethyl branch, 3, is formed.

However, a comparison of PVC with different amounts of 1 and 2
indicated that neither 1 nor 2 has a major influence on the thermal
stability of PVC. In a recent paper, van den Heuvel and Weber (15)
showed that these structures are stable at 180°C.

The ^{13}C-NMR spectra of U-PVC reduced with tributyltin hydride
(Bu$_3$SnH) showed that PVC contains butyl and long chain branches to
an increasing extent at decreasing values of P/P_o (7). For samples
obtained at P/P_o = 0.6 contents of about 3-4 and 1-1.5 per 1000
monomer units (1000 VC), respectively, were determined. In the
fractions of ordinary PVC values between 0.5 and 1 per 1000 VC were
found for the combined content of the two branch structures.
Reductions with tributyltin deuteride (Bu$_3$SnD) showed that the
microstructure of the butyl branches is 2,4-dichlorobutyl with a
chlorine attached to the branch carbon, i.e. a tertiary chlorine.
It was also shown that the major part of the long chain branches (LCB)
also contain tertiary chlorine but the presence of hydrogen at the
LCB branch carbon could not be excluded.

By following the changes in $\overline{M}_n$ due to oxidative cleavage of all
double bonds, we could also show that U-PVC contains increased
amounts of internal double bonds (8). In the S-PVC fractions only

0.05-0.3 internal double bonds per 1000 VC were found while the content had increased to about 1 in samples obtained at $P/P_0 = 0.6$.

Formation of butyl branches takes place by a back-biting mechanism via a six-membered ring, while transfer to polymer from macroradicals is a reasonable source to LCB with tertiary chlorine (7, 9). We have suggested an alternative mechanism which also explains the formation of internal double bonds and LCB with tertiary hydrogen (7, 8). This mechanism is based on transfer to polymer from chlorine atoms produced in the mechanism for transfer to monomer:

$$\sim CH_2-CH-CH_2-CH\sim \quad + \quad Cl\cdot$$
$$\qquad\quad | \qquad\qquad |$$
$$\qquad\quad Cl \qquad\quad Cl$$

-HCl

$$\sim CH_2-\overset{\cdot}{C}-CH_2 \sim \qquad\qquad \sim CH-\overset{\cdot}{C}H-CH \sim$$
$$\qquad\quad | \qquad\qquad\qquad\qquad\qquad | \qquad\quad |$$
$$\qquad\quad Cl \qquad\qquad\qquad\qquad\qquad Cl \qquad Cl$$

| VC   VC /   -Cl·

$$Cl \qquad\qquad\qquad\qquad H$$
$$\quad | \qquad\qquad\qquad\qquad\quad |$$
$$\sim CH_2-C-CH_2 \sim \quad \sim CH-C-CH \sim \qquad \sim CH=CH-CH \sim$$
$$\qquad\quad | \qquad\qquad\qquad | \quad | \quad | \qquad\qquad\qquad\qquad\qquad |$$
$$\qquad\quad CH_2 \qquad\qquad\quad Cl \; CH_2 \; Cl \qquad\qquad\qquad\qquad Cl$$
$$\qquad\quad \wr \qquad\qquad\qquad\qquad \wr$$

The rate of dehydrochlorination in nitrogen at 190°C showed linear relations to the content of both tertiary chlorine and internal double bonds (correlation and coefficients 0.97 and 0.88, respectively). With reference to the better correlation obtained with tertiary chlorine and to the fact that its concentration is roughly 5 times higher than that of internal double bonds, we considered tertiary chlorine to be the most important labile structure in PVC (7, 8).

The results discussed so far were obtained with polymers prepared at 55°C. We have now extended this investigation to U-PVC prepared at different temperatures. We report herein the thermal stability of U-PVC obtained in the temperature range 45-80°C and with the relative monomer pressure P/P_0 ranging from 0.61 to 0.97. The structural characterization includes: molecular weight by GPC/viscometry, branches by ^{13}C-NMR measurements on samples reduced with Bu_3SnH and internal double bonds by ozonolysis.

Experimental

Polymers. The samples of subsaturation PVC were obtained by the polymerization technique described earlier (10, 16). To maintain a constant monomer pressure, vinyl chloride was continuously charged as

vapor from a storage vessel kept at lower temperature than the
reactor. Distilled water (2 l) was used as suspending medium,
ammonium peroxydisulphate (0.5 g) as initiator and uncoagulated
emulsion PVC latex as seed (100 g, dry content 25%, primary particle
diameter 0.03 μm). Vinyl chloride of polymerization grade was
kindly supplied by KemaNord AB, Sweden.

To ensure that the polymerizations were not influenced by
diffusion, a sufficient agitation was used (impeller, 1500 rpm).
Furthermore, the polymerizations were stopped before any agglomeration of the primary particles occurred. Insufficient agitation
and particle agglomeration would give a diffusion controlled system
with decreased monomer concentration in the polymer gel and correspondingly higher content of defects (17). The U-PVC samples studied
earlier (6-8, 11) were obtained under conditions which partly resulted in diffusion control. The polymerization temperatures and
pressures are given in Table 1 together with the molecular weight
data.

Molecular Weight Distribution. Gel chromatography (GPC) and
viscometry were used for determination of molecular weight distribution (MWD). Details of the GPC analysis and viscometry measurements
have been given earlier (18). Intrinsic viscosity was determined in
tetrahydrofuran at 25°C with an Ubbelohde viscometer. No correction
for kinetic energy losses was necessary. A Waters Associates GPC
Model 200 operating at 25°C with tetrahydrofuran as solvent was used.
The column combination consisted of five Styragel columns with
permeabilities ranging from 10^3 to 10^7 A, giving good separation in
the molecular weight range of interest. To calculate MWD and
molecular weight averages the computer program devised by Drott and
Mendelson (19) was used, assuming trifunctional long chain branch
points. The program corrects the MWD and LCB and gives a measure of
the content of LCB. The calibration for linear PVC was obtained via
the universal calibration curve as described earlier (18).

Thermal Stability. Thermal stability was measured by
dehydrochlorination. Experimental details have been given earlier
(20). Bulk samples (100 mg) were treated at 190°C in pure nitrogen
atmosphere and HCl was measured by conductometry. The rate of
dehydrochlorination is expresed as evolved HCl (in percent of the
theoretical amount) per minute. The linear part of the conversion
curve between 0.1 and 0.3% was used.

Determination of Branches. Reductive dechlorinations were performed
with Bu$_3$SnH as reducing agent. We have modifid the original two-
step method given by Starnes et. al. (21) to a one-step method (22).
To avoid precipitation, a mixture of tetrahydrofuran and xylene is
used as solvent. At high conversion, the concentration of xylene
and the temperature is increased. By this procedure, a chlorine
content less than 0.1% is obtained in 6 h. The experimental details
are as previously given (22).
Proton-decoupled ^{13}C-NMR spectra were obtained with a Varian
XL-200 spectrometer equipped with a "High-sens" probe. Free induction decays with spectra windows of 8000 Hz were stored in 16K
computer locations with 32 bits wordlengths. The acquisition time
was 1 s, the tip angle 60° and the pulse interval 10 s. The reduced

Table 1. Molecular weight data

Sample	Pol.temp. $^{\circ}C$	P/P_o	$M_n \cdot 10^{-3}$	$M_w \cdot 10^{-3}$	M_w/M_n	$\lambda \cdot 10^{6}$ [a]
A1	45	0.97	59.8	161	2.69	12
A2		0.92	50.4	161	3.19	16
A3		0.85	46.8	153	3.27	21
A4		0.76	32.0	117	3.66	22
A5		0.70	26.5	98.2	3.71	30
A6		0.61	24.1	81.1	3.37	26
B1	55	0.97	48.6	117	2.41	12
B2		0.92	45.0	118	2.63	11
B3		0.85	41.0	110	2.68	13
B4		0.76	37.7	106	2.81	18
B5		0.70	33.1	101	3.05	19
B6		0.61	29.3	94.7	3.23	23
B7		0.53	25.1	91.9	3.66	41
C1	65	0.97	37.1	93.9	2.53	20
C2		0.92	35.2	93.4	2.65	22
C3		0.85	34.1	92.4	2.71	25
C4		0.76	31.6	90.0	2.88	23
C5		0.70	30.3	96.7	3.19	35
C6		0.61	25.2	79.5	3.15	40
D1	80	0.97	26.7	66.9	2.51	30
D2		0.92	25.3	65.0	2.57	38
D3		0.85	26.0	76.6	2.95	45
D4		0.76	20.9	66.3	3.17	44
D5		0.70	20.2	64.2	3.18	51
D6		0.61	19.2	67.0	3.49	53

a) λ is the number of long chain branches per molecular weight unit.

samples were observed at 115°C as 10-15% (w/v) solutions in 1,2,4-
trichlorobenzene with 20% benzene-d$_6$ to provide the deuterium lock.
The number of scans accumulated was 8-15000. The Fourier transformations were performed with floating point arithmetic. As reference, the main methylene peak was used; 30.0 ppm versus TMS.

Determination of Internal Double Bonds. The number of internal
double bonds was determined by following the changes in $\overline{M}_n$ due to
oxidative cleavage by ozone of all double bonds. The ozonolysis was
performed according to Michel et. al. (23) and the experimental
details have been given earlier (8). The reaction was carried out
at -20°C in tetrachloroethane solution with a small amount of
methanol added in order to facilitate the cleavage of the formed
ozonoide. The reaction time was 2 h and the polymer was recovered
by precipitation in methanol and was dried in vacuum for 24 h.

The content of internal double bonds ($C=C_{int}$) was calculated
from the number average molecular weight before and after the
oxidative treatment:

$$C=C_{int}/1000 \ VC = (1/\overline{M}_n - 1/\overline{M}_{n,o}) \cdot 62 \ 500$$

where $\overline{M}_{n,o}$ is the original number average molecular weight and $\overline{M}_n$
the number average molecular weight after ozonation.

Results and Discussion

The rate of dehydrochlorination of most commercial PVC samples fall
within a rather narrow interval. A typical scatter of stability
data can be found in ref. 24 where the results from 11 commercial
samples are given. The rate of dehydrochlorination at 190°C in
nitrogen varied in the range of $1-2.5 \cdot 10^{-2}$ % per minute. If the
labile structures should be the major reason to the instability of
PVC, the content of such structures should also vary within a correspondingly narrow interval.

There are two main reasons for the similarity in
dehydrochlorination behavior among different PVC samples. First,
vinyl chloride is most often polymerized in a rather narrow temperature interval, 40-75°C, because of the high tendency to chain
transfer to monomer which determines the molecular weight (25).
Second, the monomer concentration at the reaction site is constant
during the major part of the polymerization because PVC is insoluble
in vinyl chloride and the propagation mainly occurs in the polymeric
phase (26). At 50°C, the gel contains 30 g vinyl chloride per 100 g
PVC (27, 28) as long as a separate liquid monomer exists. In
practice, the "pressure drop" occurs at about 70% conversion and
thereafter the monomer concentration decreases in the gel. However,
only a minor part of the batch is formed under subsaturation
conditions.

The content of tertiary and internal allylic chlorine is low
in normal PVC: about 1 (7, 9) and 0.1-0.3 (8, 24, 29, 30) per
1000 VC, respectively. At these concentration levels the accuracy
of the analytical techniques is not too good. In combination with
the expected narrow concentration range, correlations between
thermal stability and labile structures are uncertain when ordinary
PVC is used.

In a subsaturation polymerization, the monomer concentration can be kept constant at any given level below the saturation level. We have, therefore, used this kind of polymerization in order to obtain polymers with decreased stability. The degradation rate (190°C, N_2) of the U-PVC samples used in our previous investigations (6-8) covered a much wider interval than ordinary PVC; i.e. 1.7-5.5 • 10^{-2} % per minute. The identification was therefore considerably simplified. For the first time, a definite relation could be proposed between the rate of dehydrochlorination and the most frequent labile structures, tertiary and internal allylic chlorine.

The results from the degradation experiments in the present investigation are given in Table 2. The rate of dehydrochlorination covers a still wider interval; 2-11 • 10^{-2} % per minute. However, it is not possible to directly compare these values with those discussed above as another degradation equipment was used. In agreement with the experience gained within the IUPAC "Working Party on PVC Defects" (31) the same ranking is obtained, although the actual values may differ. The rate of dehydrochlorination in the new equipment is 40-50% higher.

Even if the degradation had been performed with the same equipment, it would have been difficult to compare the polymers prepared at 55°C in this investigation with those studied earlier. The morphology of the latter showed that agglomeration of the primary particles had occurred (10). The polymerization after the agglomeration is diffusion-controlled, which decreases the monomer concentration (17). For comparison, a sample was prepared under the same conditions used to obtain sample B1 except that the polymerization was continued an additional 15% after the agglomeration point. The rate of dehydrochlorination was 3.4 • 10^{-2} % per minute compared to 2.2 • 10^{-2} % for sample B1. An exact comparison between the earlier polymers and the present 55°C samples is therefore not possible. However, on a relative basis the two series are similar.

Figure 1 shows the influence of P/P_o on the rate of dehydrochlorination at different polymerization temperatures. At all temperatures, the thermal stability is steadily decreasing with decreasing monomer pressure in accordance with our earlier findings (10). In the range of 55-80°C, an increase in polymerization temperature leads to decreased stability at all levels of P/P_o. In a recent paper, Hamielec et. al. (32) suggested that the temperature should be increased at the later stages of high conversion batch polymerizations of vinyl chloride in order to avoid a decrease in the thermal stability after the pressure drop. Based on free volume ideas they argued that the propagation reaction would be relatively more favored than side reactions leading to labile defects. The present experimental findings, as illustrated in Figure 1, clearly contradict that suggestion.

Apart from the highest value of P/P_o used, the rate of dehydrochlorination of the polymers prepared at 45°C is higher than for the 55°C polymers. Apparently, there is an optimum in polymerization temperature with respect to the thermal stability. As illustrated in Figure 2, this is accentuated at lower relative pressures. For ordinary batch polymerized PVC this effect should not be too large as the major part of the polymerization takes place at $P/P_o = 1$. However, we have not shown that the rate of dehydrochlorination is doubled if the conversion is pushed up from

Table 2.   Thermal stability data

Pol. temp. °C		$\dfrac{\text{deHCl}}{\text{dt}}$ $\cdot$ 10^2 a) (%/min.)						
	P/P_0	0.97	0.92	0.85	0.76	0.70	0.61	0.53
45		2.08	2.92	3.50	5.43	6.75	7.85	
55		2.21	2.44	3.26	4.28	5.41	6.38	7.89
65		3.65	4.21	5.20	6.20	7.16	8.79	
80		3.90	4.68	6.25	7.85	8.50	10.9	

a)   rate of dehydrochlorination at 190°C in nitrogen.

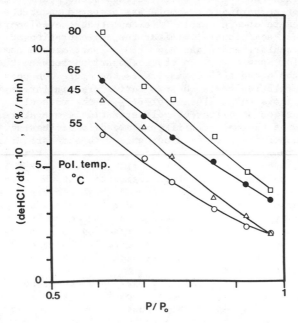

Figure 1.   The rate of dehydrochlorination at 190 °C in nitrogen as a function of relative monomer pressure (P/P_0) and polymerization temperature.

75% to 95%. For such high conversion materials, a maximum in thermal stability should be found for a polymerization temperature around 55°C. Fortunately, this temperature level coincides with the temperatures most often used in commercial production of PVC.

The maximum observed for the thermal stability can per se be explained if there are two variables with opposite influence on the formation of labile defects. As discussed in the Introduction, such structures are formed after inter- and intramolecular transfer to polymer. Reasonably, the relative frequency of these transfer reactions should increase with decreasing monomer concentration and increasing polymerization temperature. Instead of the experimental variable P/P_o, the monomer concentration at the reaction site should be considered.

According to Berens (27), the concentration of monomer in the gel is only influenced by the relative pressure in the P/P_o-range of interest. However, by using polymers prepared at 11, 50 and 90°C, Nilson et. al. (28) could show that the monomer concentration is a function of both P/P_o and the polymerization temperature. By interpolating their data, it can be estimated that an increase in polymerization temperature from 45 to 80°C should increase the monomer concentration with about 25% at saturated conditions. This will tend to counteract the expected faster formation of labile structures due to the temperature alone which qualitatively is in accordance with the observation of an optimum polymerization temperature with respect to the thermal stability.

It is also possible to use the data given in ref. 28 to calculate the monomer concentration at all P/P_o-levels. In the present calculations, no corrections have been made for the particle size or the surface tension, due to the presence of emulsifier (1 g ammoniumlaurate per 1 H_2O). If the rate of dehydrochlorination is plotted against the concentration instead of P/P_o, Figure 3 is obtained. Compared to Figure 1, the 45°C series now becomes almost identical with the polymers at 55°C, except at the lowest monomer concentration. The relations between the rate of dehydrochlorination and polymerization temperature at constant monomer concentration will therefore show a plateau below 55°C, see Figure 4. A minimum is only indicated for monomer concentrations below about 10 g VC per 100 g PVC. At temperatures above 55°C, the thermal stability decreases with increasing polymerization temperature.

If the labile structures are the main reason to the low thermal stability of PVC, Figures 3 and 4 should also reflect the concentration of the defects. In our previous work (7, 8), we showed that the rate of dehydrochlorination could be related to the amounts of tertiary and internal allylic chlorine. However, it is also likely that random dehydrochlorination will contribute to a certain extent (2, 3, 33, 34). According to our estimation, random dehydrochlorination could account for 10-15% of the initiation during degradation of ordinary PVC (3). It has been suggested that the stereo-structure should influence dehydrochlorination from ordinary monomer units (33, 35-37). The present samples also cover a change in polymerization temperature, 45-80°C. A comparison between the content of labile structures and thermal stability might therefore reveal an eventual influence of the tacticity.

The amounts of internal double bonds are given in Table 3. In agreement with our previous results (8) a decrease in P/P_o results in

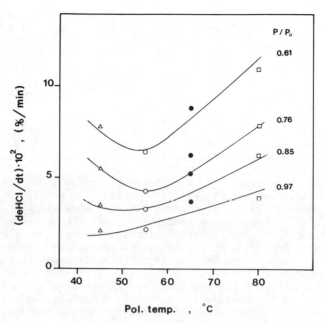

Figure 2.   Demonstration of the optimum temperature.

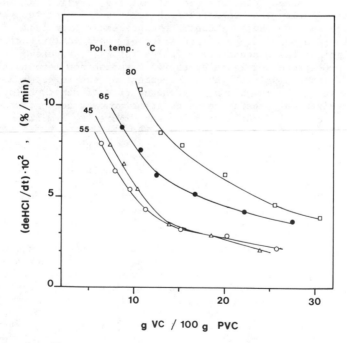

Figure 3.   The rate of dehydrochlorination as a function of monomer concentration and polymerization temperature.

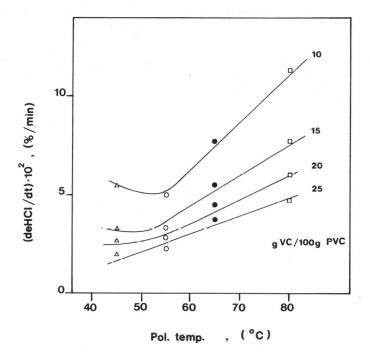

Figure 4. The influence of polymerization temperature on the degradation rate at four levels of monomer concentration.

Table 3. Content of internal double bonds

Pol. temp. °C	P/P$_o$	$C=C_{int}/1000$ VC						
		0.97	0.92	0.85	0.76	0.70	0.61	0.53
45		0.28	–	0.58	0.54	0.60	0.79	
55		0.14	0.33	0.52	0.40	0.55	0.66	0.78
65		0.23	0.25	0.32	0.44	0.63	0.55	
80		0.08	0.14	0.18	0.39	–	0.49	

higher values. In Figure 5, the content of internal double bonds is
instead plotted against the monomer concentration. It is obvious
that the monomer concentration is the most important parameter, but
the polymerization temperature may also influence the frequency of
this structure. Admittedly, there is a rather large scatter in the
data, but at increasing polymerization temperature the content of
internal double bonds tends to decrease. This can be seen from the
solid and broken lines which represent the extreme values of the
polymerization temperature, 45 and 80°C respectively.

We have suggested that internal double bonds are formed after
transfer to polymer from chlorine atoms (7, 8). Besides
unsaturation, the subsequent reactions may also result in long chain
branches:

$$Cl\cdot \ + \ \sim CHCl-CH_2-CHCl\sim$$

$$(1) \qquad\qquad -HCl \qquad (2)$$

$$\sim CH_2-\overset{\bullet}{C}Cl-CH_2\sim \qquad\qquad\qquad \sim CHCl-\overset{\bullet}{C}H-CHCl\sim$$

$$VC \ \ (3) \qquad\qquad\qquad VC \ \ (4) \qquad\qquad (5)\ \ -Cl\cdot$$

$$\sim CH_2-\overset{\overset{\displaystyle Cl}{|}}{\underset{\underset{\displaystyle \wr}{|}}{C}}-CH_2\sim \qquad \sim CHCl-\overset{\overset{\displaystyle \overset{\bullet}{H}}{|}}{\underset{\underset{\displaystyle \wr}{|}}{C}}-CHCl\sim \qquad \sim CH=CH-\overset{\overset{\displaystyle }{}}{\underset{\underset{\displaystyle Cl}{|}}{C}H}\sim$$

The chlorine atom is formed in the reaction scheme leading to
chain transfer to monomer (6, 9, 13, 15). Normally, it reacts with a
monomer molecule initiating a new polymer chain:

$$Cl\cdot \ + \ CH_2 = \underset{\underset{\displaystyle Cl}{|}}{C}H \ \ \xrightarrow{\ (6)\ } \ \ \underset{\underset{\displaystyle Cl}{|}}{C}H_2 - \underset{\underset{\displaystyle Cl}{|}}{\overset{\bullet}{C}}H$$

At decreasing monomer concentration, reactions 1 and 2 should
therefore be more favored. This can e.g. be shown by the ratio
between 1,2-dichloro alkane end groups and internal double bonds. It
can be assumed that each macromolecule contains 1 end group of this
type (6, 15). For the 65°C series, the molecular weight data in
Table 1 indicates that this ratio decreases from 8.4 at P/P_o to 4.5
at $P/P_o = 0.61$. Furthermore, the rate of formation of chlorine atoms
will increase at decreasing monomer concentration (see below). The
observed increase in the content of internal double bonds is
therefore easy to understand.

It is more difficult to explain the effect of the polymerization
temperature. As there is no mass transfer in reaction 6 in contrast

to reactions 1 and 2, the former should have lower activation energy. Accordingly, the relative frequency of reaction 6 compared to 1 and 2 should increase with decreasing temperature. The ratio discussed above decreased to 3.7 for the sample prepared at 45°C and P/P_o = 0.97, indicating a change in the opposite direction. However, the influence of the temperature on the relation between reactions 1 and 2 must be considered. Experiments with chlorination of PVC have indicated that the methylene groups are preferably attacked (38). The selectivity should, however, decrease with increasing temperature which might explain the decrease in the content of internal double bonds with increasing polymerization temperature.

Typical ^{13}C-NMR spectra of reduced samples, polymerized at P/P_o equal to 0.97 and 0.70, are shown in Figure 6. The spectrum of the sample obtained at P/P_o = 0.97 is very similar to those obtained with ordinary PVC. Besides the main peak at 30.0 ppm from undisturbed CH_2-carbons, there are some smaller peaks mainly originating from branches and normal alkane end groups. The evaluation of such spectra has been discussed in detail elsewhere (7, 9). The designation of the peaks from these structures and their chemical shifts are summarized in Figure 7 and Table 4.

As shown earlier, branches with one carbon atom are the most frequent short chain branch (4, 5, 7, 9, 13). In the polymers obtained at low values of P/P_o it is easy to observe butyl branches, although this structure can be seen also in polymers obtained close to the saturation pressure (7, 22). It is also possible to detect long chain branches in agreement with our earlier results (7, 22). The better signal-to-noise ratio obtained in the present work (due to the "High-sens" probe) has allowed us to detect the presence of ethyl branches. The existence of the latter type of branches has earlier been proven by Starnes and his coworkers (9, 39). The influence of the polymerization conditions on the content of different branches will be discussed below.

We have also been able to detect several other structures. One of them is isolated chlorine residues due to imperfect reduction. The content in the present material varies between not observable to about 1 Cl per 1000 VC. In agreement with the results published by Starnes and his coworkers (9, 40) we have also found nonallylic primary halogen, i.e. CH_2Cl- groups in long chain ends and methyl and butyl branches. For the polymers studied here it is especially chloromethyl branches which can be observed, although butyl branches with chlorine residues were observed in some polymers as well. When necessary, the content of branches (see below) has been corrected with respect to the presence of these chlorinated structures.

Starnes et. al. have shown that unsaturated structures in the starting PVC may form cyclic products during reduction with Bu_3SH. The most frequent unsaturated end groups (1) is thus converted to a 1-(ethyl)-2-(long alkyl)cyclopentane group (41) while internal double bonds are converted into internal 1,2-di(long alkyl)cyclopentane moities (42). These structures have also been observed in the present material. The most important observation is that the internal cyclopentane group can most easily be seen in polymers obtained at low values of P/P_o. This is an agreement with the measurements of internal double bonds, Table 3. These details in ^{13}C-NMR spectra of reduced PVC will be more completely dealt with in a separate paper (43). In the following discussion, it is the branching structure which is of interest.

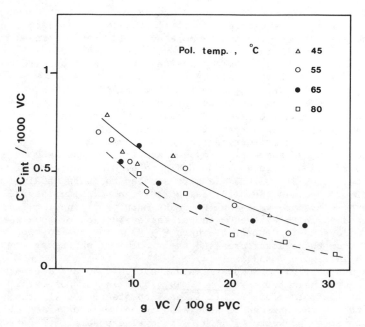

Figure 5. The content of internal double bonds ($C=C_{int}$) per 1000 VC as a function of monomer concentration and polymerization temperature. The symbols are the same as those used in Figures 1-4; (——) represents the 45 °C series; (---) represents the 80 °C series.

Table 4. ^{13}C shifts for saturated chain ends and branches expressed in ppm versus TMS [a]

Structure	Carbon [b]						
	CH₃	2	3	4	bι	α	β
LE	14.08	22.88	32.17	29.58	–	–	–
Me	19.94	–	–	–	33.21	37.51	27.44
Et	11.07	26.83	–	–	39.69	34.06	27.22
Bu	14.13	23.39	29.50	34.13	38.09	34.52	27.25
L					38.12	34.52	27.25

a)   see ref. 8.
b)   according to Figure 7.

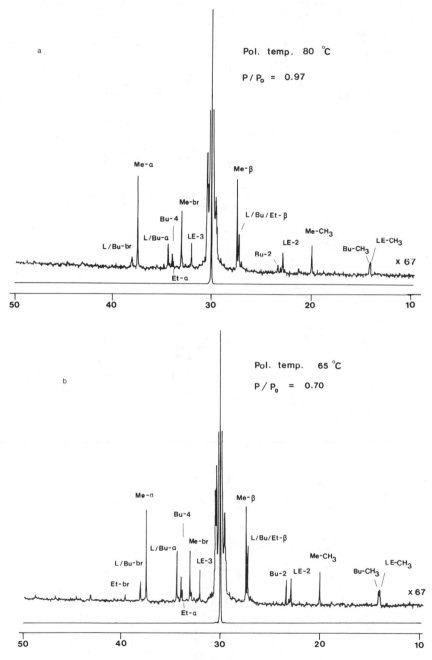

Figure 6. ^{13}C–NMR spectra of reduced U–PVC samples.   For peak assignments, see Figure 7 and Table 4.   (a) Polymerization temperature 80 °C, P/P$_o$ = 0.97.   (b) Polymerization temperature 65 °C, P/P$_o$ = 0.70.

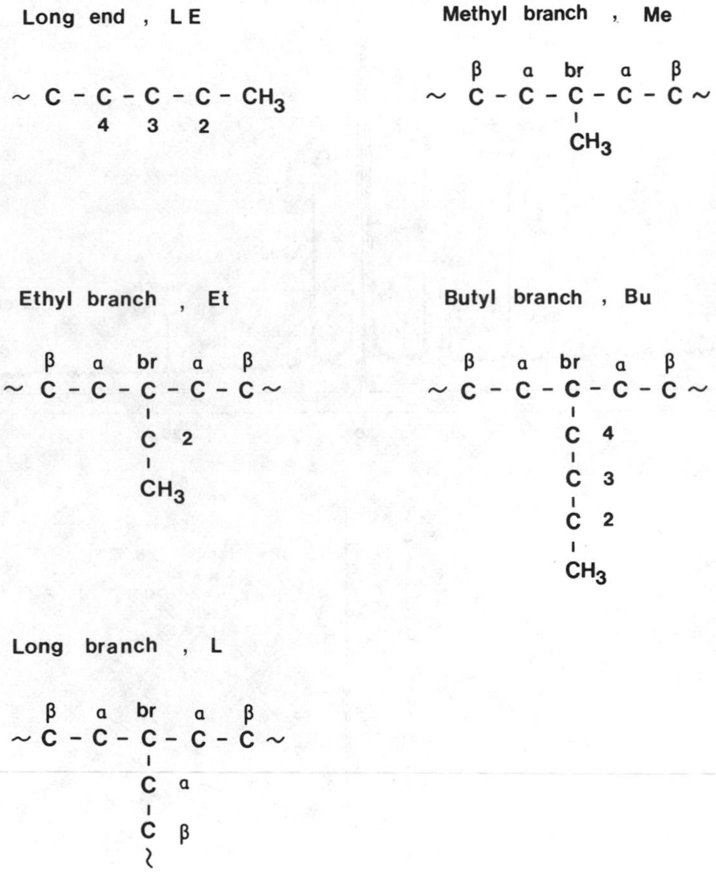

Long end , L E

$$\sim \text{C} - \underset{4}{\text{C}} - \underset{3}{\text{C}} - \underset{2}{\text{C}} - \text{CH}_3$$

Methyl branch , Me

$$\sim \overset{\beta}{\text{C}} - \overset{\alpha}{\text{C}} - \overset{\text{br}}{\underset{|}{\text{C}}} - \overset{\alpha}{\text{C}} - \overset{\beta}{\text{C}} \sim$$
$$\text{CH}_3$$

Ethyl branch , Et

$$\sim \overset{\beta}{\text{C}} - \overset{\alpha}{\text{C}} - \overset{\text{br}}{\underset{|}{\text{C}}} - \overset{\alpha}{\text{C}} - \overset{\beta}{\text{C}} \sim$$
$$\underset{|}{\text{C}} \ 2$$
$$\text{CH}_3$$

Butyl branch , Bu

$$\sim \overset{\beta}{\text{C}} - \overset{\alpha}{\text{C}} - \overset{\text{br}}{\underset{|}{\text{C}}} - \overset{\alpha}{\text{C}} - \overset{\beta}{\text{C}} \sim$$
$$\underset{|}{\text{C}} \ 4$$
$$\underset{|}{\text{C}} \ 3$$
$$\underset{|}{\text{C}} \ 2$$
$$\text{CH}_3$$

Long branch , L

$$\sim \overset{\beta}{\text{C}} - \overset{\alpha}{\text{C}} - \overset{\text{br}}{\underset{|}{\text{C}}} - \overset{\alpha}{\text{C}} - \overset{\beta}{\text{C}} \sim$$
$$\underset{|}{\text{C}} \ \alpha$$
$$\underset{\wr}{\text{C}} \ \beta$$

Figure 7.   Structures identified in Figure 6.

As shown earlier, the short chain branches have the following structures in the original PVC: chloromethyl (4, 5), 2-chloroethyl (9, 39) and 2,4-dichlorobutyl (7, 9, 44). The two latter have chlorine connected to the branch carbon. The same is valid for the main part of the long chain branches (7, 9, 44). However, the presence of tertiary hydrogen in connection with long chain branches cannot be excluded. According to the ^{13}C-NMR spectrum of a deuteride reduced U-PVC polymerized at 55°C and P/P_o = 0.59 up to one-third of the long chain branch points could contain tertiary hydrogen (7).

The content of the different branches is given in Table 5. The relation between the methyl branch frequency and the polymerization conditions is somewhat different compared to that of the others. The content of this branch structure decreases with decreasing polymerization temperature and monomer concentration, Figure 8. This is what could be expected from the mechanism for chain transfer to monomer:

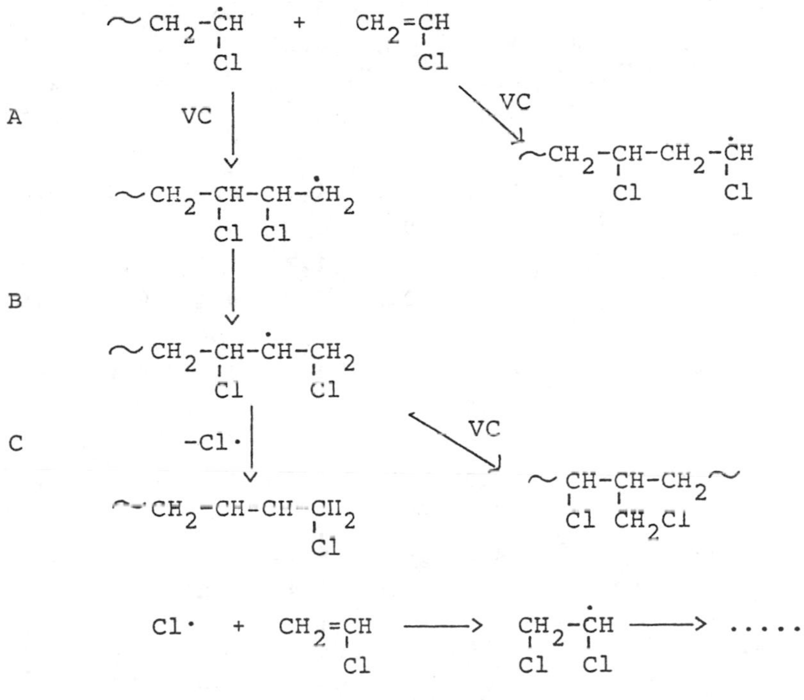

$$\sim CH_2-\overset{\cdot}{C}H \quad + \quad CH_2=CH \qquad\qquad$$
$$\qquad\quad | \qquad\qquad\qquad\quad |$$
$$\qquad\quad Cl \qquad\qquad\qquad\quad Cl$$

A

$$\qquad\quad VC \Big| \qquad\qquad\qquad \searrow VC$$

$$\qquad\qquad\qquad\qquad\qquad \sim CH_2-CH-CH_2-\overset{\cdot}{C}H$$
$$\sim CH_2-CH-CH-\overset{\cdot}{C}H_2 \qquad\qquad | \qquad\qquad |$$
$$\qquad\quad | \quad | \qquad\qquad\qquad Cl \qquad\quad Cl$$
$$\qquad\quad Cl \; Cl$$

B

$$\sim CH_2-CH-\overset{\cdot}{C}H-CH_2$$
$$\qquad\quad | \qquad\qquad |$$
$$\qquad\quad Cl \qquad\qquad Cl$$

C

$$\qquad -Cl\cdot\Big| \qquad\qquad\qquad \searrow VC$$

$$\qquad\qquad\qquad\qquad\qquad \sim CH-CH-CH_2\sim$$
$$\sim CH_2=CH-CH-CH_2 \qquad | \qquad |$$
$$\qquad\qquad\quad | \qquad\qquad Cl \; CH_2Cl$$
$$\qquad\qquad\quad Cl$$

$$Cl\cdot \quad + \quad CH_2=CH \longrightarrow CH_2-\overset{\cdot}{C}H \longrightarrow \ldots\ldots$$
$$\qquad\qquad\qquad | \qquad\qquad | \quad |$$
$$\qquad\qquad\qquad Cl \qquad\qquad Cl \; Cl$$

The changes with the polymerization temperature reflect the balance between the two reactions at level A. As expected, a decrease in polymerization temperature leads to less head-to-head additions. A change in the monomer concentration should, however, not influence this balance. Instead, the reason for the content of chloromethyl branches with decreasing monomer concentration can be found at level C. It is obvious that that balance between these reactions should change in the observed manner, i.e. less propagation. The decreasing molecular weight with decreasing P/P_o is another indication of this effect.

The content of the other branches, ethyl, butyl and long chain

Table 5.   Content of branches in the reduced PVC samples

Pol. temp. °C	Branch type	Branches/1000 VC						
		P/P_o 0.97	0.92	0.85	0.76	0.70	0.61	0.53
45	Me	3.9		3.7			3.5	
	Et	≤0.1		0.2			0.3	
	Bu	0.6		1.2			2.2	
	L	≤0.1		0.1			0.3	
55	Me	4.2	4.2	4.1	4.0	4.0	3.8	3.9
	Et	0.2	0.2	0.3	0.4	0.4	0.5	0.5
	Bu	0.7	1.1	1.2	1.6	1.7	1.9	2.2
	L	0.2	0.2	0.3	0.4	0.6	0.6	0.8
65	Me	4.6		4.5	4.5	4.4	4.2	
	Et	0.2		0.4	0.35	0.5	0.5	
	Bu	0.8		1.6	1.8	2.0	2.3	
	L	0.3		0.5	0.5	0.6	0.7	
80	Me	4.8	4.7	4.7	4.6		4.5	
	Et	0.3	0.4	0.4	0.5		0.6	
	Bu	1.4	1.5	1.7	2.1		2.4	
	L	0.3	0.5	0.6	0.6		0.8	

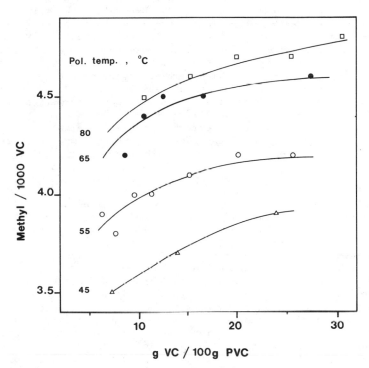

Figure 8.   The influence of polymerization conditions on the content of methyl branches.

branches, increases with increasing polymerization temperature as
well as with decreasing monomer concentration, see Figure 9. For
ethyl (9) and butyl (7, 9, 44) branches this is in agreement with the
back-biting mechanism suggested for their formation. The same
behavior should also be expected for the formation of long chain
branches by transfer to polymer from macroradicals. However, we have
also suggested that long chain branches can be formed after transfer
to polymer from chlorine atoms (7, 8). Naturally, the formation of
long chain branches via reactions 1 and 3 should increase with
decreasing monomer concentration, similar to the behavior valid for
the formation of internal double bonds (see above). On the other
hand, the balance between reaction 4 and 5 indicates that the ratio
between long chain branches with tertiary hydrogen and internal
double bonds should decrease with decreasing monomer concentration.
Obviously, the total effect leads to an increased formation of long
chain branches at low values of P/P_o.

   The discussion about the effect of the polymerization
temperature indicated that reaction 1 is favored compared to reaction
2 at increasing temperature. Accordingly, the fraction of long chain
branch points with tertiary chlorine should increase as well. To
prove this suggestion, ^{13}C-NMR spectra of polymers reduced with
Bu_3SnD must be obtained.

   In the present investigation, we have assumed that all long
chain branch points are associated with tertiary chlorine. The
content of tertiary chlorine can therefore be taken as the sum of
ethyl, butyl and long chain branches.

   Earlier, tertiary chlorine was generally considered to be less
reactive than internal double bonds. This was based on experiments
with low molecular weight model substances, see e.g. ref. 45. However, using copolymers between vinyl chloride and 2-chloro-propene,
Berens (46) stated that the presence of 1-2 tertiary chlorine per
1000 VC per se would account for the thermal lability observed in
ordinary PVC. Furthermore, our previous investigation indicated that
the thermal reactivity of internal allylic chlorine is of the same
order as that of tertiary chlorine (8). We, therefore, consider it
justifiable to use the total content of labile chlorine atoms. As
shown in Figure 10, there is a very good relation between the rate of
dehydrochlorination and the amount of labile chlorine obtained in
this way. For comparison, it can be mentioned that the degradation
rate of commercial samples is found in the interval $1.5-3.5 \cdot 10^{-2}$ %
per minute.

   In all samples the amount of tertiary is considerably higher
than that of internal allylic chlorine. This is also valid for
ordinary PVC where typical values are: 0-0.5 internal allylic and
1-2 tertiary chlorine per 1000 VC. We have, therefore, suggested
that the latter structure is the most important labile structure in
PVC (8). In a recent paper, Ivan et. al. (47) have instead claimed
internal allylic chlorine to be most important. For a commercial
sample, the content of this structure was given to 0.1 per 1000 VC
and the rate constant of the dehydrochlorination to about 10^{-2} min^{-1}.
They did not measure the amount of tertiary chlorine but suggested
the presence of an unidentified labile structure. Its rate constant
of degradation should be only somewhat less than that of internal
allylic chlorine. Furthermore, the content of the unknown structure
was claimed to be about four times higher than the content of

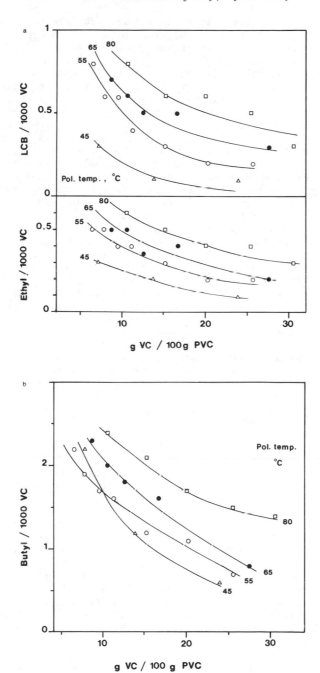

Figure 9. The influence of polymerization conditions on the content of branches. (a) Ethyl and long chain branches. (b) Butyl branches.

internal allylic chlorine. It is therefore plausible to assume the
unidentified structure to be tertiary chlorine and to conclude that
this defect is the most important for the degradation of PVC.

In a series of papers, Minsker et al. (see, e.g., ref. 48) have
claimed that all actual labile structure should be ketoallylic
(◡(C=O)-CH=CH-CHCl◠). The formation of this group is suggested to
occur by oxidation of allylic methylene groups during the production
and storage of PVC. As an alternative, Svetly et. al. (49-51) have
suggested that this structure should exert its adverse effect by
catalysis. As discussed in the preceding paper in this series (8)
¹H-NMR spectra of U-PVC samples with a high content of internal
double bonds did not show any evidence of ketoallylic groups.
Furthermore, using an appropriate model compound, Starnes et. al.
have recently shown that this structure, in fact, is relatively
thermally stable (52) and that it does not have any catalytic
influence on the dehydrochlorination (53).

If Figure 10 is inspected in more detail, it can be seen that
the 55°C series tends to have the lowest rate of dehydrochlorination
at a given level of labile chlorine. This effect is more easily seen
if the content of labile chlorine is related to the polymerization
conditions. The content of the structure at different levels of
monomer concentration can be obtained from Figures 5, 8 and 9. The
influence of the polymerization temperature is given for four
different monomer concentrations in Figure 11. Obviously, the defect
concentration increases at all monomer levels. This behavior is
different from that of the degradation rate given in Figure 4. A
comparison between the two figures indicates that the degradation
rate of the 45°C series is too high in relation to its content of
labile chlorine.

At present, we consider two possible explanations of this
difference. First, the data given in Figures 5, 8 and 9 indicates
that the relation between tertiary and internal allylic chlorine
decreases with decreasing temperature. In the 80°C series, this
ratio is 7 or higher while it is only 2 to 3.5 in the 45°C series.
If the lability of internal allylic chlorine is in fact higher than
that of tertiary chlorine, the low stability of the 45°C series can
be explained qualitatively. A statistical evaluation of the data
given here might reveal a difference in reactivity. This will be
discussed in another paper.

It must also be remembered that the rate of dehydrochlorination
is influenced by the initiation as well as by the length of the
polyene sequence. It is known that the polyene sequence length
decreases with increasing syndiotacticity (2, 36, 54, 55), i.e. with
decreasing polymerization temperature. This effect will also tend to
decrease the overall stability of the 45°C series in relation to the
amount of labile chlorine. Determination of the polyene sequence
length at low degrees of degradation should show if this factor is of
importance for the temperature range, 45-80°C, used in this
investigation.

Another observation made in Figure 10 is that the relation
between degradation rate and content of labile chlorine is nonlinear. This may be an effect of higher stationary concentrations of
HCl in samples with a high rate of dehydrochlorination. This would
give a too high value of degradation rate due to HCl-catalysis (see
e.g. ref. 56). It is e.g. known that the presence of HCl lengthens

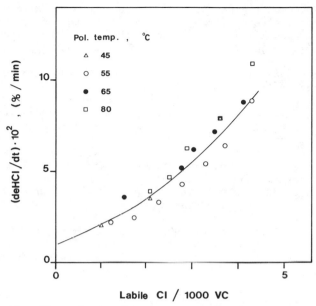

Figure 10.   The relation between rate of dehydrochlorination and the content of labile chlorine per 1000 VC.

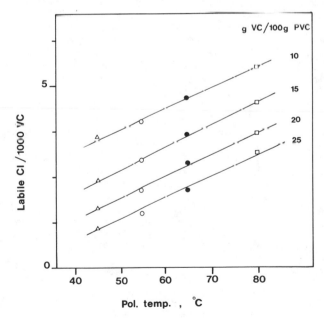

Figure 11.   The influence of the polymerization conditions on the content of labile chlorine per 1000 VC.

the polyene sequence length (56, 57). Furthermore, we have suggested
that that primary effect of HCl is to catalyze random dehydrochlorination (56). Degradation in solution instead of in the solid state
would give a better possibility to avoid HCl-catalysis (54) even in
low stability samples.

According to earlier opinion, initiation should not occur at the
normal PVC-units below 200°C. This conclusion was mainly based on
experiments with low molecular weight model substances (see e.g. ref.
45). The extrapolation to zero content of labile chlorine in
Figure 10 gives an indication of the importance of random
dehydrochlorination.

Using the data published by Ivan et. al. (47) and our own
conclusions, the relative rate constants for initiation from internal
allylic, tertiary and normal secondary chlorine càn approximately be
given as 1, 1, and $2 \cdot 10^{-4}$, respectively. On the other hand, typical relative concentrations of these structures are 2, 10 and 10^4,
respectively, in an ordinary PVC. With respect to all uncertainties
this shows that the total contribution from random dehydrochlorination is of the same order as that from internal allylic chlorine. In
agreement with our earlier conclusion (3, 8) the dominating influence
of initiation from tertiary is obvious.

Conclusions

The minimum in degradation rate found for subsaturation PVC obtained
around 55°C becomes less obvious if the monomer concentration at the
reaction site is used as variable instead of the relative monomer
pressure, P/P_o. The observed behavior is mainly due to the influence
of the polymerization conditions on the formation of thermally labile
chlorine, i.e. tertiary chlorine and internal allylic chlorine.
Tertiary chlorine is associated with ethyl, butyl and long chain
branches. The labile structures are formed after different inter-
and intramolecular transfer reactions. Generally, the content
increases with decreasing monomer concentration and increasing
temperature in accordance with the proposed mechanisms. The content
of internal double bonds instead decreases with increasing
temperatures.

The content of labile chlorine can be calculated as the sum of
tertiary and internal allylic chlorine. The relation between this
measure and the rate of dehydrochlorination is very good.
Extrapolation ·to zero content indicates the presence of random
dehydrochlorination. The total contribution from this type of initiation is of the same order as that from internal allylic chlorine.
However, tertiary chlorine must be considered as the most important
labile structure in PVC.

Acknowledgments

Financial support from the Swedish Board for Technical Development
is gratefully acknowledged. The authors also which to express their
thanks to Anne Wendel for thorough experimental work.

References

1. D. Braun in "Developments in Polymer Degradation 3", N. Grassie, Ed., Applied Science, London, 1981, p. 101.
2. W. H. Starnes, ibid., p. 135.
3. T. Hjertberg and E. M. Sorvik in "Thermal Degradation of PVC", E. D. Owen, Ed., Applied Science, London, 1984.
4. K. B. Abbas, F. A. Bovey and F. C. Schilling, Makromol. Chem., Suppl., 1, 227 (1975).
5. F. A. Bovey, K. B. Abbas, F. C. Schilling and W. H. Starnes, Macromolecules, 8, 437 (1975).
6. T. Hjertberg and E. M. Sorvik, J. Macromol. Sci., Chem., A17, 983 (1982).
7. T. Hjertberg and E. M. Sorvik, Polymer, 24, 673 (1983).
8. T. Hjertberg and E. M. Sorvik, Polymer, 24, 685 (1983).
9. W. H. Starnes, F. C. Schilling, I. M. Plitz, R. E. Cais, D. J. Freed, R. L. Hartless, and F. A. Bovey, Macromolecules, 16, 790 (1978).
10. T. Hjertberg and E. M. Sorvik, J. Polym. Sci. Chem., Ed., 16, 645 (1978).
11. T. Hjertberg, E. M. Sorvik and A. Wendel, Makromol. Chem. Rapid Commun., 4, 175 (1983).
12. R. Petiaud and Q-T Pham, Makromol. Chem., 178, 741 (1977).
13. W. H. Starnes, F. C. Schilling, K. B. Abbas, R. E. Cais, and F. A. Bovey, Macromolecules, 12, 556 (1979).
14. P. Hildebrand, W. Ahrens, F. Brandstetter, and P. Simak, J. Macromol. Sci. Chem., A17, 1093 (1982).
15. C. J. M. van den Heuvel and A. J. M. Weber, Makromol. Chem., 184, 2261 (1977).
16. E. M. Sorvik and T. Hjertberg, J. Macromol. Sci. Chem., A11, 1349 (1977).
17. T. Hjertberg and E. M. Sorvik, submitted for publication.
18. K. B. Abbas and E. M. Sorvik, J. Appl. Polym. Sci., 19, 2991 (1975).
19. E. E. Drott and R. A. Mendelson, J. Polym. Sci., A-2, 8, 1361 and 1373 (1970).
20. K. B. Abbas and E. M. Sorvik, J. Appl. Polym. Sci., 17, 3567 (1973).
21. W. H. Starnes, R. L. Hartless, F. C. Schilling, and F. A. Bovey, Adv. Chem. Ser., 169, 324 (1978).
22. T. Hjertberg and A. Wendel, Polymer, 23, 1641 (1982).
23. A. Michel, G. Schmidt and A. Guyot, Polym. Prepr., 14, 665 (1973).
24. K. B. Abbas and E. M. Sorvik, J. Appl. Polym. Sci., 20, 2395 (1976).
25. Encyclopedia of PVC, Vol. 1, L. I. Nass, Ed., Marcel Dekker, New York, 1976.
26. J. Ugelstad, H. Flogstad, T. Hertzberg, and E. Sund, Makromol. Chem., 164, 171 (1973).
27. A. R. Berens, Angew Makromol. Chem., 47, 97 (1975).
28. H. Nilsson, C. Silvergren and B. Tornell, Eur. Polym. J., 14, 737 (1978).
29. P. Burille, Diss., University Claude Bernard, Lyon, 1979.
30. D. Braun, A. Michel and D. Sonderhof, Eur. Polym. J., 17, 49 (1981).

31. IUPAC Working Party on PVC Defects, Report on
     Dehydrochlorination Tests, December 1983, compiled by J. Millan,
     to be published.
32. A. E. Hamielec, R. Gomez-Vaillard and F. L. Marten, J. Macromol.
     Sci. Chem., A17, 1005 (1982).
33. W. H. Starnes, R. C. Haddon, D. C. Hische, I. M. Plitz,
     C. L. Schosser, F. C. Schilling, and D. J. Freed, Polym. Prepr.,
     21 (2), 176 (1980).
34. F. Tudos and T. Kelen in "Macromolecular Chemistry", K. Saarela,
     Ed., Butterworths, London, 1973, Vol. 8, p. 393.
35. J. Millan, G. Martinez and C. Mijangos, Polym. Bull., 5, 407
     (1981).
36. G. Martinez, C. Mijangos and J. Millan, J. Macromol. Sci. Chem.,
     A17, 1129 (1982).
37. G. Martinez, C. Mijangos and J. Millan, J. Appl. Polym. Sci.,
     28, 33 (1983).
38. R. A. Komorski, R. G. Parker and M. H. Lehr, Macromoleculas, 15,
     844 (1982).
39. W. H. Starnes, F. C. Schilling, I. M. Plitz, R. E. Cais,
     D. J. Freed, and F. A. Bovey, Paper presented at the Third
     International Symposium on Polyvinylchloride, August 1980,
     Cleveland, USA, Preprints p. 58.
40. W. H. Starnes, G. M. Villacorta, F. C. Schilling, and
     I. M. Plitz, "Proceeding, 28th Macromolecular Symposium of the
     International Union of Pure and Applied Chemistry", Amherst,
     Mass., July 1982, p. 226.
41. W. H. Starnes, G. M. Villacorta and F. C. Schilling, Polym.
     Prepr., 22 (2), 307 (1981).
42. W. H. Starnes, G. M. Villacorta, F. C. Schilling, G. S. Park,
     and A. H. Saremi, Polym. Prepr., 24 (1), 253 (1983).
43. T. Hjertberg, to be published.
44. W. H. Starnes, F. C. Schilling, I. M. Plitz, R. E. Cais, and
     F. A. Bovey, Polym. Bull., 4, 555 (1981).
45. M. Asahina and M. Onozuka, J. Polym. Sci., A, 2, 3505 and 3515
     (1964).
46. A. R. Berens, Polym. Eng. Sci., 14, 318 (1974).
47. B. Ivan, J. P. Kennedy, T. Kelen, F. Tudos, T. T. Nagy, and
     B. Turcsanyi, J. Polym. Sci. Polym. Chem. Ed., 21, 2177 (1983).
48. K. S. Minsker, V. V. Lisitsky, S. V. Kolesov, and G. E. Zaikov,
     J. Macromol. Sci. - Rev. Macromol. Chem., C20(2), 243 (1981).
49. J. Svetly, R. Lukas and M. Kolinsky, Makromol. Chem., 180, 1363
     (1979).
50. J. Svetly, R. Lukas, J. Michalcova, and M. Kolinsky, Makromol.
     Chem., Rapid Commun., 1, 247 (1980).
51. J. Svetly, R. Lukas, S. Pokorny, and M. Kolinsky, Makromol.
     Chem., Rapid Commun., 2, 149 (1981).
52. S. L. Haynie, G. M. Villacorta, I. M. Plitz, and W. H. Starnes,
     Polym. Prepr., 24(2), 3 (1983).
53. W. H. Starnes, S. L. Haynie, H. E. Katz, I. M. Plitz, and
     G. M. Villacorta, Paper presented at "International Symposium on
     the Degradation and Stabilization of Polymers", American
     Chemical Society, Div. Polym. Chem., St. Louis, April 1984,
     Polym. Prepr., 25(1), 71, (1984).

RECEIVED December 7, 1984

Degradation of Poly(vinyl chloride) According to Non-Steady-State Kinetics

JOSEPH D. DANFORTH[1]

Grinnell College, Grinnell, IA 50112

Poly(vinyl chloride) has been shown to degrade thermally
by a chain mechanism that is best represented by non-
steady-state kinetics (NSSK). Degradations are initiated
at a very few sites within a chain, and the subsequent
zip reaction which accounts for substantially all of the
evolved hydrogen chloride is confined to a single chain.
The initiation reaction is temperamental. It is influenced by surfaces, impurities, and hydrogen chloride
pressure. The number of initiations that result in a
zip chain influences the average length of a zip chain,
which in turn establishes the time at which the maximum
degradation rate is attained. Although the uncertainty
of the initiation reaction can significantly alter
degradation patterns and make reproducibility under
presumably identical conditions difficult to achieve,
the NSSK model gives excellent agreement of observed
and calculated data over the entire range of a degradation.

The degradation of PVC has been described in terms of non-steady-
state kinetics (NSSK) which were developed on the basis of the zipper
mechanism (1,2). Observed degradation patterns were altered significantly by the presence of hydrogen chloride, inert surfaces, and intentionally added impurities (3). These degradation patterns were
reproduced by NSSK using the best fit values of three parameters
(previously named less descriptively): k_i, the fraction of chains
starting degradation per sec; k_z, the rate of unzipping of degrading
chains expressed as the fraction of a chain unzipping per sec; and
k_T, a chain terminating constant that was based on certain
assumptions about the way that impurities prevented the initiation of
degradation chains. The initiation and zip constants were primarily
responsible for degradation patterns. The terminating constant, k_T,
reduced the values of α, the fraction dehydrochlorinated, as a
function of time. Its primary function was to obtain best fit parameters when degradations did not go to completion. All degradation
reactions for all samples were acceleratory. That is, the evolution

[1] Deceased.

0097–6156/85/0280–0285$06.00/0
© 1985 American Chemical Society

of hydrogen chloride was slow at the beginning, increased through a
maximum and then decreased. The initiation reaction accounted for
only negligible amounts of hydrogen chloride. The zip reaction
accounted for greater than 99% of the evolved hydrogen chloride (3).
Alpha-time curves calculated from best fit values of the parameters
could be superimposed upon the data curves. Excellent agreement of
theory and data was also obtained in the more critical comparison in
which degradation rates were plotted as a function of time.

Individual degradation curves for nine samples of PVC from
three different suppliers have been observed at many temperatures in
the range 210-240°C. Each degradation was effectively reproduced by
the proper assignment of best fit parameters, but the duplication of
degradation curves under presumably identical run conditions was
uncertain. Duplication was often attained for runs made in sequence,
but a run made at another time under presumably identical conditions
would occasionally give an altered degradation pattern and significantly different values of the parameters at best fit.

It is now apparent that difficult reproducibility is a built-in
characteristic of zipper kinetics. Even though reproducibility is a
recognized problem, the values of the parameters k_i and k_z have
theoretical and practical significance and can be directly related
to the processes that occur in any specific degradation.

Kinetic models of polymer degradation based on chain processes
have been suggested (4-8). These models and other models for polymer degradation have normally invoked the steady-state hypothesis
(9). Barron and Boucher (10) describe a kinetic model for zip kinetics but assume an apparent first order behavior and do not account
effectively for the acceleratory phase of the degradation.
MacCallum has questioned the validity of the steady-state hypothesis
(11) but did not develop a practical kinetic model for NSSK. It is
the non-steady-state feature that is responsible for the successful
application of this model to the zip degradation of PVC.

Experimental

The apparatus which measures precisely the rates of gas evolution as
a function of time has been described (2, 12-14). PVC samples from
Goodrich and Scientific Polymer Products have been characterized by
the name of the supplier and the average degree of polymerization.

Curve fitting techniques and the computer generation of degradation curves for single and mixed parameters have been described
(15). The equations representing NSSK when initiation is first
order have been derived for chain termination that is first order in
impurities (16).

Other initiation and zip orders have been evaluated. All non-
steady-state models appear to be markedly superior to models that
invoke the steady-state hypothesis, but none is appreciably better
than the simple assumption of first order initiation and zero order
zip. The derivation and evaluation of other non-steady-state models
will be the subject of a later publication.

Results and Discussion

The equations representing the behavior of acceleratory degradations which have been previously described ($\underline{1},\underline{2},\underline{16}$) are summarized in Table I. Though complex in appearance, non-steady-state equations are based on very simple assumptions about initiation and propagation. With computer facilities available it is no longer necessary to invoke steady-state assumptions in order to solve the problems. The ideal equation represents expected behaviors when there is no chain termination by surfaces and impurities. When chain termination occurs during the initial states of a degradation, only a portion of the chains that are thermally initiated actually produces a zip reaction. Reaction of an initiated chain with surface or impurities prevents the zip reaction for that initiation. The non-ideal equation properly accounts for initiations that do not result in a zip reaction.

Table I. Equations for Non-Steady-State Kinetics

Ideal Equation - No Chain Termination

Acceleratory Phase, $t = o$ to $t = 1/k_z$

$$\alpha = k_z t + (k_z/k_i) \exp (-k_i t) - k_z/k_i$$

Deceleratory Phase, $t = 1/k_z$ to $t = \infty$

$$\alpha = 1 + (k_z/k_i) \exp (- k_i t) [(1 - \exp (k_i/k_z)]$$

Equation with Chain Termination

Acceleratory Phase, $t = o$ to $t = 1/k_z$

$$\alpha = k_z t + k_z/k_i \exp (- k_i t + (k_z k_T/k_i) \exp (-k_i t)$$

Deceleratory Phase, $t = 1/k_z$ to $t = \infty$

$$\alpha = (k_z/k_i) \exp (- k_i t) (1 - \exp (k_i/k_z))$$

$$- (k_z k_T/2k_i) \exp (- 2k_i t) (1 - \exp (k_i/k_z)) + 1$$

$$- (k_z k_T/2k_i) (1 - \exp (- k_i/k_z))$$

The α vs. time curves of Figure 1 have been generated from assigned values of k_i, k_z, and k_T to illustrate how the values of the parameters influence degradation patterns. For a zip rate, $k_z = 0.0005$ sec^{-1}, the inflection point of the resulting α-t curves will fall at 2000 sec because the maximum rate will appear at $t = 1/k_z$. At constant k_z the degradation patterns are significantly altered by the values of k_i and to a lesser extent by the values assigned to k_T.

Figure 2 presents generated rate data as $\Delta\alpha/100$ sec vs. t. The middle curve of Figure 2 represents the rate data for the case where $k_i = k_z = 0.0005$ and $k_T = 0$. The maximum rate appears at 2000 sec, which is the inflection point of the corresponding α-t curve of

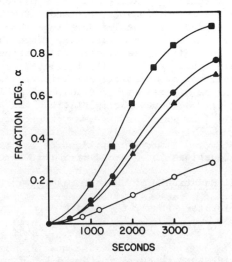

Figure 1. Calculated alpha-time curves for assigned values of k_i, sec^{-1}; k_z, sec^{-1}; and k_T. (●) k_i = 0.0005, k_T = 0; (▲) k_i = 0.0005, k_z = 0.0005, k_T = 0.2; (■) k_i = 0.00010, k_z = 0.0005, k_T = 0; (○) k_i = 0.0001, k_z = 0.0010, k_T = 0.

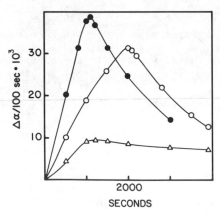

Figure 2. Calculated rate curves for changing k_i, sec^{-1} and k_z, sec^{-1}. (●) k_i = 0.0005, k_z = 0.0010; (○) k_i = k_z = 0.0005; (△) k_i = 0.0001, k_z = 0.0010.

Figure 1. The two upper curves of Figure 2 represent degradation
behaviors at constant initiation (k_i = 0.0005 sec^{-1}) for k_z values
of 0.001 sec^{-1} (upper curve) and 0.0005 sec^{-1} (middle curve). The
lower rate curve has the same k_z value as the upper curve but the
value of k_i has been reduced ten-fold to 0.0001 sec^{-1}. These generated rate curves illustrate the drastic changes in degradation
patterns that can be expected from changes in the initiation and zip
constants. At values within the range normally encountered the chain
terminating constant lowers somewhat the initial portions of a
degradation curve but does not significantly change degradation
patterns at longer times and higher conversions.

In addition to the three parameters, k_i, k_z, and k_T, there is a
time shift, t_s. The time shift is necessary to account for induction
periods. When chains are immediately terminated during an induction
period, the kinetic model based on zip kinetics obviously does not
apply. The time shift gives an initial time that depends on the
behavior of the degradation curve after chain termination has become
almost negligible. It eliminates the induction period for purposes
of curve fitting but does not alter the general shape of the degradation curve.

It has been shown in earlier work that zip kinetics give best
fit parameters that will reproduce α-t and rate-time curves over the
entire range of a degradation (3,16). The quantitative significance
of the initiation and zip constants was implied but not emphasized
because different degradation patterns and correspondingly different
parameter values were often obtained for the same sample under presumably the same conditions.

It is now apparent that difficult reproducibility is an expected
characteristic of zip kinetics. The nature of zip kinetics that
makes reproduction of data difficult will be described and the quantitative characterization of sample behavior in terms of initiation
and zip parameters will be demonstrated.

It is useful for the better understanding of zip kinetics to
illustrate how approximate degradation data corresponding to observed
data can be calculated for two hypothetical samples. Each is 100 mg;
their chain lengths are 500 and 1000 vinyl chloride units. The term,
"approximate", is appropriate because of the way $\Delta\alpha/\Delta$ more closely
approaches $d\alpha/dt$, and the term, "approximate", would be unnecessary
for the calculated data. These calculated data represent an ideal
situation in which there is no mixture of chain lengths and in which
no premature termination of zip chains occurs. Actual samples will
always have some variations in chain length, and starting characteristics may also show variations due to premature chain termination.
Although it has been shown that single valued parameters give good
representation of data generated from samples in which chain lengths
and starting characteristics have been purposely mixed (3,15), the
term "approximate" will always be appropriate when single valued
parameters are used to represent actual degradation in which there
must be chains of different lengths.

Calculation of Degradation Data for Hypothetical Samples of
Different Degrees of Polymerization. One hundred milligrams of PVC,
regardless of chain length, will contain $\dfrac{0.100 \cdot DP}{MW}$ moles hydrogen

chloride where DP is the vinyl chloride units per chain, and MW is
the number average molecular weight. If the weight of the sample is
doubled the rate of hydrogen chloride evolution expressed as moles
hydrogen chloride $\cdot$ sec^{-1} will double, but the rate expressed as
$\Delta\alpha$sec^{-1} will not change with the sample size. The rate expressed as
moles HCl $\cdot$ sec^{-1} will depend upon the number of chains that are unzipping and the intrinsic rate of unzipping expressed as moles HCl $\cdot$
chain^{-1} $\cdot$ sec^{-1}. In the zipper mechanism the intrinsic rate of unzipping is assumed to be the same for all samples regardless of
chain length. However k_z, the fraction of a chain unzipping per sec,
will depend upon the chain length. During the acceleratory phase of
a degradation the moles hydrogen chloride evolved per sec at any time
will be the moles of degrading chains multiplied by the intrinsic
rate. The moles of degrading chains during acceleration can be estimated at any time as the cumulative sum of chains that have started
during preceding intervals.

The chains that start during an interval can be estimated from
the sample weight, the degree of polymerization, the first order
initiation constant, k_i, and the length of the interval. Thus,
0.10 g samples will contain initially 1.60 $\cdot$ 10^{-6} (DP 1000) and 3.2
x 10^{-6} (DP 500) moles of polymer chains. If k_i is assigned the
reasonable value, 0.0005 sec^{-1}, the moles of polymer chains unzipping
after 100 sec will be (moles of initial polymer) $\cdot$ 0.0005 sec^{-1} $\cdot$
100 sec. The moles of polymer chains starting during the first 100
sec can be subtracted from the number of polymer chains present initially to give the number of unstarted chains present at the start
of the interval, 100–200 sec. The number of chains starting during
the 100–200 sec interval can then be calculated. During the accelerating period the cumulative sum of the average number of chains
starting per interval multiplied by the intrinsic zip rate will give
the moles HCl per interval. When t = 1/k_z the first started chains
are completely decomposed. At t = 1/k_z + 100 the number of producing chains calculated as the cumulative sum must be corrected by
subtracting the chains that have terminated. The subtracted number
will be the average number of chains producing from 0–100 sec, and
so on for subsequent intervals. The average number of producing
chains multiplied by the intrinsic rate (DP $\cdot$ k_z) gives the rate of
hydrogen chloride evolution as moles HCl per interval. This value
divided by the moles hydrogen chloride in the original samples gives
$\Delta\alpha$/interval. The calculated moles hydrogen chloride evolved per
interval can be entered in the same way that observed peak areas of
an actual run are entered and the best fit values of the parameters
obtained. The best fit values of the parameters from the calculated
data, as would be expected, are essentially the same as those values
that were used in making the calculations. The ideal equation shown
in Table 1 describes more simply the kinetic behavior that has been
estimated by the preceding stepwise calculations. The stepwise calculation is similar to the way run data are obtained since in a run
the peak area recorded represents $\Delta\alpha$/interval.

There is one critical point that requires additional comment.
When the intrinsic rate is assigned a value, DP $\cdot$ k_z, the unwarranted
assumption that a zip chain length is the same as the length of a
polymer chain has been made. This assumption would be true if there
were one initiation site per polymer chain. If there were two ini-

tiation sites per polymer chain the zip chain would be one-half as
long but the fraction unzipping per sec would be doubled. The intrinsic zip rate as initially assumed for zip kinetics would not
depend upon chain length. However, the number of starts per chain
will certainly influence both k_i and k_z. The starts per chain must
be a small number if the zipper mechanism is to apply. A reasonable
model for PVC would assume that each sample contained an average
number of potential starting positions per chain characteristic of
that sample. The actual number of starts that led to a zip reaction
would depend upon the impurities present and the number of surface
contacts that were chain terminating. The partial pressure of hydrogen chloride, which has been shown to influence the initial reaction,
would also influence not only the rate of initiation but also the
fraction of potential starting positions that actually initiates a
zip reaction. Since the buildup of hydrogen chloride at the beginning of a degradation depends on zip chains already present and weak
linkages that slowly decompose thermally, an erratic buildup of
hydrogen chloride contributes to difficult reproducibility in the
same way that chain terminating impurities and surfaces contribute.
Thus, initiation comprises only an insignificant part of degradation
but influences significantly the number of chains that will be producing at any time and the avergage chain length that those chains
will have. The average chain length is directly related to the time
at which the maximum rate is attained ($t_{max} = 1/k_z$) so those substances or run conditions that alter starting characteristics will
also alter somewhat the value of k_z. Changes in k_i and k_z, as illustrated in Figures 1 and 2, significantly influence degradation patterns. Thus, the presence of Chromsorb or a piece of a paper clip,
preheating a sample, the sample size which influences the ratio of
sample to surface contacts, the partial pressure of hydrogen chloride,
the rate at which hydrogen chloride forms in the sample, and chain
terminating impurities already present (3) will alter the initiation
characteristics and these in turn will influence the average chain
length.

Since it has already been established that PVC data from the
beginning to the end of a thermal degradation can be reproduced from
NSSK, (2,3,16) there seems to be no reason to burden the literature
with the many degradation curves and best fit parameters that have
been obtained for over 600 individual runs using nine different
samples and a variety of run conditions. It does seem appropriate to
illustrate how data can be reproduced under very carefully controlled
conditions and to show how occasional samples deviate from their expected behavior. It will also be suggested that for eight of the
nine samples studied the chain length of the polymer has an overriding influence on degradation patterns. The behavior of one sample
that did not follow the degradation pattern that was expected on the
basis of its chain length will be considered. The influence of preheating on degradation patterns will also be described.

The Characterization of PVC Samples in Terms of Their Kinetic
Parameters. Under carefully controlled conditions runs that are made
in sequence may give reproducible data and illustrate that difficulties of reproducibility are not caused by shortcomings of the techniques and apparatus used in the measurements. The sample Goodrich

684 has been chosen to illustrate the reproducibility of data and
also to illustrate that occasional runs do not duplicate the expected
degradation pattern. Figure 3 shows α-t curves for identical runs at
225°C using samples of Goodrich 684 in the 70-80 mg range. Three
samples demonstrate reproducibility. The other sample is significantly different. Rate curves for these samples are reproduced in
Figure 4. Table II gives the values of the best fit parameters
obtained by minimizing the differences between the observed and calculated values of α over the entire range of the degradation. In
these runs the sample was removed shortly after the degradation
reached the deceleratory phase.

Table II. Best Fit Parameters for Goodrich 684 at 225°C

Run	k_i (sec$^{-1} \cdot 10^3$)	k_i (sec$^{-1} \cdot 10^3$)	k_T	$t_{s,sec}$	$\Delta,\%$ dif per point
2D46A	1.37	0.65	0.36	189	0.73
2D25A	2.24	0.82	0.78	131	0.97
2D24A	2.15	0.83	0.59	197	0.94
2D49A	2.13	0.76	0.48	211	1.17

Using the non-ideal equation of Table I and the best fit values
of the parameters, calculated α-t and rate data agree with observed
data, and the calculated time of the maximum rate falls very close to
the time of the observed maximum rate. Each set of degradation data
is effectively represented by the best fit parameters, yet there is a
significant variation of parameter values for run 2D46A which was
presumably run under conditions identical with the other runs. Run
2D46A displays the characteristics that are always encountered in
runs that misbehave. The values of k_i and k_z are lower than those of
the "good" runs which invariably show more rapid acceleration. The
justification for difficult reproducibility and decrease in k_i and k_z
for runs that misbehave is of course inherent in the mechanism. The
initiation reaction represents only an insignificant part of the overall reaction. It is extremely sensitive to impurities, to surface
contacts, and to the partial pressure of hydrogen chloride. The data
of Figures 1 and 2 illustrate the dramatic changes in degradation
patterns that are to be expected for changes in the fraction of
chains starting per sec even if the chain length remained unchanged.
However, when the fraction initiated per sec is decreased the fraction of actual starts per chain is also somewhat decreased. Even
though the potential starts per chain are initially identical for a
given sample, the actual number of zip chains started per polymer
chain is less than normal under poor starting conditions. The fewer
the number starts per chain the longer will be the average length of
a zip chain, and this is reflected in a lower value of k_z and a
longer time to achieve the maximum rate.

Before the implications of the zipper mechanism in terms of reproducible data were fully appreciated, efforts to attain reproducibility commensurate with the quality of the apparatus and the method
of operation led to hundreds of degradation runs. Some of these runs

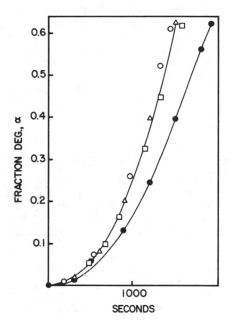

Figure 3. Fraction degraded vs time for identical run conditions:
Goodrich 684 at 225°C. (●) 2D46A, (○) 2D25A, (△) 2D24A,
(□) 2D49A.

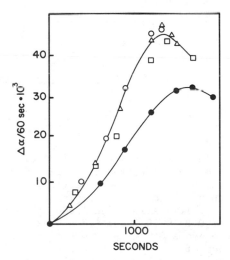

Figure 4. Rate-time curves for Goodrich 684 at 225°C, (●) 2D46A,
(○) 2D25A, (△) 2D24A, (□) 2D49 A.

gave good initial acceleration and were intuitively considered to be "good runs". Others did not "take off" and although they gave best fit parameters that would duplicate the data for that run, the parameter values differed somewhat from those obtained for other runs under presumably the same conditions. From many hundreds of runs on a variety of samples there are a number of degradation behaviors that can be quantitatively expressed in terms of parameter values. From the very large amounts of data available some of the sample characteristics and run variables that are reflected in the values of the parameters will be described.

Changes in the Number of Potential Starts Per Chain Can Influence the Degradation Pattern. The sample SP 917 has previously been shown to accelerate more rapidly and to attain its maximum rate in less time than SP 634 and SP 1381 (3). Samples SP 634 and SP 1381 follow approximately the pattern of all Goodrich samples in which at comparable conditions long chain samples accelerated slowly and required a longer time to achieve their maximum rates (16). Figure 5 shows representative plots of $\Delta\alpha/60$ sec as a function of time for SP 634, SP 917 and SP 1381. All runs were made by the direct introduction of the sample into the reaction chamber after the temperature of the run had been attained. Best fit values of the parameters for each of the three runs and the interpolated values expected for SP 917 on the basis of the length of the chain are recorded in Table III.

Table III.   Best Fit Values of Parameters for the Degradations of SP 634, SP 917 and SP 1381 at 225°C
Comparable Conditions

Sample	Run	k_{i-1} (sec^{-1}·10^3)	k_i (sec^{-1}·10^3)	k_T	$t_{s,sec}$	Δ,% dif per point
SP 634	C110	1.12	0.68	0.6	171	0.33
SP 917	C109	1.73	0.85	1.0	46	0.39
SP 1381	C108	0.58	0.32	1.0	164	0.44
Interp. 917	–	0.88	0.52	–	–	–

The parameter values of SP 634 and SP 1381 are in the range of values expected from degradation studies on five Goodrich samples in which the time to achieve the maximum rate $(1/k_z)$ increased with increasing chain length of the sample. Sample SP 917 is completely out of line, giving k_i, 1.73 x 10^{-3} sec^{-1} and k_z, 0.85 x 10^{-3} sec^{-1} vs expected values of 0.88 x 10^{-3} sec^{-1} and 0.52 x 10^{-3} sec^{-1}, respectively. On the basis of the previous discussions of mechanism, the anomalous behavior of SP 917 can be understood if that sample, either in preparation or in subsequent handling, had significantly more potential starting positions per chain. The fraction starting per sec and the number of actual zips per chain are significantly greater than would be expected for that chain length. Although no technical information is available concerning the stability characteristics of SP 917, it would be expected that the sample would be less stable than the other samples that have been studied.

Thermal Pretreatment Influences the Starting Characteristics and the Degradation Pattern. Even though all PVC samples except SP 917 demonstrate that there is a uniform dependence of the degradation pattern on the chain length of the sample, there is a significant influence of the thermal pretreatment of the sample on the degradation pattern. Figure 6 shows a comparison of degradation patterns for SP 1381 after thermal pretreatments at 235°C of 0, 2, and 3 minutes. Thermal pretreatment seems definitely to increase the number of potential starts per chain and the number of zip chains actually started. Because the number of zip chains started is increased, the average length of a zip chain is reduced, so k_z, the fraction of a chain unzipping per sec, is increased.

The behavior of samples after preheating at 235°C illustrates the problem that one encounters when runs at different temperatures are compared. During the initial period of a run the bulk of the sample is preheated at the run temperature. As has been shown, the time, and presumably the temperature at which the preheating occurs, influence the degradation pattern of the sample. Thus, even though runs at different temperatures give degradation patterns that can be precisely accounted for by the best fit parameters of NSSK, there is no assurance that the parameters at the two different temperatures are directly comparable.

This problem was recognized and has been more fully discussed in a publication that describes the special precautions that are necessary in order to obtain meaningful activation energies and frequency factors for the acceleratory degradation of PVC (17).

Mechanistic Implications of NSSK. The effective reproduction of degradation patterns over the entire degradation range by best fit values of parameters based on NSSK leaves little doubt that degradations are well represented by the NSSK model. The difficult reproducibility is expected for chain processes in which a very minor amount of an initiation process is responsible for a large amount of product. The average length of the zip chain is directly related to the inflection point of the α-t curves, which occurs at the time of the maximum rate shown in the rate-time curves. Except for one, all samples show a degradation behavior that is directly related to the average chain length of the PVC. This observation is consistent with a model in which the zip chain is primarily confined to a single polymer chain. Each chain presumably contains a limited number of potential starting positions. Under a given set of conditions, number of starts resulting in the propagation of a zip chain is influenced by hydrogen chloride pressure and chain terminating substances. The number of potential starting positions per chain is presumably established by the method of preparation and subsequent treatment of the sample. Preheating a sample for brief times at elevated temperatures before making a degradation run at a lower temperature increases the number of potential starting positions and the number of zip chains actually starting. The preheating also renders chain terminating substances inactive and eliminates excessive induction periods. The greater the number of zip chains starting per polymer chain, the shorter will be the average zip length.

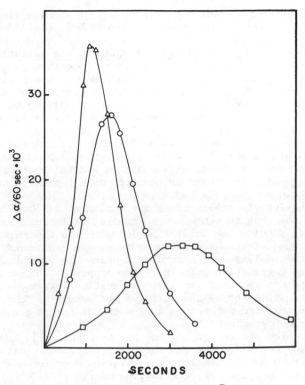

Figure 5. Rate-time curves for SP 634 (◯), SP 917 (△), and
SP 1381 (☐) at 225°C.

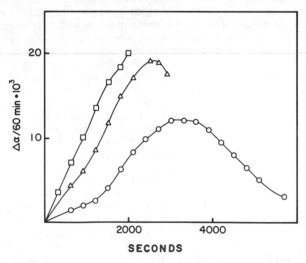

Figure 6. Rate-time curves for SP 1381 at 225°C after preheating at 235°C for: (◯) 0 min, (△) 2 min, (☐) 3 min.

In addition to accounting in a reasonable way for the unusual
behavior shown in the thermal degradation of PVC, the NSSK based on
a chain mechanism raises serious doubts about the reproducibility
and the interpretation of data obtained during incipient degradation.
It has already been pointed out that even at very low conversions
the bulk of the hydrogen chloride evolved represents the zip reaction
rather than initiation. Furthermore, the number of chains unzipping
during the initial stages of a degradation is influenced by traces of
impurities, surface contacts, and hydrogen chloride pressures, variables which cannot be controlled in any simple manner. The accelerating reaction does not get underway until initiation sites have
reacted with and rendered inactive the chain terminating impurities
and surfaces. At low temperatures induction periods are often encountered. During an induction period all initiation sites react
with terminating substances. Acceleration begins only after most of
the chain terminating impurities have been rendered inactive.
     Studies of PVC degradation have often been focused especially
on incipient degradations with the implication that it is this range
that holds the key to understanding the complex behavior. This range
has now been shown to represent a range where reproducible data are
difficult if not impossible to obtain by techniques normally applied
to degradation.
     The degradation behavior of PVC seems related to chain mechan-
isms that have been postulated for the spoiling of vegetable oils
and the formation of gum in gasolines. Some of the techniques of
study that have been applied to evaluating inhibitors for these
materials may also be applicable to studies of the stabilization of
PVC. Gasoline inhibitors have traditionally been evaluated according to the time required at elevated temperatures to form significant quantities of gum deposits. Inhibitors for edible oils have
been evaluated by the time required at elevated temperatures to
develop rancidity when air is bubbled through the oil. The time
required at elevated temperatures for acceleratory degradation of
PVC to begin should represent an appropriate parameter for evaluating stabilizers for PVC. Such a technique would evaluate inhibitors
for the initiation reaction only and would not be especially useful
in the evaluation of substances that would prevent zip chains
already present from continuing their degradation.

Ackowledgments

Acknowledgment is made to the donors of the Petroleuum Research Fund,
administered by the ACS, for support of this work.

Literature Cited

1.   Danforth, J. D.; Takeuchi, T.   J. Polym. Sci.,Polym. Chem. Ed.
     1973, 11, 2083-2090.
2.   Danforth, J. D.; Takeuchi, T.   J. Polym. Sci.,Polym. Chem. Ed.
     1973, 11, 2091-8.
3.   Danforth, J. D., J. Macromol. Sci., Chem. 1983, A-19, 897-917.
4.   Simha, R.; Wall, L.A.; Blatzy, P.K.   J. Polym. Sci. 1950, 5,
     615.

5.   Simha, R.; Wall, L.A.   J. Phys. Chem. 1952, 56, 707.
6.   Boyd, R.H.   J. Chem. Phys. 1959, 31, 321.
7.   Boyd, R.H.; Lin, T.P.   J. Chem. Phys. 1959, 31, 773.
8.   Boyd, R.H.   J. Polym. Sci. A-1, 1967, 5, 1573.
9.   Jellinek, H.H.G.   "Aspects of Degradation and Stabilization
     of Polymers"; Elsevier Scientific Publishing Co.: New York,
     1978; p.9.
10. Barron, T.H.K.;Boucher, E.A., Proceedings of the Seventh Inter-
     national Symposium on Reactivity of Solids, 1972, pp. 472-505.
11. Atkinson, J.; MacCallum,J.R.   J. Macromol. Sci., Chem. 1971,
     A5, 945.
12. Danforth, J.D.   U.S. Patent 3 431 077, 1969.
13. Danforth, J.D.; Dix, J.   Inorg. Chem. 1971, 10, 1623.
14. Danforth, J.D.; Dix, J.   J. Am. Chem. Soc. 1971, 93, 6843.
15. Danforth, J.D.   In "Computer Applications in Applied Polymer
     Science"; Provder, Theodore, Ed.; ACS SYMPOSIUM SERIES No. 197,
     American Chemical Society: Washington, D.C, 1982; pp. 377-384.
16. Danforth, J.D.; Spiegel, J.; Bloom, J.   J. Macromol. Sci., Chem.
     1982, A17, 1107.
17. Danforth, J.D.; Indiveri, J.   J. Phys. Chem. 1983, 87, 5376.

RECEIVED December 7, 1984

Effect of Antioxidants on the Thermooxidative Stabilities of Ultraviolet-Cured Coatings

I. P. HEYWARD[1], M. G. CHAN[1], and A. G. LUDWICK[2]

[1] AT&T Bell Laboratories, Murray Hill, NJ 07974
[2] Department of Chemistry, Tuskegee Institute, Tuskegee, AL 36088

The thermal stabilities of several UV-cured acrylate coatings were examined and it was determined that these materials can undergo oxidative reactions at moderate temperatures (140°C or greater). If these materials are to be exposed to elevated temperatures, thermal stabilizers will be necessary to impart good long-term stability. A number of phenolic antioxidants and hindered amine light stabilizers were evaluated in an epoxy diacrylate coating, and three major factors were found to influence their efficiency. The stabilizers must: 1) be sufficiently soluble in the monomers 2) be stable enough to UV light to survive the cure and 3) have minimal interactions with the photoinitiator. Several stabilizers which meet these criteria have been identified, and thermal analysis data indicate that they can suppress oxidation of the coating. Among these additives, the hindered amine light stabilizers are of particular interest.

Highly crosslinked, UV-cured acrylates are used extensively as protective coatings on the glass fibers (1) used for optical lightwave transmission systems. The acrylates cure rapidly during fiber drawing, provide good abrasion resistance, protect against microbending, and have optical properties that make them ideally suited as coatings for the optical fibers used in lightwave communications (2).

Coatings are applied to fibers during the fabrication process. Optical fibers are manufactured from a solid rod or "preform," consisting of a core made from oxides of germanium, silicon and boron surrounded by a high purity quartz cladding of lower refractive index. This construction confines the transmitted light to the core where it is propagated by internal reflection. Fibers are drawn from the preform on a draw apparatus.

The preform is suspended at the top of the apparatus and is introduced into a high temperature zirconia furnace where it is softened and drawn to a thin fiber. Coating material, contained in an applicator cup, is applied to the fiber and then cured by a UV-lamp system. This requires the coating to be cured rapidly. The coated fiber goes around a fast response capstan and is taken up on a reel. The coating material is critical to the overall performance of the fiber since damage to the fragile glass can severely hamper light transmission.

Many recent studies have probed the polymerization mechanisms (3,4), UV stabilities (5) and physical properties (6) of UV-cured coatings. However little attention has been given to the aging properties of these materials and their effectiveness in long-term applications. We

0097–6156/85/0280–0299$06.00/0

have begun a series of studies to determine the stabilities of fiber coatings in a variety of environments. The initial work emphasizes the oxidative stabilities of UV-cured acrylate coatings and the effect of antioxidants on coating performance.

To be useful, an antioxidant must be soluble in the uncured resin components, must have sufficient UV stability to survive the curing process, and must not adversely affect the efficiency of the photoinitiator. We have studied a number of commercial stabilizers to determine if they have the requisite properties, and have examined their effectiveness in suppressing the oxidation of an epoxy diacrylate.

EXPERIMENTAL

Materials. A high viscosity epoxy diacrylate resin and an acrylate diluent were used to prepare the UV-cured films. The photoinitiator was 2,2-dimethoxy-2-phenylacetophenone, DMPA. Stabilizers are identified in Table I. All materials were from commercial sources and used as received from the suppliers.

Solution Studies. The solubilities of antioxidants in the diluent were determined at 1 wt% concentration by visual examination after mild heating ($<60°C$).

Photolysis studies were performed on freshly prepared solutions of the photoinitiator, (4×10^{-3} g/mL in hexane), the stabilizers (1×10^{-3} g/mL hexane), and combinations of the photoinitiator and stabilizer solutions. Solutions were degassed with dried N_2 and photolyzed for 27 s with a broad band, high pressure Hg lamp (200-400 nm 0.030 J/s·cm²). The photolyzed samples were analyzed within 4 h on a DuPont Model 850 HPLC System equipped with a CN column. The solvent systems, $CHCl_3$/hexane or IPA/hexane, were pumped at 1.5 mL/min. Products were detected at 280 nm with a minimum of five injections used for each analysis.

Film Studies. Resin mixtures were prepared by first dissolving the DMPA and the desired antioxidant in the acrylate diluent, then adding the diacrylate resin with stirring and mild heating ($<60°C$). Films (6 mil) were made by placing the mixtures on quartz plates or release paper and irradiating for 8-9 s with a medium pressure Hg lamp (200-400 nm, 0.108 J/s·cm²).

Oxidative stabilities of film samples (0.05-0.1 g) were determined in quadruplicate at $139 \pm 1°C$ using a standard oxygen absorption method (7). Stabilities of the films at temperatures above 139°C were measured on a DuPont 990 Thermal Analyzer equipped with a DSC cell.

Microtensile samples were conditioned overnight (22°C, 60% RH) and tested on an Instron at a crosshead speed of 5.1 mm/min.

Equilibrium moduli were measured on a Rheometrics Dynamic Mechanical Spectrometer, Model RD 7700, equipped with an S-7 tension-compression fixture.

Infrared measurements were recorded on a Perkin-Elmer Model 580B Spectrophotometer. Absorption data were obtained on NaCl crystals. Standard instrument conditions were used and a computer driven data station was used to assist in analyzing the spectra.

RESULTS

Oxidation Studies. Oxygen absorption studies of representative acrylate coatings show that these materials can absorb oxygen at elevated temperatures. Figure 1 shows data at $139 \pm 1°C$ for an epoxy diacrylate coating, A, and two urethane diacrylates, B and C. All materials undergo some degree of oxidation, but the epoxy diacrylate is the most unstable.

Dynamic thermal analysis confirm these findings, where the epoxy diacrylate material begins to oxidize at a considerably lower temperature than the urethane diacrylate (Table II). The instability of the epoxy diacrylate is also confirmed by UV spectra of films heated in oxygen to 200°C on quartz plates (Figure 2). Absorptions at 282 and 278 nm in the initial spectrum are

Table I. Solubility Parameters of Stabilizers

Identifier	Chemical Name	Soluble[1]
	Hydrogen Donors	
HD-1	Tetrakis[methylene-3(3',5'-di-tert-butyl-4'-hydroxyphenyl)]propionate methane	yes[2]
HD-2	Di-n-octadecyl-3,5-di-tert-butyl-4-hydroxy benzyl phosphonate	yes[2]
HD-3	Octadecyl 3,5-di-tert-butyl-4-hydroxy hydrocinnamate	yes
	Hindered Amine Light Stabilizers	
HALS-1	Bis(2,2,6,6-tetramethyl-4-piperidinyl) sebacate	yes
HALS-2	Bis(1,2,2,6,6-pentamethyl-4-piperidinyl) sebacate	yes
HALS-3	Bis(1,2,2,6,6-pentamethyl-4-piperidinyl)-2-n-butyl-2-(3,5-di-tert-butyl-4-hydroxy-benzyl)malonate	no
HALS-4	Polymeric hindered amine light stabilizer	no
	Peroxide Decomposers	
PD-1	Nickel di-n-butyl dithocarbamate	yes[2]
PD-2	Dilaurylthiodipropionate	yes
PD-3	Trioctadecyl phosphite	yes
PD-4	Substituted triphenyl phosphite	no

1 At 1% by weight in diluent.

2 Heated to 60°C.

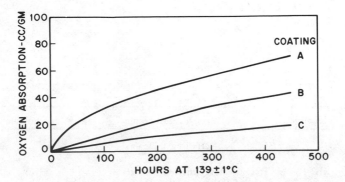

Figure 1.   Oxygen absorption data for unstabilized acrylate coating resins aged at 140°C.

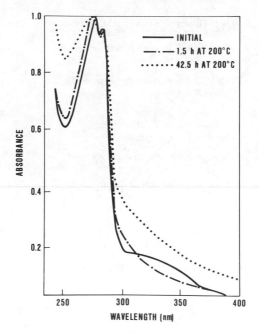

Figure 2.   Ultraviolet spectra of the epoxy diacrylate resin before and after aging at 200°C.

due to the polymer backbone. Unreacted photoinitiator is indicated by longer wavelength absorptions (330-340 nm). Upon heating, an initial decrease in the longer wavelength absorption is observed indicating heat catalyzed decomposition of the DMPA. At longer exposure times an increase in those absorptions indicative of yellowing occurs. Yellowing of epoxy resins is not uncommon (8) and in the case of UV-cured epoxy-acrylates initial yellowing is presumed to occur by residual photoinitiator decomposition (9), followed by yellowing due to the appearance of oxidation products (10). Slight shifts in the 270-280 region in addition to enhanced absorptions below 260 nm are also noted. These chemical changes cause significant changes in physical properties of the epoxy diacrylate resin (Figure 3). Physical tests on films before and after aging at 200°C in air indicate a substantial increase in modulus and tensile strength with a subsequent dramatic decrease in elongation. Changes in physical properties are known to occur in other resins upon oxidation (11-13). In addition to oxidation, in uv-cured materials other phenomena such as post cure reactions and volatilization of low molecular weight components can occur. Stabilizers were incorporated into the epoxy acrylate formulation to suppress oxidative degradation.

Table II. Onset of Oxidation of Coating Materials

Coating	T_1 °C
A	169
B	236
C	221

Solution Studies. The solubility of additives within an optical fiber coating material is of prime importance since particulates can damage the surface of the glass (14) in addition to causing pinholes or tears in the coating. Solubility data for a number of commercial stabilizers are summarized in Table I. The most soluble stabilizers in the acrylate diluent are monomeric materials, particularly those with long alkyl chains.

In order for a stabilizer to function well it must also be sufficiently stable to UV light to survive the irradiation process. Photolysis studies were performed with a number of representative antioxidants of the hydrogen donor and peroxide decomposer type (Table III). When the stabilizers are photolyzed alone in n-hexane, conversions range from 10-22% with the hindered phenol, HD-1, the least stable and the aryl phosphite, PD-4, the most stable. When solutions containing both the antioxidant and the photoinitiator are photolyzed, the photoinitiator accelerates the decomposition of the antioxidants by a factor of about five. This results in the total decomposition of the HD-1 which can no longer be detected. The other antioxidants are not completely decomposed.

Solution photolysis studies were also used to study the effect of a number of stabilizers on the efficiency of the photoinitiator during cure. Both the amount of the photoinitiator decomposed and the photoproduct distribution were analyzed (Table IV). In the absence of antioxidant, 47% of the photoinitiator is converted to photoproducts. Several of the stabilizers (HD-2, PD-2 and PD-3) decrease the amount of DMPA converted while three other stabilizers (HD-1, PD-1 and PD-4) increase the amount of DMPA decomposed. In all cases residual photoinitiator capable of catalyzing additional crosslinking or oxidation reactions remains.

Photoproduct distribution is also affected by the stabilizers. A marked decrease in the methyl benzoate/DMPA ratio from 0.52 to less than 0.17 occurs in the presence of all of the

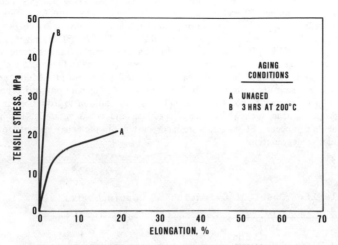

Figure 3.   Stress strain profiles at the epoxy diacrylate resin before and after aging at 200°C.

Table III.   Photolysis of Stabilizers

Sample Description	Photostability Alone		Photostability with DMPA	
	Final Conc[1] (g/mL)	Amount Converted (%)	Final Conc[1] (g/mL)	Amount Converted (%)
HD-1	7.75×10^{-5}	22	0^2	100
HD-2	8.13×10^{-5}	19	2.6×10^{-5}	74
PD-1	8.45×10^{-5}	15	3.25×10^{-5}	67
PD-4	8.9×10^{-5}	10	3.0×10^{-5}	70

1 Initial additive concentration 1.0×10^{-4} (g/mL).

2 Not observed under limits of detection, 5×10^{-6} (g/mL).

Table IV. Effect of Stabilizer upon DMPA Photochemistry

| Additive | Amount Converted DMPA (%) | Mole Ratio Products Produced/DMPA Consumed | | |
		$\overset{\text{O}}{\underset{\text{PhCOCH}_3}{\parallel}}$	$\overset{\text{O}}{\underset{\text{PhCOH}}{\parallel}}$	$\overset{\text{OO}}{\underset{\text{PhCCPh}}{\parallel\parallel}}$
None	47	.52	.34	.03
HD-1	61	.17	--	.01
HD-2	28	.50	~.25–.30[1]	.03
PD-1	59	.08	--	.01
PD-2	30	.14	.09	.01
PD-3	31	.14	.06	.01
PD-4	68	.12	.08	.01

$\overset{\text{O}}{\underset{}{\parallel}}$
1 PhCOH was not quantified due to broadness under separation conditions.

additives studied with the exception of the phosphonate, HD-2. A similar trend is observed for benzoic acid. The HD-2 does not substantially affect the product ratios indicating that it alone does not participate in the product formation reactions and that it can be assumed to have a minimal effect upon the curing process.

It should be noted that the products of the photodecomposition of DMPA in air differ from those observed in nitrogen (15—18). In air, a new product, benzoic acid, is formed at a relatively high yield (30-35%) at the expense of benzaldehyde and benzil.

Stabilized Film Studies. Differential Scanning Calorimetry (DSC) was used to determine the temperature at which a steady state oxidation of the coating is observed. This proved to be an effective method for screening stabilizers. Data for stabilized formulations are summarized in Table V. All the stabilizers show antioxidant activity as indicated by an increase in the temperature at which oxidation is observed. The sample containing the secondary amine, HALS-1, exhibits the greatest antioxidant activity and is considerably more stable than the film containing the fully substituted HALS-2. Apparently, the labile hydrogen present in the HALS-1 is significant in the overall stabilization mechanism. Lower temperature isothermal aging studies using the standard oxygen absorption method support the dynamic data indicating that the stabilizers suppress oxidation and again HALS-1 is the most effective antioxidant (Figure 4).

Table V. Onset of Oxidation of Stabilized Films

Sample	Temperature °C
Unstabilized	196
HD-2	229
HALS-1	238
HALS-2	213
HD-3	234

Isothermal DSC measurements at higher temperatures (T > 200°C) gave interesting results. The unstabilized resin had oxidative induction times at 180°C of approximately 2 minutes, while the samples containing HD-2 or HD-3 had enhanced stabilities of approximately 11 minutes at 210°C. Samples containing HALS-1 had an indication of reaction occurring almost immediately with increasing exothermicity at the isothermal temperatures ranging from 210-250°C, after which no reactions could be detected. A typical example of this behavior is shown in Figure 5. After this initial reaction no exotherm was observed after 3 hours at 210°C. These results are somewhat puzzling since dynamic thermal data indicate that the onset of oxidation of the film containing HALS-1 to be 238°C. Similar results were obtained for HALS-2.

Analysis of Cure. Infrared spectroscopy was utilized to examine film cure and the effect of stabilizers upon the cure of the epoxy acrylate. Pre-cured material has absorptions at 1635, 1410 and 810 cm^{-1} which clearly diminish after the resin is UV-irradiated (Figure 6). The 1635 cm^{-1} absorption can be assigned to the carbon-carbon stretch of an olefinic bond in conjugation with a carbonyl group; the 1410 cm^{-1} can be assigned to the CH$_2$ in-plane deformation of a vinylic group. The 810 cm^{-1} can also be assigned to some aspect of vinylic

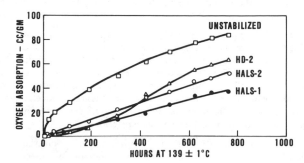

Figure 4. Oxygen absorption data for stabilized acrylate coating resin aged at 140°C.

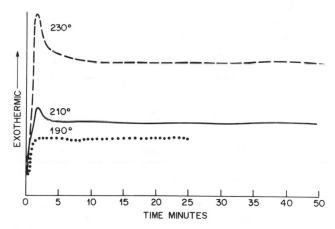

Figure 5. Isothermal DSC measurements at the epoxy diacrylate resin stabilized with HALS-1 at various temperatures.

vibrational transitions, perhaps a CH_2 out-of-plane deformation (19). Hence IR spectroscopy clearly shows that most of this structural feature has reacted during the UV cure.

Films stabilized with HD-2, HD-3 or HALS-1 were also examined for residual vinyl unsaturation after cure. All the stabilized films had non-tacky surfaces and had essentially equivalent vinyl content to the unstabilized film indicating that the stabilizers have a minimal effect upon cure reactions.

The effect of the stabilizers upon post-cure reactions was also examined. Films coated on NaCl discs were aged at room temperature in a desiccator and in an 80°C oven. Spectra of samples aged at room temperature for 7 and 14 days showed no significant changes in vinyl content indicating that additional cure reactions are minimal. At 80°C, however, changes in the vinylic absorption at 1410 cm^{-1} were observed. Figure 7 illustrates this change for an unstabilized film and demonstrates that post-cure reactions do occur at elevated temperatures. Similar results were obtained for the films containing stabilizers demonstrating that the presence of the stabilizers has no significant effect on post-cure reactions.

Physical test data support the IR findings. Tensile strength, elongation and equilibrium moduli are tabulated in Table VI. All of the physical properties examined are indicators of crosslink density and can reflect differences in cure. Table VI shows that these physical properties are essentially equivalent for stabilized and unstabilized films.

Table VI. Physical Properties of Films

Sample	E, MPa*	Tensile Strength MPa	Elongation (%)
Unstabilized	25.6 ± 1.2	17.0 ± 1.4	3.8 ± 0.6
HD-2	24.6 ± 4.2	19.4 ± 1.0	7.0 ± 0.6
HALS-1	25.3 ± 1.0	19.0 ± 1.3	7.0 ± 0.3
HD-3	28.7 ± 2.8	22.3 ± 1.0	5.4 ± 0.5

* At 85°C.

** Based on original amount in polymer.

DISCUSSION

The mechanistic interpretations of the superior stabilizing ability of the hindered amines compared with hindered phenols are highly speculative since experiments in this study have not been designed to elucidate these sequences. Some general conclusions can be made. In these photocured systems radicals capable of propagating oxidation could be formed by the decomposition of residual photoinitiator or the subsequent thermal oxidation of the resin. Hindered phenols are presumed to function as chain terminators in these systems via their mechanism in other polymeric systems (20). In these stabilizing sequences the phenol is sacrificed and thereby no longer able to function as a hydrogen donor. The mechanism of action of the amines is much more complex.

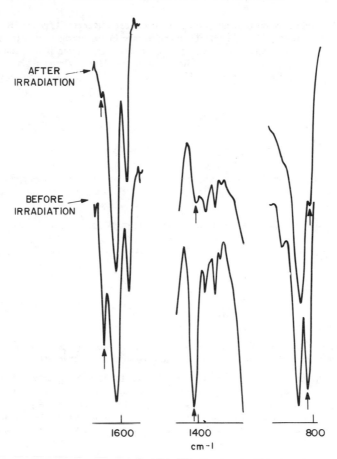

Figure 6.   Infared spectra of vinyl unsaturation absorptions of the epoxy diacrylate resin before and after a 0.8 J/cm² irradiation dose.

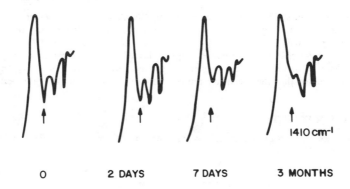

Figure 7.   Decrease in infared intensity at 1410 cm⁻¹ for the unstabilized epoxy diacrylate resin after subsequent aging at 80°C.

Hindered amines have been found to function as radical scavengers (21) and to interact with species such as hydroperoxides (22,23). One key feature in the mechanism is the regenerative nature of this class of compounds, where the parent secondary amine is first converted to an active radical scavenger, the nitroxide radical, which then can undergo reactions leading to hydroxylamine and hydroxylamine ether. These species can then be reconverted to the nitroxide by (21)

$$\text{>NH} \xrightarrow[\text{medium}]{\text{oxidizing}} \text{>NO}\cdot$$

$$\text{>NO}\cdot + \text{R}\cdot \longrightarrow \text{>NOR} \xrightarrow{\Delta} \text{>NOH}$$

$$\text{>NOR} + \text{RO}_2\cdot \longrightarrow \text{>NO}\cdot + \text{ROOR}$$

and the sequence is therefore cyclic. In the presence of oxidizing species tertiary amines are converted first to secondary amines and then follow the sequences above (21).

In solid polymers, secondary amines have also been found to complex with hydroperoxides and it is felt that the deactivation of hydroperoxides is a significant factor in the ability of secondary amines to stabilize resins. Tertiary amines do not react with hydroperoxides (23). This fact, along with the necessary conversions of tertiary amines to secondary amines before radical scavenging occurs, could well explain the differences observed between these stabilizers in photo-cured resins.

CONCLUSIONS

UV-cured acrylate coatings oxidize upon exposure to elevated temperatures ($\geqslant 139°\text{C}$). The oxidation can be retarded by the addition to the coating of suitable antioxidants.

A number of antioxidants were examined for solubility in the monomer, effect upon the photoinitiator, UV stability, effect upon physical properties of the cured coating and their effectiveness as stabilizers. The addition of hydrogen donors, HD-2 or HD-3, or a hindered amine light stabilizer, HALS-1, improves the stability of a cured epoxy acrylate system with a minimal change in the physical properties of the resin. DSC and oxygen absorption measurements show that the secondary hindered amine, HALS-1, imparts the best stability to the resin.

ACKNOWLEDGEMENT

The authors wish to gratefully acknowledge D. Simoff for obtaining the physical test data shown in Figure 3.

REFERENCES

1. L. L. Blyler, Jr., *Polym. News,* **8**, 6 (1981).
2. U. C. Paek and C. M. Schroeder, *Appl. Optics,* **20**, No. 23 (1981).
3. D. R. Terrell, *Polymer,* **23**, 1045-1050 (1982).
4. J. P. Fouassier and D. J. Lougnot, *Polym. Photochem.,* **3**, 79-95 (1983).
5. G. P. Cunningham and C. M. Hansen, *J. Coatings Tech.,* **53**, No. 682, 39, (1981).
6. M. Koshiba, K. S. Hwang, S. K. Foley, D. J. Yarusso and S. L. Cooper, *J. Mat. Sci.,* **17**, 1447-1458 (1982).
7. W. L. Hawkins, R. H. Hansen, W. Matreyek and F. H. Winslow, *J. Appl. Polym. Sci.,* **1**, 37 (1959).
8. C. S. Schollenberger and F. D. Stewart, *Advances in Polyurethane Science and Technology*, K. C. Frish and S. L. Reegen, eds., Technomic Publishing, Connecticut, Vol. 2, p. 71 (1973) and Vol. 4, p. 68 (1976).
9. C. Decker and T. Bendaikha, *Polym. Preprints,* **25**, No. 1, p. 42 (1984).
10. J. Gardette and J. Lemaire, *Polym. Deg. and Stab.,* **6**, p. 143 (1984).
11. N. Grassie, D. H. Grant, *Polymer,* Lond. **1**, 445 (1960).
12. F. H. Winslow, M. Y. Hellman, W. Matreyck, S. M. Stills, *Polym. Eng. Sci,* **6** (3), 1 (1966).
13. C. S. Schollenberger, K. Dinbergs, *SPE Trans.,* **2**, 31 (1961).
14. L. L. Blyler, Jr. and C. J. Aloisio, in *Appl. Polym. Sci.,* R. W. Tess and G. W. Poehleim eds., ACS Division of Organic Coatings and Plastics Chemistry, in press.
15. A. Medin and J. P. Fouassier, *J. Polym. Sci.,* Polym. Letters ed., **17**, 709 (1979).
16. J. Faure, J. P. Fouassier, D. J. Lougnot and R. Savin, *Eur. Polym. J.,* **13**, 891 (1977).
17. M. R. Sandner and C. L. Osborn, *Tetrahedron Letters,* **5**, 415 (1974).
18. A. Merlin and J. P. Fouassier, *Macromol. Chem.,* **181**, 1307 (1980).
19. L. J. Bellamy, *The Infra-Red Spectra of Complex Molecules*, Vol. 1, 3rd Ed., Chapman and Hall, London, 1975.
20. J. Kovařová-Lerchová, J. Pospisil, *Europ. Polym. J.,* **13**, 975 (1977).
21. D. W. Gratton, D. J. Carlsson, D. M. Wiles, *Polym. Degr. Stab.,* **1**, 69-84 (1979).
22. G. Scott, *Pure and Appl. Chem.,* **52**, 365 (1980).
23. D. M. Wiles, J. P. Torborg Jensen, D. J. Carlsson, *Pure and Appl. Chem.,* **55**, No. 10, 1658 (1983).

RECEIVED October 15, 1984

Photooxidation of Poly(phenylene oxide) Polymer

JAMES E. PICKETT

General Electric Corporate Research and Development Center, Schenectady, NY 12301

Poly(2,6-dimethyl-1,4-phenylene oxide) has been found to undergo photooxidation not at the benzylic methyl groups but rather across the aromatic ring. The mechanism is an electron-transfer reaction leading to a polymer radical cation ($Ar^{\dagger}$) and superoxide (O_2^{-}) which combine to give oxidatively degraded polymer. Sensitization, quenching, radical trapping and flash photolysis experiments support this mechanism.

The photodegradation of poly(2,6-dimethyl-1,4-phenylene oxide), $\underset{\sim}{1}$, has received considerable attention both in industrial and in academic laboratories. Workers have observed that when poly-(phenylene oxide) films are exposed to light of wavelengths greater than 300 nm in the presence of oxygen, considerable discoloration and crosslinking occur accompanied by the appearance of carbonyl and hydroxyl bands in the infrared spectrum (2-5). Most workers in the field have ascribed these results to a hydroperoxide-mediated free radical oxidation of the benzylic methyl groups (Scheme I). In this mechanism, an initiator species abstracts a hydrogen atom to give a benzylic radical, $\underset{\sim}{2}$, which reacts with oxygen to yield a hydroperoxy radical, $\underset{\sim}{3}$. This radical would abstract a hydrogen atom from another repeating unit to give the hydroperoxide, $\underset{\sim}{4}$. The hydroperoxide then could undergo photolysis to give more initiating species or decompose to aldehydes and esters. These schemes have been described in detail elsewhere (6).

We have recently found that this free radical oxidation of the methyl groups is in fact not a major pathway in the photooxidation of poly(phenylene oxide). Instead, the oxidation apparently occurs through an electron-transfer mechanism on the backbone of the polymer not chemically involving the methyl groups at all. In this paper, we present evidence inconsistent with the free radical mechanism and supporting this novel pathway for polymer photooxidation.

0097-6156/85/0280-0313$06.00/0

Results and Discussion

Evidence Against The Free Radical Mechanism. An essential feature
of the free radical mechanism is that the methyl groups are the
reactive sites. As the photooxidation proceeds, the methyl groups
should disappear. Table I shows the loss of methyl groups as determined by proton NMR of poly(phenylene oxide) exposed to Pyrex-
filtered mercury lamps in 2% benzene solution or as a 1 mil cast
film. Oxygen consumption was measured by a gas buret at 1 atm.
When 0.5 eq. of O_2 per repeating unit has been consumed, a 25% loss
of the methyl groups is expected if all of the oxidation occurs
there. In fact, the methyl group loss was found to be only one
tenth that amount. Most of the oxidation therefore must occur on
the aromatic ring. Similar results were obtained when the surface
of a photooxidized film was examined by attenuated total reflectance
infrared spectroscopy (Figure 1). The band at 960 cm^{-1} is due
solely to the aromatic rings whereas the band at 1380 cm^{-1} is due
to the methyl groups. Comparison of the areas under these peaks
shows that while the methyl band is broadened and decreases slightly
after irradiation, the aromatic band is reduced by 60%. These
results are consistent with the work of Dilks and Clark (7, 8) who
examined poly(phenylene oxide) films by ESCA. Contrary to the predictions of the free radical mechanism, the aromatic rings of poly-
(phenylene oxide) undergo photooxidation much faster than the methyl
groups.

Table I. Loss of Methyl Groups Upon Photooxidation

Extent of Oxidation (Equiv. O_2/mer)	% Lost (by nmr)	% Loss Predicted for Methyl Oxidation
0	0	–
0.12 (solution)	0	6
0.52 (solution)	2.3	26
0.48 (film)	3.4	24

An important measure of polymer degradation is chain scission.
However, study of the photooxidative chain scission in solid polymer
is complicated by accompanying crosslinking. This problem can be
avoided by carrying out the photooxidation in dilute solution where
crosslinking is negligible. The effect of photooxidation (2%
solution in benzene, Pyrex-filtered Hg lamp) on the number average
and weight average molecular weights as determined by gel permeation
chromatography is shown in Figure 2. Both the weight average and
number average molecular weights decrease rapidly upon photooxidation. A plot of the intrinsic viscosity (Figure 3) also shows
the rapid decrease in molecular weight as the oxidation proceeds.
The decreasing molecular weight cannot be explained by the oxidation
of methyl groups alone since this does not result in chain scission.
Some other mechanism must account for the chain scission.

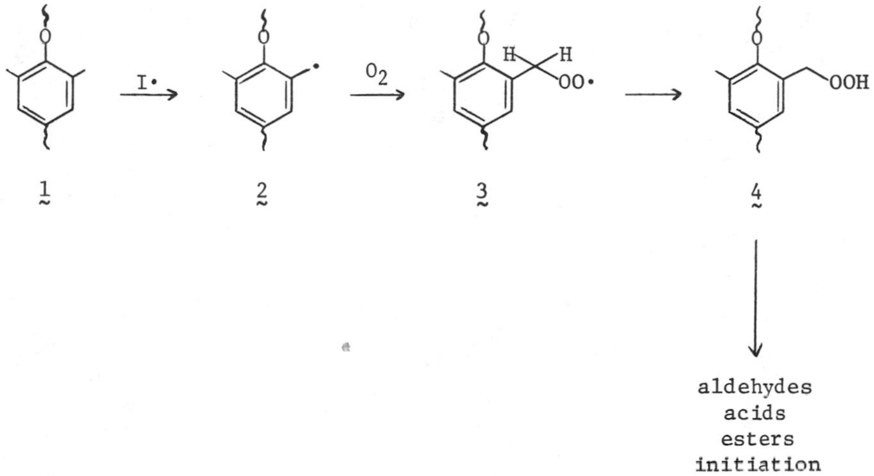

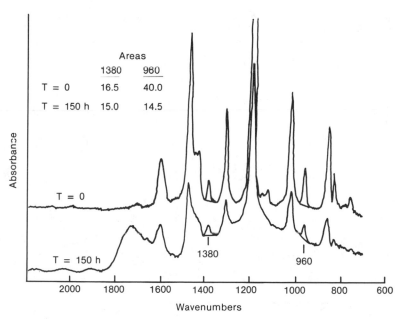

Scheme I. Hydroperoxide—mediated free radical oxidation of the benzylic methyl groups.

Figure 1. Attenuated total reflectance IR spectra of poly-(phenylene oxide) film before and after Pyrex—filtered Hg lamp irradiation.

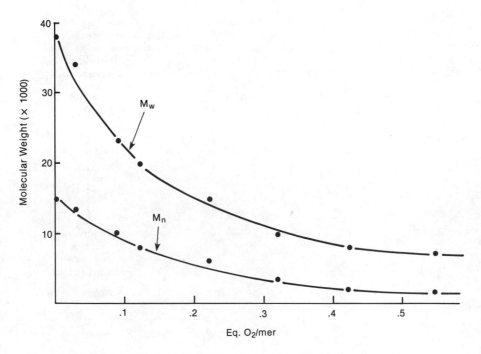

Figure 2. Loss of molecular weight upon photooxidation in benzene solution.

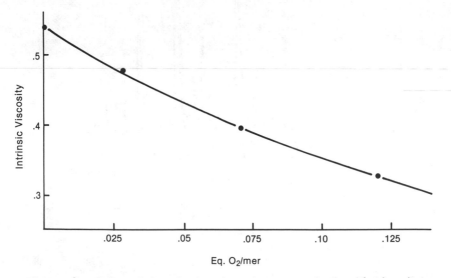

Figure 3. Loss of intrinsic viscosity upon photooxidation in benzene solution.

In the absence of oxygen, the molecular weights are not reduced
and in fact increase by approximately 10% upon irradiation with a
Pyrex-filtered Hg lamp as shown in Table II. This phenomenon occurs
even in the presence of a hydrogen donor such as isopropyl alcohol
thus ruling out the possibility of direct photolytic cleavage as
outlined in Scheme II. This type of cleavage apparently occurs
upon 254 nm irradiation (9) but is not important in real-world
polymer degradation where there is no light of wavelengths shorter
than ca. 290 nm. Since direct photolytic cleavage does not occur,
and methyl group oxidation cannot account for chain scission,
another mechanism such as oxidative cleavage of the aromatic rings
appears to be the major pathway for poly(phenylene oxide) photodegradation (8).

Table II.   Changes in Molecular Weight
Upon Photolysis

	I.V.	M_n	M_w
No Irradiation	0.523	25,000	53,000
5 Days*, N_2, iPrOH	0.618	27,000	63,000
5 Days*, O_2	0.33	13,000	30,000

* Pyrex filtered Mercury lamp.

Finally, free radical oxidation of the methyl groups would be
expected to be initiated by peroxides. The polymer has been found
to contain approximately 1 micromole of active oxygen species per
gram (4), and this has been invoked to explain the initiation of
photooxidation in poly(phenylene oxide). If this were so, one would
expect that higher concentrations of peroxide would cause more rapid
photooxidation. To test this, a benzene solution of 2.4 g of polymer (20 mmol of repeating units) and 200 micromoles of cumene hydroperoxide was exposed to a Pyrex-filtered mercury lamp. The rate of
oxygen uptake, after correcting for that due to the cumene impurity,
was only 1.1 times the rate of a peroxide-free polymer solution.
Similarly, a cast film containing 100 micromoles of cumene hydroperoxide per gram of polymer showed nearly the same rate of carbonyl
formation and yellowing as a control film when exposed to a Pyrex-
filtered mercury lamp (Figure 4). Thus, in contrast to the pre-
diction of the methyl group oxidation theory, added peroxide has no
significant effect on the photooxidation of poly(phenylene oxide).

Model Compounds.   Compounds 5 and 6 were prepared according to
standard Ullmann procedures. When these compounds were subjected
to photooxidation conditions (10 mmol in 100 mL of benzene, Pyrex-
filtered mercury lamp) both consumed oxygen at a rate only slightly
less than the equivalent amount of polymer (Figure 5). That the
rate was slower for the model compounds suggests that some sensitizing impurities are present in the polymer. However, the simple
backbone oxidation alone is sufficient to account for approximately
2/3 of the oxidation rate, and the presence or absence of methyl
groups has little effect on the rate. Interestingly, the rate of
disappearance of starting material was only about 1/6 the rate of

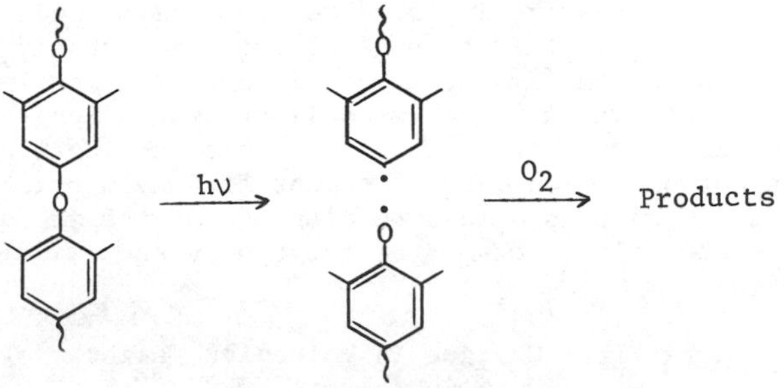

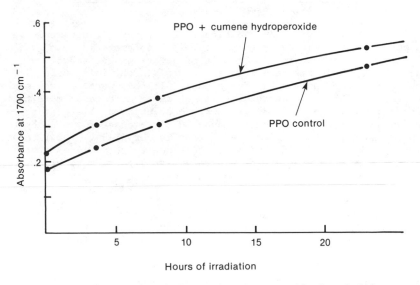

Scheme II.   Direct photolytic cleavage.

Figure 4.   Carbonyl formation in peroxide-doped film.

oxygen uptake. This indicates that the primary oxidation products
are much more susceptible to photooxidation than the starting material. Once a ring suffers oxidation, it becomes the site of several
more oxidations until only small fragments remain. Liquid chromatography showed more than forty reaction products all in minor
amounts. None as yet has been isolated and identified. Attempts to
isolate the primary products through low-temperature, controlled
oxidation conditions have also been unsuccessful.

The photooxidation of these compounds can be contrasted to the
thermal oxidation of 5 at 175°C (Table III). Three major products
were isolated, all arising from methyl group oxidation (10). One
molecule of starting material disappears per mole of oxygen consumed
and compound 6, having no methyl groups, is stable at 175°C. It is
therefore clear that the mechanisms of thermal and photooxidations
are different.

Table III. Oxidation of

	Thermal Oxidation 175°C	Photooxidation ($\lambda > 300$ nm)
Moles O_2 / Moles Consumed	~ 1	~ 6
Products	3 major	> 40
Site of Oxidation	Methyl Groups	Aromatic Ring
Absence of Methyls (Compound 6)	No reaction	Little effect on rate

Electron-Transfer Mechanism. A mechanism consistent with our
results is shown in Scheme III. In this mechanism an excited polymer repeating unit undergoes electron transfer with another unit
to generate a radical cation and radical anion pair (7 and 8).
Oxygen reacts rapidly with the radical anion to generate superoxide,
$O_2^{\cdot -}$. The superoxide and radical cation then combine to form an
unstable, presumably endoperoxidic product that undergoes further
reactions to give the degraded products. This mechanism can be
considered to be a self-sensitized version of the cyanoanthracene-
sensitized photooxidations recently described by Foote, et. al.
(11, 12, 13) and others (14, 15, 16) (Scheme IV).

The first evidence for the electron-transfer mechanism is that
known electron-transfer sensitizers promote the photooxidation of
poly(phenylene oxide). Thus, 9-cyanoanthracene, 9,10-dicyano-
anthracene (12) and methylene blue (13) cause a great increase in
the rate of polymer photooxidation both in solution and in cast
films. Polystyrene-bound Rose Bengal (17), which is an efficient
generator of singlet oxygen but a poor electron-transfer sensitizer,
does not sensitize the reaction. Singlet oxygen therefore may be
ruled out (20). The sensitizers accelerate the rate of oxygen

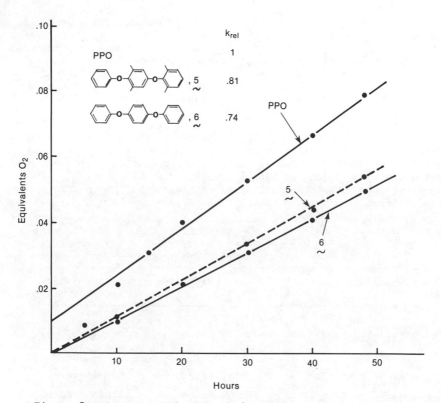

Figure 5. Oxygen uptake of poly(phenylene oxide) and model compounds. Pyrex-filtered Hg lamp, 1-M benzene solution.

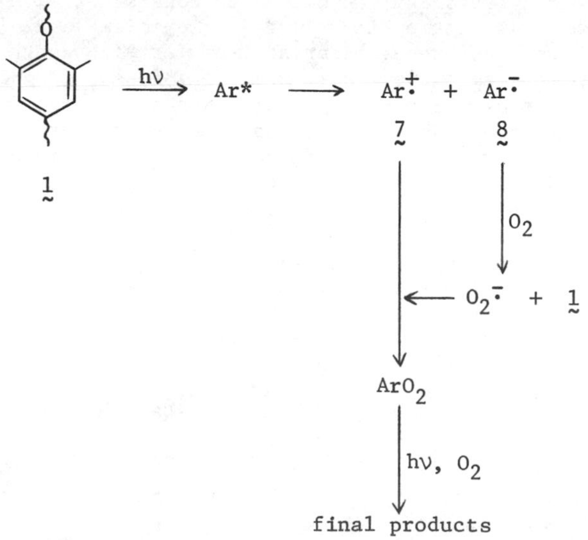

Scheme III. Electron-transfer mechanism.

uptake in the model compounds as well. It is clear therefore that $Ar^{\ddagger}$ and $O_2^{\overline{\cdot}}$, if formed, will react to give degraded polymer.

Direct evidence for the formation of $O_2^{\overline{\cdot}}$ was obtained through a nitrone-trapping experiment. A radical trapping agent, 5,5-dimethyl-1-pyrrolidine-N-oxide, 9, reacts rapidly with superoxide ion in the presence of a proton source to give the radical hydroperoxide 10 (Scheme V) (18, 19). This compound is not thermally or photochemically stable and is converted to the N-oxy pyrrolidone 11. The latter compound has a carbonyl in the infrared spectrum at 1780 cm^{-1}. The electron paramagnetic resonance (epr) spectra of these compounds are shown in Figure 6. The resolution limits of the instrument used do not permit the observation of the expected fine structure, but the four-line spectrum going to the three-line spectrum with the attendant formation of a carbonyl in the infrared spectrum should be diagnostic for superoxide.

To guard against the possibility of trapping polymer-bound radicals, a two phase system was employed. A cast film of the polymer was placed in a Pyrex NMR tube and covered with a methanol solution of the radical trap 9. This was irradiated at 0°C for several minutes with a mercury lamp. The methanol solution was then decanted and the epr spectrum taken. The results are shown in Figure 7. The initial spectrum showed the four-line signal with the three-line spectrum beginning to appear. After standing overnight in the dark at room temperature, only the three-line spectrum of 11 was seen. An infrared spectrum showed a weak carbonyl peak at ca. 1780 cm^{-1}. An epr spectrum of the washed polymer film showed no radical signals at all. The nitrone alone under these conditions produces no detectable radicals. Thus, the nitrone is trapping a photochemically-generated, mobile radical species that has epr and ir spectral characteristics consistent with superoxide.

Chemical evidence for the radical cation, 7, can be obtained from quenching experiments. A known electron transfer quencher, 1,2,4-trimethoxybenzene (12), was found to reduce the rate of oxygen uptake by 50% (Figure 8). Other compounds, diazabicyclo-2.2.2-octane (DABCO) and bis(1,1,6,6-tetramethyl-4-piperidinyl) sebacate, 12, also quenched 50% of the reaction. These compounds presumably act by donating an electron to the radical cation, 7, and then reacting reversibly with the superoxide to return all the species to their ground states and disperse the energy thermally (Scheme VI). Compounds with higher oxidation potentials (and thereby less likely to transfer an electron), such as 1,3,5-trimethoxybenzene, 1,4-dimethoxybenzene, and trans-stilbene, do not quench the reaction. Some quenching was observed in cast films but it was significant only at very high concentrations of quencher. Thus, compounds expected to divert the path of oxidation by donating an electron to the radical cation reduce the rate of oxidation. The question of why only 50% of the oxidation can be stopped, even at very high concentrations of quencher, remains open and will be discussed below.

Spectroscopic evidence for the radical cation and radical anion was obtained from flash photolysis experiments (20). Transient spectra of a 2% polymer solution in benzene 10 microseconds after a Pyrex-filtered Xenon flash are shown in Figure 9. There are two major transients of interest, one centered at 450 nm and another at 510-560 nm. Approximate lifetimes of these transients, obtained by

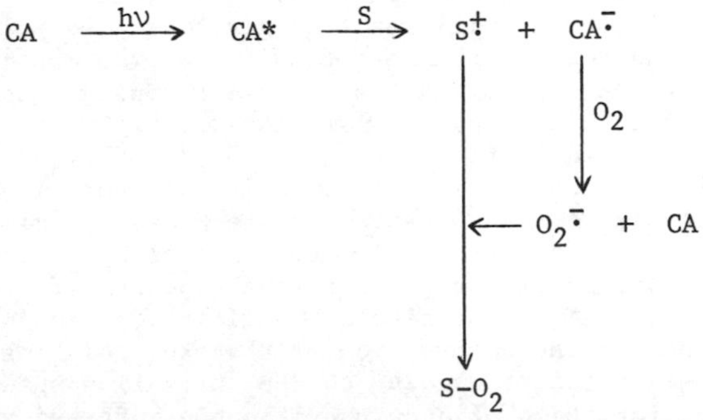

$$CA \xrightarrow{\quad h\nu \quad} CA^* \xrightarrow{\quad S \quad} S^{\dagger} + CA^{\overline{\cdot}}$$

$$\downarrow O_2$$

$$\longleftarrow O_2^{\overline{\cdot}} + CA$$

$$S-O_2$$

CA = cyanoanthracene

S = electron-rich molecule

Scheme IV.   Cyanoanthracene-sensitized photooxidations.

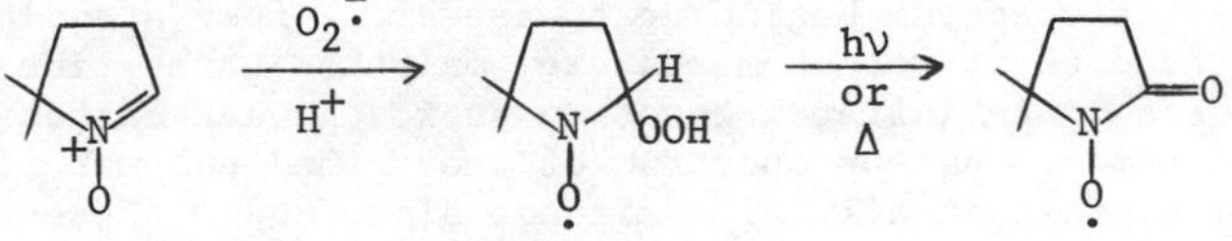

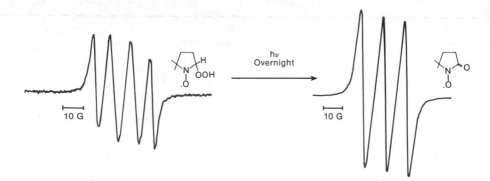

$$\underset{9}{} \xrightarrow[H^+]{O_2^{\overline{\cdot}}} \underset{10}{} \xrightarrow[\Delta]{h\nu \atop or} \underset{11}{}$$

Scheme V.   Reaction of 5,5-dimethyl-1-pyrrolidine-N-oxide with superoxide ion in the presence of a proton source to give the radical hydroperoxide 10.

Figure 6.   EPR spectra of authentic trapped $O_2^{\overline{\cdot}}$.

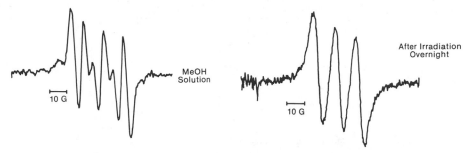

Figure 7.   EPR spectra of polymer–generated O_2^{-}.

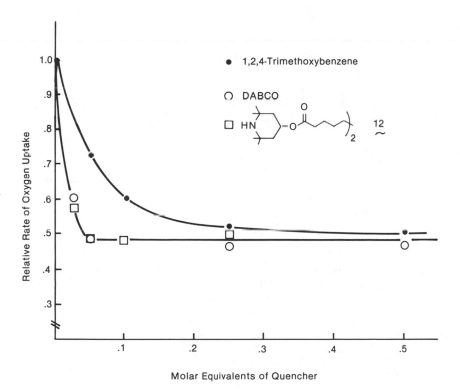

Figure 8.   Effect of electron–transfer quenchers on oxygen uptake.

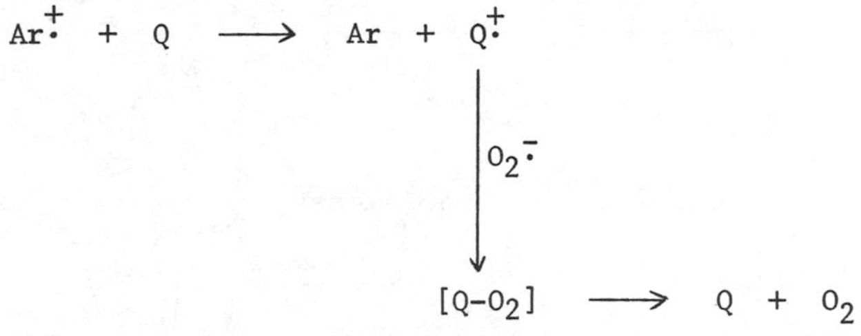

$$Ar \overset{+}{\cdot} + Q \longrightarrow Ar + Q \overset{+}{\cdot}$$

$$\Big\downarrow O_2 \overline{\cdot}$$

$$[Q\text{--}O_2] \longrightarrow Q + O_2$$

Scheme VI.   Mechanism of quenchers.

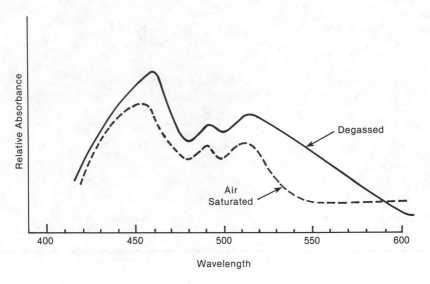

Figure 9.   Poly(phenylene oxide) transient spectra 10 μsec after Xenon flash in benzene solution.

kinetic flash spectroscopy, are shown in Table IV (21). The 450 nm
transient has been assigned to the radical cation $\underline{7}$. It is relatively insensitive to oxygen and its lifetime is reduced by the
addition of electron-transfer quenchers. The 510–560 nm transient
is believed to be the radical anion. Its lifetime is less than
50 μsec in the presence of air, presumably due to rapid reaction
with oxygen to give O_2^-. In degassed solvent, the lifetime of this
transient is increased slightly in the presence of the radical
cation quenchers. These results are consistent with the behavior of
radical cations and radical anions photolytically produced in other
electron-transfer photooxidations (12). Thus, evidence has been
obtained for all three of the proposed intermediates of Scheme III.

Table IV.   Transient Lifetimes (Microseconds) After Xenon Flash
of 1.6-$\underline{M}$ Poly(Phenylene Oxide) Solution in Benzene

Additive	450 nm	510 nm
None		
Degassed	220	100
Air	180	< 50
1,2,4-Trimethoxybenzene (0.6-$\underline{M}$)		
Degassed	140	140
Air	120	< 50
DABCO (0.8-$\underline{M}$)		
Degassed	140	120
Air	140	< 50
Hindered Amine 12 (0.6-$\underline{M}$)		
Degassed	< 50	140
Air	< 50	< 50

We have been unable thus far to isolate the primary oxidation
product Ar-O_2 even at low temperatures for either the polymer or
the model compounds. Since electron-transfer oxidations often
mimic singlet oxygen reactions, one expects that the primary products will be the result of 1,4 or 1,2 addition across the aromatic
ring (Scheme VII) (12, 22). None of these primary products would
be expected to be photostable and would eventually lead to a complex
final product mixture.

Scheme III is not, however, the only possible electron-transfer
mechanism. In the presence of oxygen, an electron may be transferred directly from the excited polymer repeating unit to oxygen,
generating the radical cation and superoxide without the intermediate radical anion (Scheme VIII). Our flash photolysis experiments cannot rule out this possibility because the transients formed

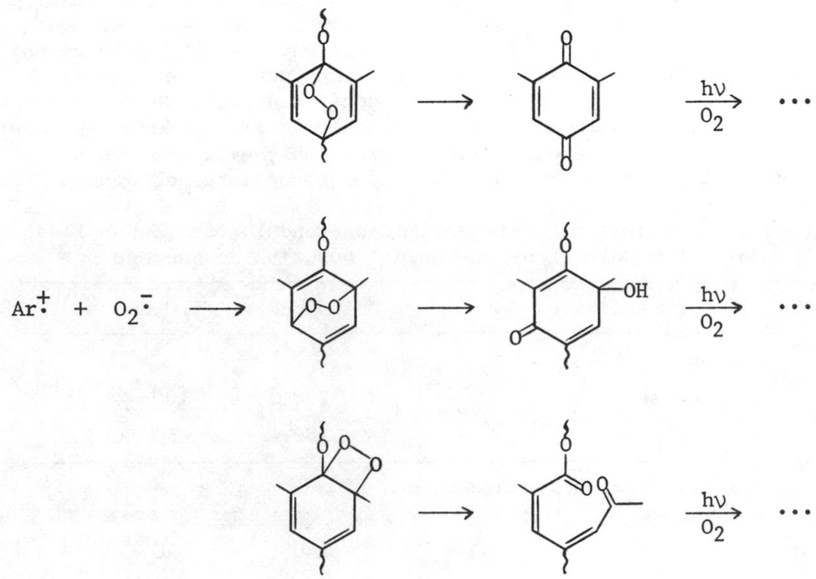

Scheme VII. 1,4- or 1,2-addition across the aromatic ring.

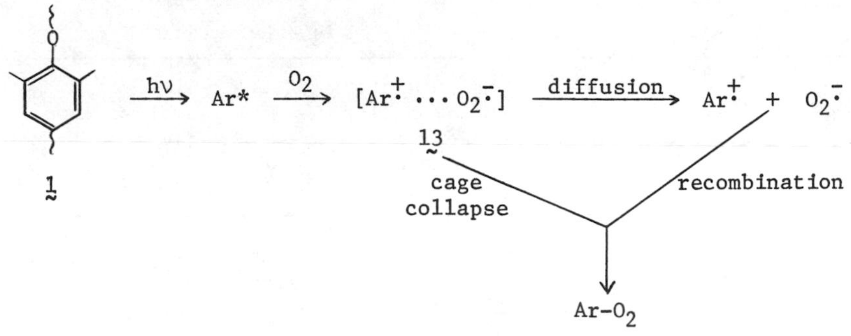

Scheme VIII. Electron-transfer mechanism without the intermediate radical anion.

from this mechanism may have a lifetime too short to be observed by our Xenon flash photolysis. The quenching experiments shown in Figure 8 suggest that the mechanism of Scheme VIII may in fact be operating in the presence of oxygen. The rate of oxygen uptake can be reduced by only about 50% even in the presence of a large amount of quencher. The non-quenchable portion may be an oxidation process that takes place within the solvent cage, 13, and is thus inaccessible to quenchers unless a quenching molecule is already part of that solvent cage. A trimolecular process such as in Scheme III is unlikely to be so limited. However, it is possible that an oxygen molecule might accept an electron from Ar* and immediately react to give products. Such a process could not be quenched. Those cases in which the immediate products did diffuse apart would be subject to quenching, but could not be distinguished from Ar^{+} and O_2^{-} generated as in Scheme III. The role of the radical anion is therefore unclear, but the net result is the formation of radical cations and O_2^{-} in any case.

Experimental

The poly(phenylene oxide) polymer was commercial resin obtained from Noryl Products Division, General Electric Company, Selkirk, NY. Solvents were spectroscopic grade. All other reagents were obtained from commercial sources and were used as received.

The solution photooxidations were performed on 2% solutions of polymer in benzene at 25°C under pure O_2 irradiated with a Pyrex-filtered high pressure mercury lamp. Oxygen consumption was monitored continuously by a self-leveling gas buret maintained at atmospheric pressure. Typically, 2.4 g of polymer (20 mmol of repeating units) would consume 0.7 cc of O_2 per hour and an experiment would run 24-48 hours. Aliquots could be removed and analyzed by standard methods.

Film experiments were performed on 1 mil films cast from chloroform solutions and air-dried for 48 hours. Flash photolysis experiments were carried out in the laboratories of Sir George Porter at the Royal Institution on an Applied Photophysics Xenon flash tube apparatus for kinetic experiments and from a Xenon flash spectrograph for spectroscopic experiments.

Acknowledgments

I wish to acknowledge the support of my colleagues at the General Electric Company, especially A. Factor and G. Davis for their helpful comments, S. Valenty and J. Chera for the infrared spectroscopy, and J. Carnahan, P. Gundlach, and S. Weisman for their analytical services. I also thank Professor George Porter and Dr. A. Harriman of the Royal Institution, London, for their hospitality and assistance during the flash photolysis experiments.

Literature Cited

1.   This paper was first presented at the Fifth International Conference on Advances in the Stabilization and Controlled Degradation of Polymers, Zurich, Switzerland, June 1-3, 1983. Preliminary reports: Pickett, J. E., *Polymer Preprints*, 1984, 25 (1), 48.

2. Kelleher, P. G.; Jassie, L. B.; Gesner, B.D., J. Appl. Polym. Sci., 1967, 11, 137.
3. Jerussi, R. A., J. Polym. Sci., 1971, A-1, 9, 2009.
4. Slama, Z.; Svajova, E.; Majer, J., Makromol. Chem., 1980, 181, 2449.
5. Petrůj, J.; Slamã, Z., Makromol. Chem., 1980, 181, 2461.
6. Rånby, B.; Rabek, J. F., "Photodegradation, Photooxidation and Photostabilization of Polymers"; John Wiley & Sons: London, 1975, p. 215.
7. Dilks, A.; Clark, D. T., J. Polym. Sci., Polym. Chem. Ed., 1981, 19, 2847.
8. Peeling, J.; Clark, D. T., J. Appl. Polym. Sci., 1981, 26, 3761.
9. Wandelt, B.; Jachowicz, J.; Kryszewski, M., Acta Polymerica, 1981, 32, 637.
10. Davis, G. C., General Electric Corporate Research & Development Center, unpublished results. See also Reference 3.
11. Eriksen, J.; Foote, C. S.; Parker, T. L., J. Am. Chem. Soc., 1977, 99, 6455.
12. Spada, L. T.; Foote, C. S., J. Am. Chem. Soc., 1980, 102, 391.
13. Manring, L. E.; Eriksen, J.; Foote, C. S., J. Am. Chem. Soc., 1980, 102, 4275.
14. Schaap, A. P.; Zaklika, K. A.; Kashir, B.; Fung, L. W-M., J. Am. Chem. Soc., 1980, 102, 389.
15. Mattes, S. L.; Farid, S., JCS. Chem. Comm., 1980, 457.
16. Santamaria, J., Tet. Lett., 1981, 22, 4511.
17. Schaap, A. P.; Thayer, A. L.; Blossey, E. C.; Neckers, D. C., J. Am. Chem. Soc., 1975, 97, 3741.
18. Harbour, J. R.; Chow, V.; Bolton, J. R., Can. J. Chem., 1974, 52, 3549.
19. Harbour, J. R.; Bolton, J. R., Biochem. Biophys. Res. Comm., 1975, 64, 803.
20. Jensen, J. P. Tovborg; Kops, J., J. Polym. Sci., Polym. Chem. Ed., 1980, 18, 2737.
21. There was also a very long-lined transient (∼ 0.6 sec) that may be due to diphenoquinone species that are always present in low concentrations in uncapped PPO solutions.
22. Saito, I.; Matsuura, T., In "Singlet Oxygen"; Wasserman, H. H.; Murray, R. W., Eds.; Academic Press: New York, 1979, pp. 511-550.

RECEIVED October 26, 1984

Photoaging of Polycarbonate: Effects of Selected Variables on Degradation Pathways

C. A. PRYDE

AT&T Bell Laboratories, Murray Hill, NJ 07974

Films of bisphenol A polycarbonate have been exposed to photo-aging at 25°C with the following variables controlled: 1) minimum wavelength of light, 2) O_2 or N_2 atmosphere, 3) presence or absence of humidity, 4) degree of prior hydrolysis. Changes in the samples have been followed using ultraviolet and infrared spectroscopy as well as gel permeation and liquid chromatography. Observed changes include cross-linking, chain scission and further reactions of scission fragments. No evidence for photo-Fries products (the salicylate or dihydroxybenzophenone derivatives) was seen for samples exposed to wavelengths normally present in outdoor light. Although some differences are apparent in samples exposed in oxygen as opposed to those exposed in nitrogen, many other changes are similar. During exposure under most conditions, a combination of cross-linking and scission occurs. Cross-linking is favored at lower wavelengths (below 315-320 nm) and by the absence of oxygen or prior hydrolysis. Chain scission appears to be nonrandom: large amounts of low-molecular-weight materials, including bisphenol A, are formed. The bisphenol A undergoes further reactions. The presence of bisphenol A, which is also a product of hydrolysis, significantly changes the degradation patterns seen in oxygen. Possible mechanistic pathways for degradation are discussed.

Because of its excellent physical properties, polycarbonate is specified for a number of engineering uses. Some of these may involve significant exposure to light. As an aromatic polymer, polycarbonate absorbs significantly in the near UV and is thus susceptible to photodegradation.

The chemical processes involved in polycarbonate photodegradation are not well defined. Much early work concentrated on the photo-Fries reaction, which involves rearrangement to

0097–6156/85/0280–0329$07.00/0
© 1985 American Chemical Society

the salicylate product (a), which can further rearrange to the dihydroxybenzophenone (b). An early paper by Gesner and Kelleher (1) showed that oxidation is also involved, although polycarbonate is relatively stable to oxidation compared to the other aromatic polymers they studied. More recently, a paper by Factor and Chu (2) showed photo-oxidative processes to be much more important than the photo-Fries reaction in the degradation of the solid polymer. Recent studies by Moore (3) and Clark and Munro (4) have similarly stressed the importance of processes other than the photo-Fries reaction.

A number of variables are involved in polycarbonate photo-aging (5). For instance, the degree of cross-linking has been found to be dependent on wavelength (2,5), as have the amounts of the photo-Fries products (6). Changes in the near UV (300-360 nm) have been reported to be dependent on the relative humidity (7,8) and on the presence or absence of oxygen (8).

In this laboratory apparatus has been constructed to examine the photodegradation of polycarbonate with simultaneous control of the following variables: 1) minimum wavelength of light, 2) oxygen or nitrogen atmospheres, 3) relative humidity, and 4) state of prior hydrolysis of the sample. In addition, the temperature is controlled at 22-25°C.

Samples exposed under selected combinations of the above variables have been analyzed by a variety of techniques including UV and IR spectroscopy and liquid and gel permeation chromotography. The goal of this survey has been to elucidate not only the chemical nature of the reactions occurring but also to determine where they occur in the polymer chain, since this has a primary importance in determining the degradation in physical properties due to weathering.

EXPERIMENTAL

Materials. The polycarbonate used in this study was Lexan 105-111 (Lot J11Z) obtained from General Electric. This is a powdered research material containing no additives. Before use the polymer was dissolved in methylene chloride, filtered through Filter-Cel and precipitated with acetone. This process removed any particulate contaminants present as well as any very low-moleculer-weight material in the polymer. The values of Mn and Mw as determined by GPC were 15,500 and 33,200 respectively, based on the use of General Electric ND62 as a polycarbonate standard.

Instruments. UV and visible spectra were recorded on a Cary 219 spectrometer. IR spectra were taken on a Perkin Elmer 580B spectrometer. Liquid chromatography (HPLC) analyses were done on a Du Pont 850 Liquid Chromatograph using a 60Å size exclusion column (Zorbax PSM-60). Gel permeation chromatography (GPC) analyses were performed on a Waters High Speed GPC (model 244) equipped with microstyragel columns. Traces were obtained using a refractive index (RI) detector.

Mass Spectral Analysis. MS analyses were performed by Oneida Research Services, New Hartford, NY.

Films. Films for exposure include very thin films (1-4 μ) and moderately thick ones (1-3 mils). Thicknesses are generally chosen so that, depending on the filter selected (see below), there is no significant difference initially in the light reaching the front and back surfaces. Exceptions to this were made for exposures to the shortest wavelengths. After prolonged exposure degradation compounds may absorb a significant amount of the lower wavelength light. The thin films were cast onto quartz plates from 6-14% (w/v) solutions of the polymer in methylene or ethylene chloride by flooding the plates and then letting them drain in a vertical position. Thicker films were prepared by pouring methylene chloride solutions (12-15%) into a flat bottomed dish and then evaporating the solvent at about 50°C. These films were then detached from the dish and baked under vacuum at ~140°C for 1 hr to ensure complete removal of the solvent.

Apparatus. (See Figure 1.) The source used in these exposures is a Hanovia 1000W mercury-xenon arc lamp housed in an Oriel illuminator with an ellipsoidal reflector. On exiting the housing the beam is passed through an Oriel integrator. This results in an essentially uniform distribution of light at distances greater than 8 in. Immediately after the integration, the beam is passed through a water filter which serves both to remove infrared radiation and to keep the UV cut-off filters which are placed in it at a constant temperature.

The exposure cell is constructed of aluminum. The back plate has channels to provide for circulation of cooling water. The side piece has nozzles which provide entrance and exit ports for the exposure atmospheres. The face plate is made from 1/8 in. quartz. All aluminum parts were subjected to a "hard-coat" anodizing process to minimize corrosion. The back plate was covered with a one mil thick sheet of teflon to protect the samples from contamination.

Filters. The UV filters used here are Schott glass, colorless, sharp cut-off filters obtained from Melles Griot. The glass type (see Table I) gives the approximate wavelength for 50% transmission while the cut-off is the wavelength at which the transmission becomes insignificant. (The actual transmissions at the cut-off wavelengths are listed under %T.) Intensity measurements (at 11 in. from the exit port) were made using an Eppley thermopile and a calibrated Eplab model 901 readout meter. The measurements were made through the water filter and a UV transmitting black glass filter (Schott type UG5) was used to screen out visible radiation. The readings thus obtained give a rough measurement of the light intensity between the wavelength defined by the cut-off filter and about 400 nm (the approximate cut-off for the black glass filter). Visible light is present in the actual exposure, but is assumed in this work to be essentially non-actinic, at least for the undegraded, ground-state polymer.

Exposures. Samples to be exposed were placed in the cell and the cell sealed and flushed for at least one hour with the atmosphere to be used. Samples to be exposed to dry atmospheres were heated gently ($\sim$100°C') in the cell to remove adsorbed water. For samples to be run in saturated atmospheres, the gas was bubbled through a fritted glass filter submerged in distilled water to give gases with $\sim$97% R.H.

RESULTS

The polycarbonate films exposed in this survey have received varying total irradiations. With the exception of films exposed to only the longest wavelengths ($\lambda > 337$ nm), all the exposed films show obvious changes in their physical and spectral characteristics indicating moderate to extensive damage. The physical changes are discussed below and summarized in Table II.

Embrittlement is seen in many of the samples. It is most obvious in those exposed in oxygen, but is also evident in films exposed in nitrogen to shorter wavelengths. The solubility properties of the polymer are also changed on exposure. Thin films exposed for 2-3 hr to the shortest wavelengths ($\lambda > 250$ nm) show considerable amounts of a gel which is insoluble in chloroform or tetrahydrofuran. As the minimum wavelength increases ($\lambda > 279$ nm and $\lambda > 300$ nm) the amount of gel appears to decrease. More gel is found in samples exposed in nitrogen. For $\lambda > 314$ nm traces of gel are found after relatively long exposures (720 hrs). No gel was seen after extensive irradiation (60-70 hrs) at the longest wavelengths ($\lambda > 337$).

Accompanying cross-linking is the formation of small, polar molecules which may be separated from the polymer by precipitation of its chloroform solutions with alcohol. In general, films exposed in oxygen are less readily soluble in chloroform and more soluble in THF or in chloroform with a trace of alcohol, suggesting the presence of appreciable amounts of oxidized materials.

Irradiations under most conditions result in a yellowing of the film. Even when yellowing is not obvious, there are other changes in the UV-visible spectra of the films. These changes have provided the primary means for following the photodegradation. The results from UV-Vis spectroscopy will be divided into sections on changes above and below 300 nm; both sections will include relevant IR data. A short section on results from GPC data will follow. Finally, addition of bisphenol A (BPA), which is a product of polycarbonate degradation, causes marked changes in the course of photodegradation. Some initial results from analysis of films containing added BPA will be given.

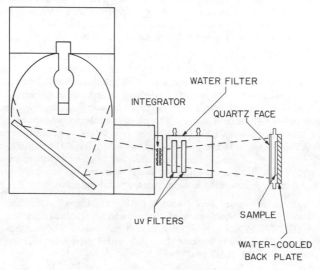

1. Schematic of apparatus.

Table I.   Spectral Characteristics of UV Filters

Schott Glass Type	Cut-off (nm)	%T	Intentsity (mW/cm^2)
WG 280	250	~1.6	160
WG 295	279	1.0	130
WG 305	287	0.5	120
WG 320	300	0.3	110
WG 335	314	0.2	85
WG 360	337	0.2	50

Table II.   General Observations

N$_2$	O$_2$
Brittleness	
very brittle for shortest wavelengths; less at moderate ones	more embrittlement than in N$_2$ at all wavelengths
Gel Formation	
large amount at shortest wavelengths decreasing to trace for $\lambda > 314$ nm.	less than in N$_2$ at all wavelengths
Solubility	
non-gel portion readily soluble in chloroform.	some non-gel material is insoluble in chloroform, dissolves in alcohol or THF.
Yellowing	
intense yellow-brown at shortest wavelengths decreases to faint yellow at $\lambda > 314$ nm.	less yellowing than in N$_2$ for short times at shortest wavelengths; yellowing still significant for $\lambda > 314$ nm.

I. Films Without Added BPA

A. Changes in the Near-UV and Visible Region — Photoyellowing and Photo-Fries Products

1) The effect of wavelength in dry atmospheres. Yellowing is reflected in a gradual rise in the absorption spectra starting as far out as 500 nm It is not significant on exposures using the longest wavelength light [λ > 337 nm (see Table I)] even after moderately long (~70 hr or 3500 mW·h/cm²) exposures.

Photoyellowing is clearly influenced by all the variables studied here. At the shortest wavelengths, the yellowing is pronounced after only 1 or 2 hours exposure (160–320 mW·h/cm²). When short wavelengths are employed the photoyellowing appears to be stronger in nitrogen for short exposures. The yellow has a brownish cast in this case. Typical differences for short wavelength exposures are shown in Figure 2, which shows the Vis-near-UV spectra of thick (~1.5 mil) films exposed for 4 hours to λ > 279 nm. On longer exposures the yellowing in oxygen increases and can become stronger than that in nitrogen. As the minimum wavelength increases, the relative importance of the yellowing in oxygen increases. For films exposed to λ > 300 nm, the yellowing in nitrogen is still somewhat stronger than in oxygen for moderate exposures (700-3500 mW·h/cm²). But for samples exposed to λ > 314 nm (5000–6000 mW·h/cm²), the results are reversed: The sample exposed in oxygen is markedly more yellow and its near-UV absorbance is higher (see Figure 3). As noted before, yellowing is not significant in samples exposed to light of λ > 337 nm. Samples exposed to only these long wavelengths in either oxygen or nitrogen show an anomalous peak near 320 nm. This will be discussed in the section on the photo-Fries reaction.

Yellowing in thin film samples (2-4 μ) appears to be uniform, but yellowing in thick samples shows significant spot-to-spot variations. The areas of strongest yellowing correspond to the presence of cloudiness in the unexposed film, suggesting partial crystallization of the polymer. Solution spectra of different areas confirm that there are marked differences between cloudy and clear areas after exposure.

2) The effects of humid atmospheres and prior hydrolysis. One series of tests was done to measure the effects of oxygen, humidity and hydrolysis with the minimum wavelength held constant (at λ > 287 nm). The films used were 3.5 ± .4μ thick. Unless noted otherwise, samples were exposed for 15 hr (at 120 mW/cm²). The results are given in Figure 4, which shows the ratios of initial-to-final absorbance at selected wavelengths. The major observation in the changes occurring above 300 nm is that the presence of water vapor slows the yellowing of samples. This is indicated in Figure 4 by the readings at 360 nm. A change is reflected also in the lower absorbances at 320 nm for samples exposed in a saturated atmosphere. The differences in oxygen are small, but those in nitrogen are significantly larger. When samples have been hydrolyzed prior to photolysis, this effect is enhanced: The differences between exposures in wet and dry atmospheres are somewhat larger. Again, the effect is stronger for samples exposed in nitrogen. Finally, prior hydrolysis increases the yellowing in samples exposed in oxygen so that it is stronger than that in nitrogen.

3) Appearance of photo-Fries products. Yellowing in exposed polycarbonate is often attributed to the presence of photo-Fries products. Model compounds for the photo-Fries products, phenyl salicylate and 2,2'-dihydroxybenzophenone, have broad peaks centered at ~312 and 355 nm, respectively, when they are present in low concentrations in polycarbonate film (see Figure 5). Similar groups attached to the polymer chain might be expected to have their absorbances shifted 2-4 nm toward longer wavelengths. In films exposed in nitrogen at the lowest wavelengths (Figure 5), there do appear to be slight increases in the absorbances in these regions, suggestive of small bands superimposed on a rising curve. The peaks are most obvious in the sample shown, where the absorbance at 355 nm suggests that the benzophenone could be present in concentrations of up to ~0.4%.

In the IR phenyl salicylate absorbs at 1690 cm⁻¹. The band can still be faintly distinguished when the compound is present in a film at 0.5% concentration. (The absorption bands near 1630 cm⁻¹ associated with the o-hydroxy benzophenones are less easily distinguished in the polymer.) The exposed films showed a very faint broadening of the carbonyl peak (1775 cm⁻¹) for samples exposed in oxygen, but the band at 1690 cm⁻¹ did not appear to be

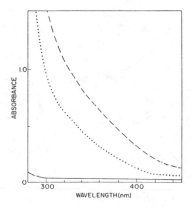

2.   Films exposed to λ > 279 nm: unexposed (———), exposed in nitrogen (– – –), exposed in oxygen (···).

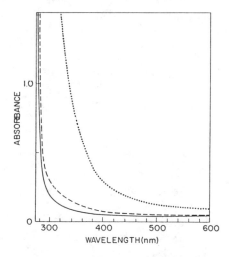

3.   Films exposed to λ > 314 nm: unexposed (———), exposed in nitrogen (– – –), exposed in oxygen (···).

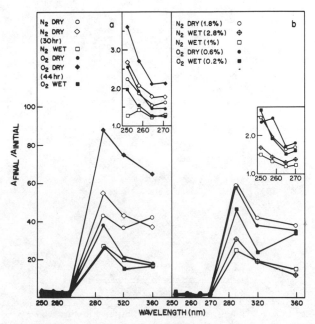

4. Films exposed 15 hr (unless otherwise noted) to λ > 287 nm. a) Standard Films. b) Films which have undergone prior hydrolysis. (Degree of hydrolysis listed in parentheses.)

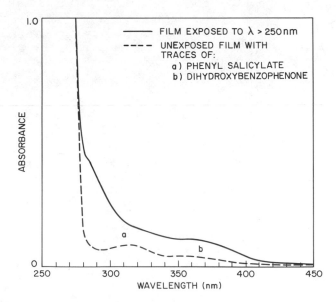

5. Comparison of PC film exposed in nitrogen with films containing 0.2% each of model compounds for photo-Fries products.

present in any of the exposed films. Thus the substituted salicylate is present, if at all, only in low concentrations.

As mentioned before, exposure of films to only the longest wavelength light ($\lambda > 337$ nm) in either oxygen or nitrogen results in the formation of a small but readily distinguishable peak at ~320 nm (see Figure 6). The peak appears to reach a maximum size after relatively short exposures (700-1000 mW·h/cm^2). When a sample exhibiting this peak was exposed to shorter wavelengths ($\lambda > 314$ nm), the peak became indistinguishable in a slowly rising absorption curve. Although the absorbance strongly suggests a substituted phenyl salicylate, this is unlikely to be the product. No phenyl salicylate was isolated following exposure of diphenyl carbonate in polycarbonate, indicating that the first step of the photo-Fries rearrangement does not occur at such long wavelengths.

One possibility is that the peak is due to a substituted 4-hydroxystilbene. A model compound, *trans*-4-hydroxystilbene, has peaks at 315 and 325 nm (Figure 7). On exposure to moderate wavelength light (e.g. $\lambda > 300$ nm), these peaks are replaced by a steadily rising curve.

B. Changes in the 220-300 nm Region

Below 300 nm a number of changes are evident in the UV spectra of exposed films. Most obvious is the overall increase in the intensity of the envelope of peaks in the 260-280 nm region. For samples exposed to $\lambda > 287$, this may be seen in Figure 8 and in the ratios of the final-to-initial A_{265} readings shown in Figure 4. The ratio continues to increase as exposure time is lengthened, particularly for samples exposed in oxygen. The samples exposed in wet atmospheres show significantly smaller increases. Prior hydrolysis increases these absorption ratios, particularly in oxygen.

At the same time as the overall absorbance increases, the characteristic shape of the polycarbonate absorption changes. The relative absorption increases around 285 nm and also in the 250-260 nm region. In samples exposed in nitrogen, a relatively clear shoulder is formed at 285 nm, consistent with the formation of phenolic end groups. In samples exposed in the presence of oxygen, this shoulder is generally not readily apparent. The IR spectra of exposed samples all show growth in the region 3300-3500 cm^{-1}, indicating the presence of hydroxyl groups. The alcohol-soluble extracts of samples exposed in nitrogen strongly suggest the presence of bisphenol A and oligomers (see Figure 9), although they clearly contain other materials as well (photo-yellowing products and products that absorb strongly in the 250-260 nm region). The UV spectra of the alcohol-soluble extracts from films exposed in oxygen vary with the exposure conditions. The spectra of some extracts are consistent with the presence of BPA, as is the case with the sample shown in Figure 9. In other cases they are rather featureless, with high absorption below ~230 decreasing to quite low absorption in the region 260-280 nm where the polymer itself or other aromatic products absorb strongly. Other analyses for BPA in exposed films will be discussed later.

In the region 250-260 nm, the samples listed in Table III suggest several trends. First, changes in this region appear to be favored by shorter wavelengths: The relative increases are quite small for the samples exposed to only long wavelengths. The growth in these regions generally appears to reach a maximum rather quickly (compare, for instance 15 and 44 hrs with $\lambda < 287$). Second, the changes in the 250-260 nm region are primarily associated with the small, alcohol-soluble molecules. Ong and Bair (9) have shown similar results in spectra of films exposed in air and then extracted with alcohol. The lack of clear relationship between the size of the 252 nm and 258 nm peaks suggest that two or more different chromophores are contributing to the changes in this area. Finally, the growth in this region is always greater for samples exposed in oxygen, suggesting that some of the absorbing species are oxidation products.

C. Gel Permeation Chromatography (GPC)

Typical GPC traces are shown in Figure 10. Results from a series of exposures analyzed by GPC are given in Table IV. The listed $\overline{M}w$ and $\overline{M}n$ values are based on main-peak data and do not include either gel or the sometimes considerable amount of low-molecular-weight materials listed in the last column of Table III. Decreases in polymer $\overline{M}w$ on exposure are generally quite

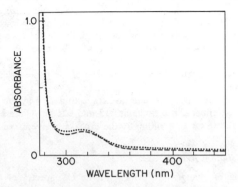

6. Anomalous peak in films exposed to λ > 337 nm in nitrogen (– – –) or oxygen (···).

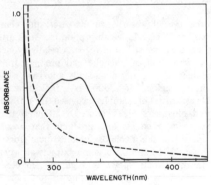

7. Films containing 1% *trans*-4-hydroxystilbene before (——) and after (– – –) irradiation with λ > 300 nm.

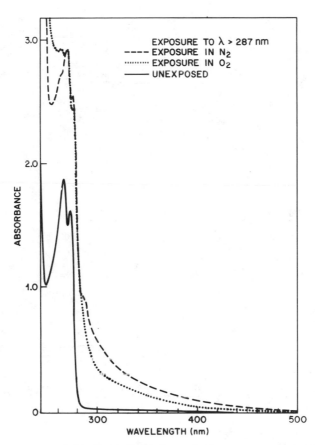

8.   UV-Vis spectra of thin (3.5μ) films before and after exposure to λ > 287 nm.

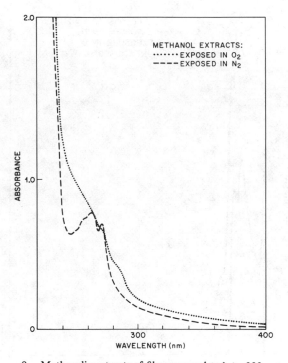

9.   Methanolic extracts of films exposed to $\lambda > 300$ nm.

Table III. UV Data for Alcohol Soluble Portions of Degraded PC

Minimum Wavelength (nm)	Exp. Conditions	Time (hr)	Sample Type	Absorbances Relative to A_{265}	
				252	258
287	N₂, dry	15	Film	0.85	0.93
287	N₂, wet	15	"	0.79	0.88
287	O₂, dry	15	"	1.01	1.0
287	O₂, dry	44	"	1.01	1.0
287	O₂, wet	15	"	0.92	0.96
300	N₂, dry	17	CHCl₃ sol'n	0.70	0.89
300	N₂, dry	17	EtOH extracts	1.13	1.06
300	O₂, dry	41	CHCl₃ sol'n	0.85	0.92
300	O₂, dry	41	MEOH extracts	1.34	1.22
314	N₂, dry	64	CHCl₃ sol'n	0.63	0.83
314	N₂, dry	64	MeOH extracts	1.15	1.10
314	O₂, dry	67	CHCl₃ sol'n	0.79	0.93
314	O₂, dry	67	MeOH extracts	1.2	1.1
337	N₂, dry	68	CHCl₃ sol'n	0.62	0.81
337	N₂, dry	"	EtOH extracts	0.82	0.91
337	O₂, dry	71	CHCl₃ sol'n	0.73	0.87
337	O₂, dry	"	MeOH extracts	1.21	1.09
Unexposed	—	—	Film	0.58	0.79

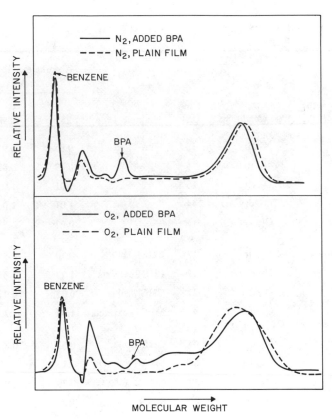

10. Effects of added BPA on molecular weight distribution for PC exposed to
    λ > 314 nm.

Table IV. Molecular Weight Changes from GPC Studies

Minimum Wavelength (nm)	ATM.	Time (hr.)	$\overline{M_n}^*$ ($\times 10^3$)	$\overline{M_w}^*$	Low-molecular-wt products[**]
UNEXPOSED			15.5	33.2	
250[**]	N_2	4	8.23	28.7	48%
250[**]	N_2	15	7.79	40.1	50%
279	N_2	4	13.4	34.8	4%
279	O_2	4	8.64	24.6	4%
300	N_2	34	16.1	32.0	6%
300	O_2	35	11.6	26.0	4%
314	N_2	64	16.8	34.1	10%
314	O_2	68	6.14	27.6	2%
337	N_2	68	17.0	32.7	1%
337	O_2	71	19.1	35.4	20%

[*] Low MW products not included

[**] number given corresponds to area under low-molecular-weight ($Mw < 600$) peaks as percent of total area (from RI traces).

small. This is particularly true for those exposures done in nitrogen, where the values even appear to increase significantly for long exposures to very short wavelength light. For exposures in oxygen, the values of $\overline{M}w$ tend to decrease a small but significant amount.

The values of $\overline{M}n$ show larger decreases than do those of $\overline{M}w$. Here again the largest decreases are seen in those samples exposed in oxygen. For longer wavelengths, ($\lambda > 300$ nm) the $\overline{M}n$ values of samples exposed in nitrogen do not decrease at all. (The values listed are slightly higher than those of the starting material, but it is not clear that they are significantly different.) For the longest wavelength exposures ($\lambda > 337$ nm) the results are anomalous, since both $\overline{M}n$ and $\overline{M}w$ appear to be slightly higher in the oxygen sample.

II. Films with Added Bisphenol A

The results shown in Figure 4 indicate that prior hydrolysis significantly increases the changes in the UV-Vis spectra, especially for samples exposed in oxygen. Because the monomer, BPA, is known to be a major product of hydrolysis (11), its addition to the polymer would be expected to provide a readily controlled way of modelling the effects of hydrolysis. At the same time, BPA and other phenolic molecules are commonly found in samples exposed to nitrogen, but are not always so apparent in those exposed in oxygen. This suggests that BPA is formed on photolysis, as indicated by Ong and Bair (9), but is then destroyed on further irradiation in the presence of oxygen. In order to determine both the fate of the added BPA and its effect on polycarbonate degradation, a series of photolyses were done with films containing added BPA. The results are summarized here and will be discussed in more detail in a later paper (10).

In the UV spectra, the addition of BPA to samples exposed to $\lambda > 287$ nm caused changes very similar to, but larger than, those seen in the samples which had undergone prior hydrolysis. Samples exposed to shorter wavelengths did not exhibit significant differences between films exposed with or without added BPA. Measurable differences were seen for longer wavelengths, with both the increase in yellowing and the other spectroscopic changes larger in samples exposed in oxygen.

Comparisons of the GPC traces of polycarbonate exposed to $\lambda > 314$ nm with and without 10% added BPA show that the added monomer significantly affects the degradation of the polymer. This may be seen in Figure 10, which shows the chromatograms of films exposed to $\lambda > 314$ nm. All the samples contain appreciable amounts of one low-molecular-weight component which is also found after irradiations under other conditions. The peak attributed to BPA is larger in the samples with added BPA, particularly in the sample exposed in nitrogen. In nitrogen the addition of BPA results in a small decrease in $\overline{M}w$ (to 28,000). In oxygen the addition of the monomer enhances the shoulder on the low-molecular-weight side of the main peak, resulting in decreases in $\overline{M}w$ (to 16,000), even though the maximum of the main peak is higher than that of the plain film (28,000).

The alcohol-soluble extracts of samples exposed to $\lambda > 314$ nm were analyzed via high pressure liquid chromotography. Comparison of the chromatograms with that of a control obtained from the extracts of an unexposed film showed that the BPA was largely destroyed in each case after 70-72 hrs. About twice as much (35%) survived in the nitrogen atmosphere as in oxygen ($\sim$16%). Several new peaks appeared in the chromatograms. One peak, which is also present in trace amounts in samples exposed in the absence of added BPA, was trapped and analyzed following exposure in nitrogen. It is nonaromatic (from UV, IR and both C^{13} and proton NMR) but it may contain vinylic protons (NMR). The IR indicates a weak carbonyl but a very strong hydroxyl peak. Mass spectral data were consistent with a molecule containing several hydroxyl groups. Work on the alternate preparation and identification of this material is continuing.

DISCUSSION

The results from the previous sections confirm that all the variables surveyed in this study have measurable effects on the photodegradation of bisphenol A polycarbonate. The first part of this discussion will summarize these effects. The second part will examine possible reaction pathways to the products seen in this and other works and suggest how some of the variables could affect these reactions.

I. Effects of Variables

A. *Minimum Wavelength*

Wavelength is clearly an important factor in photo-aging. The minimum wavelengths present affect the relative amounts of oxidative and nonoxidative yellowing; the amounts of cross-linking; and the formation of alcohol-soluble, low-molecular-weight material. Finally, shorter wavelengths may cause further reaction of some products formed at longer wavelengths, as was seen in the case of the anomalous peak at ~320 nm (4-hydroxystilbene?) which was found after irradiation at long wavelengths, but disappeared in subsequent irradiation with shorter wavelength light.

The importance of wavelength in cross-linking has been noted in a number of investigations. Previous workers (2,5) have suggested that only wavelengths <280-290 nm are actinic in cross-linking. The results from the experiments described here, which have allowed a variety of minimum wavelengths to be selected while keeping other variables constant, suggest that relatively long wavelength light (up to 315–320 nm) can cause cross-linking.

The degree of oxidative versus nonoxidative yellowing as a function of wavelength has not been extensively documented. These results indicate that nonoxidative yellowing is significant only below --314 nm.

The light sources used in artificial weathering are rarely very closely matched to actual sunlight. With the source and filters used here, the distribution differs from sunlight in that the relative amount of near UV radiation (~300-350 nm) is considerably higher. A Xenon lamp would give a closer approximation to sunlight in this region. For removing wavelengths not admitted by the atmosphere, the WG320 filter ($\lambda > 300$ nm) probably gives the closest match for general outdoor weathering. If samples need to be tested for extensive summer exposure or exposure at high altitude, where the lower wavelengths are not as thoroughly filtered out, use of the WG305 filter ($\lambda > 287$ nm) would provide a rigorous test. In general the strong wavelength dependence of the degradation patterns seen in this study shows that accelerating aging studies by use of shorter wavelengths is very likely to lead to significant errors.

B. *Presence of Oxygen*

The presence or absence of oxygen also affects a number of reactions in the photodegradation. At short wavelenghts the photoyellowing appears, at least initially, to be greater in the absence of oxygen. At long exposure times the yellowing in oxygen becomes greater. These results are consistent with work reported by Moore (3), who showed that a decrease occurred initially in the absorption above 300 nm on exposure in air: On longer exposure, the absorbance rose. It was suggested that the initial decrease was due to "photobleaching." If this is an oxidation process, it could explain the initial low rate of yellowing seen here. For normal outdoor exposure, photoyellowing involving oxygen is probably a significant factor. Without determining the differences between the yellowing products obtained with and without oxygen, however, we cannot be sure that nonaerobic yellowing processes are not also involved even when oxygen is present.

A second effect of oxygen is seen in the relative amounts of gel formation and scission. The results reported here indicate that these are competing processes under most conditions. The lower values of $\overline{M}n$ and $\overline{M}w$ for most wavelengths, and the slower appearance of gel at intermediate wavelengths ($\lambda > 314$ nm), are both consistent with an increased degree of scission in oxygen.

Finally, in the series done with added BPA, much larger changes are seen in both the UV spectra and the GPC traces for samples exposed in oxygen. Both the BPA and the products formed from the polymer are apparently undergoing oxidative reactions. It is also clear, however, that the BPA can undergo further reactions even in the absence of oxygen. This was shown by the appearance of significant amounts of the as yet unidentified, nonaromatic materials when samples with added BPA were exposed in either oxygen or nitrogen. Similar products, in much lesser amounts, were seen in the soluble extracts of films exposed without added monomer.

In summary, there appear to be a large number of reactions occurring in polycarbonate even when oxygen is absent. The presence of oxygen is nevertheless very important because it leads to a greater lowering of molecular weight and thus to significant loss of physical properties.

C. Relative Humidity

The presence of high relative humidity clearly decreases photoyellowing when light of $\lambda > 287$ nm is used. Clark and Munro (8) have reported a similar finding for thin films irradiated ($\lambda > 290$ nm) in air in 10% and 100% R.H. In the results reported here, the inhibiting effect of humidity was stronger in samples irradiated in nitrogen, although there was a small effect also in oxygen. Humidity was also seen to decrease the spectroscopic changes below 300 nm.

D. Effects of Prior Hydrolysis

As was seen, prior hydrolysis caused samples exposed in oxygen to $\lambda > 287$ nm to undergo more yellowing and a larger increase in the main absorption bands (272 and 265 nm). These changes are very similar to the larger ones seen in samples with added BPA, confirming that addition of BPA offers a good method for modelling the chemical effects of hydrolysis.

The patterns of molecular weight distribution were also seen to be changed by the addition of BPA when moderately long wavelength light ($\lambda > 314$ nm) was used. Here again the effects are most obvious in exposures done in the presence of oxygen, where the added monomer appears to cause not only an increase in the amount of very-low-molecular-weight material, but also an significant increase in oligomers and small polymer molecules. These changes will clearly affect the properties of the polymer.

II. Photolysis Products and Possible Mechanisms

A. Effects of Oxygen or Nitrogen Atmospheres

A series of reactions which can account for some of the observed spectral changes is shown in Schemes A and B. The primary step involves loss of CO to form the aryloxy radical I. Evidence for this species in polycarbonate photolysis has come from ESR (2) and flash photolysis (12) studies.

Radical I may couple through C or O to give the dihydroxybiphenyl (II) or hydroxydiphenyl ether (III). Alternatively, I could abstract a hydrogen from an isopropylidene linkage to form a phenol (IV). (This probably requires excitation of the aryloxy radical.) The resulting alkyl radicals (V) could couple giving cross-linked units (VI) or could rearrange as shown in Eq. 5 to give the stilbene (IX) or, on exposure to short wavelengths, the 3-hydroxy-9-methyl-9,10-dihydrophenanthrene (X).

The large growth in the main absorption envelope (265-285 nm) may be partially accounted for by the formation of phenolic products. In addition, several of the other products suggested in Scheme A should have fairly strong absorbance bands in the 260-290 nm region. These include the coupling products II and III. They are commonly accepted as reasonable products to be formed from the aryloxy radical pairs, but no specific evidence for their presence has been given in the solid-state studies discussed here. Another possible contributor to the absorbance in this region is the dihydrophenanthrene, X. The formation of the quinone methide (VIII) and its subsequent rearrangement to the stilbene (IX) were suggested by Humphrey, et. al. (12), although they found no clear evidence for the presence of IX. As was shown in this work, the stilbene would be unstable to moderate wavelength light and would react, probably giving the dihydrophenanthrene.

In the region above 300 nm, these same products II, III and X might make some contribution to the increased absorption seen after exposure. For the shortest wavelength exposures, the photo-Fries products may also be contributing to the yellowing, but there is no evidence that they contribute significantly in samples exposed at longer wavelengths ($\lambda > 287$ nm).

The final group of changes, those below 260 nm, were somewhat more pronounced in the oxygen irradiations, but still significant in nitrogen. These changes were largely associated with the alcohol-soluble products. These alcohol extracts have low absorbances in the typical

1. PHOTO-FRIES REARRANGEMENT

Scheme A.   Nonoxidative reactions.

Scheme B. Oxidative reactions.

aromatic regions but strong absorbances below 230 nm. This, along with the isolation of the unidentified aromatic products from the added BPA experiments, suggests that there could be some aromatic ring cleavages even in nitrogen. Clark and Munro have suggested ring-opening reactions, but only in oxygen (8).

In oxygen, any of the above reactions might still occur, but the various radical intermediates could instead react with oxygen (see Scheme B). The aryloxy radical in irradiated PC films for instance, has been shown to be destroyed by exposure to oxygen (2). It most likely forms the cyclohexadienone peroxyl radical (12), as in Eq. 6, which would yield the quinone reported by Factor and Chu (2). Reaction of the tertiary radical (XIV) with oxygen would yield the acetophenone (XVII) they observed (2). As suggested by the same authors, reactions of the alkyl radicals with oxygen, with or without prior rearrangement, would lead to a variety of oxidation products including ketones, acids and anhydrides.

The lack of evidence of photo-Fries products in most of the samples discussed here indicates either a) that the products are not formed in significant amounts or b) that they are formed and subsequently destroyed. Rivaton *et al.* (6) have shown that the products formed by irradiation at 253 nm are largely destroyed by subsequent irradiation at 365 nm in oxygen, whereas similar irradiation under vacuum causes only a slight shift favoring further formation of the benzophenone. Photooxidation could explain the lack of evidence for the photo-Fries products in the shortest wavelength exposures ($\lambda > 250$ nm) done here in oxygen. In the longer wavelength experiments, it seems more likely that the products were not formed in appreciable amounts, since there was no evidence for them in either oxygen or nitrogen exposures.

Overall, the scheme suggested here explains several of the spectral changes seen on irradiations in oxygen or nitrogen. It also suggests that the formation of gel is lower in oxygen because the alkyl radicals form the hydroperoxides rather than couple. Experimental evidence for the alkyl peroxide has been given by other workers (8).

B. Effects of Water Vapor and Prior Hydrolysis

Clark and Munro (8) have reported that, for exposures in air (to $\lambda > 290$ nm), higher humidity decreases yellowing. The experiments reported here similarly indicate that yellowing and other spectroscopic changes are inhibited in the presence of water. In this work, the effect of high humidity was seen most strongly in nitrogen, so nonoxidative processes are being affected by the water. Abstraction of hydrogen from water by the aryloxy radical itself is not very likely, but it could occur following excitation of the radical. Alternatively the water could interact with the excited species to decrease the initial scission to radicals.

As was discussed previously, the effects of prior hydrolysis can be approximated by the addition of BPA. The mechanism by which the BPA influences the reaction is not known. It is clear that in the ground state the BPA associates with the carbonate linkage of the polymer through hydrogen bonding. The effect of the bonding on the OH stretching frequency has been discussed before (11). The carbonyl absorption is also affected: Spectra of films containing 10% added BPA show a shoulder on the 1775 cm^{-1} carbonate band. Subtraction of the spectrum of plain polymer from the added-BPA spectrum gives a new peak at 1755 cm^{-1}. It is possible that hydrogen bonding interactions could have important effects in the excited state. Alternatively, other reactions, perhaps between the aromatic rings, could occur. No clear evidence for such interactions was seen in solution UV spectra of the polymer and BPA, but they may exist in the solid in either the ground or excited states.

III. Scission Patterns

A work by Abbas (13) suggested that on exposure in air to $\lambda > 330$ nm, polycarbonate undergoes scission which is not completely random in that formation of low-molecular-weight fragments is favored. The results of this study support this idea. The GPC data and extraction results indicate that significant amounts of low-molecular-weight material are formed even though there is little or no change in the M_w calculated from main peak data. With exposure to moderately long wavelengths ($\lambda > 315$ nm) the chromatographs may be described as tending toward a two-peak distribution. This is clearly not what would happen in totally random scission. (At shorter wavelengths, there is a three-peak distribution in which the third peak does not appear since it comprises the insoluble gel. This situation is too complex to analyze unambiguously.)

The presence of BPA and possibly other phenolic molecules could be a major factor in this non-random behavior. These species appear to be formed (and later destroyed) in the irradiations. In the regions in which they are formed there will be enhanced reaction as seen from the added BPA experiments. If these secondary reactions also involve phenol or BPA formation (as seems likely in view of the enhanced scission) then the process will repeat and regions of highly degraded material will result. As was shown previously in connection with hydrolysis (11), this situation can lead to end scission being favored even when, as in the case of hydrolysis, there is no inherent difference in reactivity between end and internal units. The present case is more complicated, of course, because the ends may absorb more light than the other units and, having absorbed it, they may react through different pathways.

CONCLUSIONS

All of the factors surveyed in this study have an effect on at least some of the reactions involved in the photodegradation of polycarbonate.

Under most conditions there is a competition between cross-linking and scission reactions. Cross-linking, as indicated by gel formation, is favored by short wavelength light. Wavelengths as long as 314 nm are actinic in cross-linking. Scission reactions are favored in oxygen.

Photoyellowing occurs in both oxygen and nitrogen, but different pathways are involved. Films degraded in nitrogen and oxygen have slightly different coloration and show significant differences in their UV spectra around 300 nm. Photoyellowing occurs very strongly in nitrogen at shorter wavelengths. It is markedly less intense for wavelengths above ~314 nm. Oxidative yellowing also occurs strongly at short wavelengths. At longer wavelengths (>314 nm) it is significantly stronger than the yellowing in nitrogen. The presence of water vapor hinders photoyellowing, particularly in exposures in nitrogen.

Prior hydrolysis of the sample was found to affect the changes in the UV spectra obtained on exposure. It enhances the difference in photoyellowing seen between films exposed in humid or dry atmospheres. Suggestions of other significant changes in the photodegradation patterns were seen in samples containing added BPA, which mimics the effects of hydrolysis. Samples containing BPA showed an increase in the amounts of both very-low-molecular-weight products and oligomers resulting in significant lowering of the molecular weight. This was seen most obviously in samples exposed in oxygen. These results indicate that prior or concurrent hydrolysis can have markedly deleterious effects on the photostability of polycarbonate.

The photo-Fries products are not present in significant amounts in solid polycarbonate degraded under normal weathering conditions.

The results from this survey indicate that a wide variety of reactions occur in the photodegradation of polycarbonate under natural weathering conditions. Both oxidative and non-oxidative processes are involved, but the oxidative reactions probably cause greater loss in physical properties. Accelerated aging studies should take into account the variables tested here. In particular, acceleration should not be accomplished by the use of very short wavelength light.

Acknowledgments

The author thanks M. Y. Hellman for the GPC data, I. P. Heyward for the LC analysis, R. S. Hutton for the NMR spectra and M. G. Chan, E. A. Chandross, R. Gooden, W. D. Reents, W. H. Starnes and G. N. Taylor for helpful discussions.

Literature Cited

1. B. D. Gesner and P. G. Kelleher, *J. Appl. Poly. Sci.* **13**, 3183 (1969).
2. A. Factor and M. L. Chu, *Poly Deg. and Stab.* **2**, 203 (1980).
3. J. E. Moore, *Polymer Preprints* **23**, (No. 1) 97 (1981).
4. D. T. Clark and H. S. Munro, *Polymer Deg. and Stab.* **4**, 441 (1982).

5.   A. Davis and J. H. Golden, *J. Macromol. Sci. — Revs. Macromol. Chem.* C3 (1) 49 (1969).
6.   A. Rivaton, D. Sallet and J. Lemaire, *Poly. Photochem.* 3, 463 (1983).
7.   A. Gupta, A. Rembaum, and J. Moacanin, *Macromolecules* 11, 1285 (1978).
8.   D. T. Clark and H. S. Munro, *Poly. Deg. and Stab.* 5, 227 (1983).
9.   E. Ong and H. E. Bair, *Polymer Preprints* 20, 945 (1979).
10. Paper in progress.
11. C. A. Pryde and M. Y. Hellman, *J. Appl. Poly. Sci.* 25, 2573 (1980).
12. J. S. Humphrey, Jr., A. R. Shultz, and D. B. G. Jaquiss, *Macromolecules* 6, 305 (1973).
13. K. B. Abbas, *J. Appl. Poly. Sci.* 35, 345 (1979).
14. A. R. Forrester, J. M. Hay and R. H. Thomson, "Organic Chemistry of Stable Free Radicals," Academic Press, NY, p 294, 1968.

RECEIVED October 26, 1984

Photooxidation and Photostabilization of Unsaturated Cross-linked Polyesters

SONG ZHONG JIAN, JULIA LUCKI, JAN F. RABEK, and BENGT RÅNBY

Department of Polymer Technology, The Royal Institute of Technology, S-10044 Stockholm, Sweden

IR-ATR and ESCA spectroscopy have been applied to the study of photo-oxidation and photo-stabilization of unsaturated polyesters crosslinked with styrene. Results obtained show that both styrene and polyester units of the cured resins are vulnerable to UV oxidative degradation and are easily photo-oxidized. The resin has been stabilized with different types of commercially available photostabilizers, and the best results have been obtained with Tinuvin P as a photostabilizer.

Unsaturated polyesters obtained by polycondensation of saturated and unsaturated acids, anhydrides and aliphatic glycols are used as coating resins and as matrices in glass-reiforced products. Styrene is usually added to unsaturated polyesters to act not only as solvent but also as a crosslinking comonomer which reacts with unsaturated groups along the polyester chain. The detailed structure of crosslinked polyesters are, in general, not well known in spite of great efforts made in this field (1-13).This is mainly due to the insolubility of the cured products. In the three dimensional polyester copolymer network, the structural rings are composed of four segments; two segments that are part of the polyester chain and two segments that are composed of polymerized styrene. In the actual polyester network itself, it is more probable that these polymeric rings of which the network is composed, contain more than four chain segments. The average length of the crosslinks depends upon both styrene concentration and the number and type of reactive double bonds along the polyester chain. The average crosslink consists of two styrene molecules.

In general crosslinked polyesters have better light and weathering resistance than uncured polyester resins. Apparently the concentration of unsaturated double bonds is a determining factor. In our previous studies (14) we have shown that under UV irradiation of unsaturated polyesters the primary photoreactions involve excitation of conjugated structures: carbonyl groups (in ester

0097-6156/85/0280-0353$06.00/0

bonds)–double bonds–phenylene rings. Secondary reactions occur by
complicated mechanisms resulting in oxidation, chain scission,
radical termination and crosslinking of structures present in the
photolyzed polyesters. Introduction of styrene segments to the
crosslinked polyester structures provides chain structures
responsible for the yellowing of UV irradiated resins. Styrene
segments are easily photolyzed with formation of different types of
strongly absorbing chromophoric groups (15–20).

Because unsaturated polyesters are widely used as material for
the mechanical, electrical and building industries, for skis, huts,
disposable tanks and various containers used outdoors, it is
important to stabilize them against sun light irradiation. In this
paper we present results of the photostabilization of cured
polyesters using commercially produced photostabilizers.

Experimental

A commercial polyester sample (CDS 2230, No 6462 from Syntes AB,
Sweden) containing maleic anhydride (1 mol), isophthalic acid
(1 mol), 1,2-propylene glycol (1 mol) and ethylene glycol (1 mol)
and hydroquinone (50 ppm) as stabilizer was used in the experiments.
Styrene stabilized with 4-tert-butyl pyrocatechol (20 ppm) without
purification was added as crosslinking agent, benzoyl peroxide
(1 wt-%) as initiator and Co-octate (1 wt-% in styrene solution) as
accelerator. The following composition for thermal curing has been
prepared: polyester 70, styrene 15, benzoyl peroxide 1 and Co-octate
solution (in styrene) 10 (all in weight parts).

The following photostabilizers were used: 2-hydroxy-benzophenone
(I) (Merck, Germany), 2,2'-dihydroxy-benzophenone (II) (Aldrich,
Belgium) and 2(2'-hydroxy-5-methylphenyl)benzotriazole (III)
(Tinuvin P, Ciba-Geigy, Switzerland), all added at 0.3 wt-%.

The samples were cured between glass plates as thin sheets at
$60^{\circ}C$ for 1 hr, at $80^{\circ}C$ for 1 hr or at $130^{\circ}C$ for 0.5 hr in the
presence of air.

Reflection IR spectra were obtained with a Perkin-Elmer
computerized spectrometer 580B using a micro MIR accessory at a
crystal angle of 45° incidence. The absorbance values have been
normalized by using the IR band for CH_2 groups at 2930 cm^{-1} as a
standard in order to overcome variations in optical density
resulting from differences in contact between the polymer films and
the ATR crystal.

ESCA core-level spectra for C_{1s} and O_{1s} were recorded with a
Leybold-Heraeus spectrometer using $AlK_{\alpha_{1,2}}$ excitation radiation.
Typical operating conditions for the X-ray gun were 13 kV and 14 mA
and a pressure of $3x10^{-8}$ mbar in the sample chamber.

The cured samples were UV irradiated in an Atlas UVCON
Weatherometer for 10,20,40,60,100 and 200 hrs.

Results and Discussion

The mechanism of photo-oxidation at the solid surface of a cured
polyester can differ from that in the bulk of the solid sample.
For instance, the bulk photo-oxidation mechanism is diffusion
controlled, while the surface is continually exposed to an abundant

oxygen supply from the air. For that reason, the photo—oxidation reactions at the surface occur much more rapidly than in the bulk. This is important from a practical point of view, because many of the properties of cured polyester resins depend specifically upon the nature of the surface of the sample.

ATR (Fig.1) and ESCA (Fig.2) measurements, which analyze a very thin surface layer of the polymer sample (400–800 nm for ATR and 5–10 nm for ESCA), show that UV exposure of the cured polyester in air gives a gradual increase in the formation of carbon—oxygen groups such as C—O—, HC=O and —O—C=O.

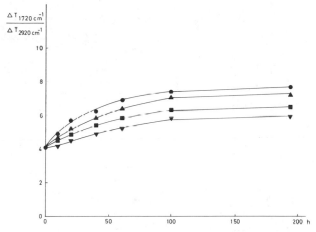

Fig.1. Kinetics of the formation of C=O band at 1720 cm^{-1} from ATR spectra.Key: (●) unstabilized sample and (▲) with I; (■) with II and (▼) with III.

Fig.2 shows the high resolution ESCA spectra of the C_{1s} (Fig.2A) and O_{1s} (Fig. 2B) bands of unexposed and photo—oxidized polyester samples. These ESCA spectra have similar character as those of photo—oxidized polystyrene, previously published (21). The C_{1s} spectrum of polystyrene consists of a single sharp peak at 285 eV binding energy with an accompanying shake—up structure, centered at around 7 eV of the higher binding energy (22). After UV irradiation, the O_{1s} signal rapidly increases in intensity relative to that in the C_{1s} region. Additional peaks at high binding energy, indicating oxidation, appear and increase in intensity with irradiation time. Carbon single bonded to one oxygen is shifted by ca. 1.5 eV to higher binding energy, while the corresponding shift for a carbon single bonded to two oxygens or double bonded to one oxygen is ca. 3 eV, these shifts being additive (21). Carbons in hydroxy, ether, peroxide and hydroperoxide groups contribute to the peak at 286.5 eV, while those at 287.9 eV and 289 eV originate from aldehydic, ketonic and ester carbons (Fig. 2A). The range of binding energies covered by the O_{1s} core levels is much smaller than for C_{1s} levels (Fig.2B). From the band it is evident that more than a single component is present (two peaks centered at 533 eV and 534.4–534.6 eV).

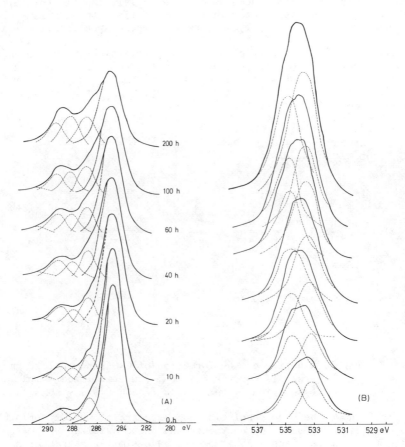

Fig.2. C_{1s} (A) and O_{1s} (B) ESCA spectra of cured polyester at different times of UV irradiation.

The lower binding energy component is assigned to double bonded oxygen and to single bonded oxygen in such groups as alcohols, ethers and peroxides. The higher binding energy is attributed to single bonded oxygen in acid, ester, hydroperoxide, carbonate, peracid or perester groups (21). Unfortunatelly, the ESCA spectroscopy is not enough sensitive technique to differenciate C_{1s} and O_{1s} signals from styrene and other units present in a complicated structure of a crosslinked polyester.

The kinetics of photo-oxidation of a crosslinked polyester were studied by ATR (Fig.1) and the ESCA (Figs 2-4) measurements. The O_{1s} / C_{1s} peak intensity ratio as a function of UV irradiation time (Fig.3) shows that photo-oxidation increases continously with time of irradiation. The subsequent leveling-off C_{1s}/ C_{1s} ESCA peak intensity ratio (Fig.4) indicates that a steady state condition in the photo-oxidation is reached after ca. 100 hrs of UV irradiation.

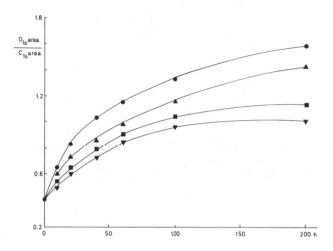

Fig.3. The ratio of the areas of O_{1s} and C_{1s} bands from
the ESCA spectra. Key: (●) unstabilized sample
and (▲) with I; (■) with II and (▼) with III.

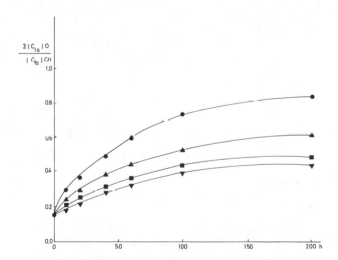

Fig. 4. The ratio of the total (Σ) C_{1s} areas for carbon
bonded to hydrogen. Key: (●) unstabilized sample
and (▲) with I; (■) with II and (▼) with III.

The IR spectra show distinctive change on photo-oxidation. A broad peak at 3200-3500 cm^{-1} appears as a result of hydroxy/ hydroperoxide group formation. The peak at 1720 cm^{-1} emerges and it can be attributed to the formation of carbonyl groups. Formation of these bands is characteristic for the photo-oxidation of both polystyrene (15,23) and polyester (14) samples.

Addition of 0.3 wt-% of the photostabilizers (I-III) causes a clear decrease in the rate of photo-oxidation (Figs 1,3,4). The stabilizers show increased effect in the order (I)< (II)< (III) which is expected from the increased UV absorption of the compounds in this order.

Results completed from our experiments show that application of ATR and ESCA techniques do not give direct answer for the question which structures in the photocrosslinked polyester with styrene are more susceptible towards photo-oxidation reactions.

Acknowledgment

These investigations are part of a research program on the role of commercial additives in the photodegradation of polymers supported by the Swedish National Board for Technical Development, which the authors gratefully acknowledge.

Literature Cited

1. W.Funke, K.Hamann and H.Gilch, Angew. Chem., 19,596 (1959).
2. W.Funke, S.Knödler and R.Feinauer, Makromol.Chem.,49,52 (1961).
3. W.Funke and H.Janssen, Makromol.Chem., 50, 188 (1961).
4. W.Funke, K.Hamann and H.Roth, Kunstoffe 51, 75 (1961).
5. W.Funke, S.Knödler and K.Hamann, Makromol.Chem.,57,192 (1962).
6. W.Funke, S.Knödler and K.Hamann, Makromol.Chem.,53,212 (1962).
7. W.Funke, W.Finus and K.Hamann, Kunstoffe 54,423 (1964).
8. W.Funke, Kolloid-Zeitschr., 197, 71 (1964).
9. W.Funke, Chimia 19, 55 (1965).
10. W.Funke, E.Gulbins and K.Hamann, Kunstoffe 55, 7 (1965).
11. W.Funke,R.Feinauer and K.Hamann, Makromol.Chem.,82,123 (1965).
12. W.Funke,R.Feinauer and K.Hamann, Makromol.Chem.,84,178 (1965).
13. W.Funke, G.Ammon and W.Pechold, Kolloid-Zeitschr.,206,9 (1965).
14. J.Lucki, J.F.Rabek,B.Rånby and C.Ekström,Europ.Polym.J.,17, 919 (1981).
15. J.F.Rabek and B.Rånby, J.Polym.Sci., A1, 12, 273 (1974).
16. J.Lucki, J.F.Rabek and B.Rånby, J.Appl.Polym.Sci.,Polym.Symp., 35, 275 (1979).
17. J.Lucki and B.Rånby, Polym.Degrad.Stabil., 1, 1 (1979).
18. J.Lucki and B.Rånby, Polym.Degrad.Stabil.,1, 165 (1979).
19. J.Lucki and B.Rånby, Polym.Degrad.Stabil.,1, 251 (1979).
20. N.A.Weir,Develop.in Polymer Degradation (Ed.N.Grassie), Appl. Science Publishers, London, 1982, p. 143.
21. J.Peeling and D.T.Clark,Polym.Degrad.Stabil.,3,97 (1980-81).
22. D.T.Clark,D.B.Adams,A.Dilks,J.Peeling and H.R.Thomas, J.Electr.Spectr.Relat.Phenom., 8, 51 (1976).
23. G.Geuskens,D.Bayens-Volant,G.Deleunois, Q.L.Ninh,W.Piret and C.David, Europ.Polym.J., 14, 291, 299, 501 (1978).

RECEIVED February 28, 1985

Polypropylene Degradation by γ-Irradiation in Air

D. J. CARLSSON, C. J. B. DOBBIN, J. P. T. JENSEN, and D. M. WILES

Division of Chemistry, National Research Council of Canada, Ottawa, Canada K1A 0R9

The γ-irradiation induced, oxidative deterioration of polypropylene film has been investigated by infrared and electron spin resonance spectroscopy and the observed changes correlated with film embrittlement. Both deterioration during γ-irradiation and during storage at 23° or 60°C after irradiation were investigated in the presence and absence of stabilizers including piperidyl derivitives, a hindered phenol and a thioester. The origin of the post-γ-irradiation deterioration was investigated by the use of a peroxyl radical trap, 2-methyl-2-nitrosopropane and selective destruction of the hydroperoxide oxidation product with reactive gases (SF_4 and SO_2). Storage oxidation was concluded to result primarily from the slow decomposition of hydroperoxide and possibly peroxide oxidation products rather than from trapped macro-radicals as proposed perviously. Ozone may also play a minor role in the oxidative process.

Polymer degradation upon exposure to γ-or other high energy radiation is important in diverse areas including radiation sterilization (medical equipment and non-wovens for hospital use), the use of plastics in the vicinity of nuclear reactor cores and the controlled modification of polymers for industrial purposes. The radiation degradation of polyolefins in both air and vacuum has already been extensively studied (1-7). Two stages to the degradation are apparent: chemical modification during the actual radiation exposure, then a slow, insidious post-irradiation deterioration during months or years of storage in air. Degradation can take the form of discolouration but is more usually shown by a progressive embrittlement.

0097–6156/85/0280–0359$06.00/0

Various theories for the post-γ-irradiation deterioration have been proposed, the most frequently cited being the presence of long-lived, macro-radicals which result from the γ-irradiation (2-4). These radicals are suggested to be trapped in the crystalline domains in polyolefins and to migrate slowly into the amorphous regions where reaction with O_2 can occur to trigger on oxidation chain. The experimental evidence for this mechanism is largely derived from electron spin resonance (e.s.r.) studies of the free radical species produced by γ-irradiation and their subsequent reaction with oxygen. Although extremely sensitive, e.s.r. can of course only detect free radical species and when used alone may give a biased view of the degradation process. We have reinvestigated the oxidative deterioration of γ-irradiated polypropylene (PP) by the use of e.s.r. together with sample characterization by infrared spectroscopy, iodometry and stress-strain analysis. Our experiments have been performed at dose rates and total doses including those used in commercial γ-sterilization equipment (~ 0.2 Mrad h^{-1}, 2.5 Mrad total dose). In this paper, some preliminary results on the respective roles of trapped radicals, unstable oxidation products and ozone in degradations are considered.

Experimental

Thin PP film (30 μm, essentially unoriented, Hercules Profax resin, 55% crystalline by differential scanning calorimetry, DSC) were freed from processing additives by acetone extraction. Film was γ-irradiated in a ^{60}Co A.E.C.L. Gamma cell (0.14 or 1.35 Mrad.h^{-1} dose rate). Films were then analyzed immediately after irradiation or after a storage period either at room temperature or at 60°C in a forced air oven. For comparison purposes some extracted PP film was photo-oxidized in a xenon arc WeatherOmeter (Atlas 6500 watt, Pyrex inner and outer filters).

A sample of commercial atactic polypropylene (Montecatini resin) was purified by extraction with hot toluene followed by precipitation from the toluene solution with methanol to remove isotactic PP contamination. The purified material showed no detectable crystalline melting endotherm by DSC (< 0.5% crystalline content). However, IR indicated the presence of some stereoblock material. The atactic PP was cast into thin films (~ 50 μm) on glass from heptane solution, vacuum dried and peeled from the glass with a sharp blade.

The e.s.r. spectra of small rolls of film (~ 30 mg) were recorded on a Varian E4 spectrometer. For film sample γ-irradiated actually in the e.s.r. tube, colour centres in the glass were carefully removed by annealing the tube end before sliding the film into this zone and insertion of the tube into the spectrometer.

Measurements were performed at a nominal power level of 0.5 mW and modulation amplitude of 5 gauss. In separate experiments signal intensity did not show saturation problems for either macro-alkyl or peroxyl radicals at this power level on our spectrometer; deviations from (power)$^{1/2}$ dependence only occurred at above ~ 0.8 mW for the macro-alkyl radicals, with no saturation apparent at < 20 mW for peroxyl radicals.

IR spectra were recorded with a Perkin Elmer 1500 Fourier Transform IR spectrophotometer with a TGS detector. Typically 200 scans were averaged for each sample to improve the signal-to-noise ratio and allow spectral subtraction at high sensitivities. Spectral subtraction was performed using the 973 cm^{-1} PP band (which is largely insensitive to variations in helical content) as the reference band to be suppressed to zero. Iodometric analysis of oxidized films were performed as described previously ([8]). Stress-strain measurements were performed on 4 mm film strips on an Instron Model 1123 at 500 % min^{-1}.

Prior to irradiation, some extracted films were immersed in hexane solutions of stabilizers for 15 h. After removal, rinsing and drying the film, the absorbed stabilizer levels were quantified by IR spectroscopy using the 1738 cm^{-1} ester band in all cases. The stabilizers employed included 1,2,2,6,6-pentamethyl-4-piperidyl octadecanoate (StNCH$_3$), 2,2,6,6-tetramethyl-4-piperidyl-N-oxyl (StNO·), octadecyl β-(3,5-di-tert.-butyl-4-hydroxyphenyl)-propionate (AOX) and di-octadecyl 3,3'-thiodipropionate (DSTDP). Subsequent to irradiation, some films were subjected to treatment by overnight immersion in a solution of a spin trap (2-methyl-2-nitrosopropane, 0.024 mol·L^{-1} in hexane) or in an atmosphere of a reactive gas (SF$_4$ or SO$_2$) to selectively derivitize the polymeric products ([8]).

Results

Immediately After γ-Exposure. Exposure of unstabilized PP film to 2-3 Mrad. of γ-irradiation in air results in oxidation which is readily observed in the IR (Figure 1) at ~ 3400 (-OH species), ~ 1715 (carboxyl species) and ~ 1200 cm^{-1}. The origin of the latter absorption is not well established, but may be indicative of C-O-C- and C-O-O linkages. From iodometric determinations, the hydroperoxide (-OOH) groups dominated at the lower dose rate, whereas alcohol -OH was much more important at the higher dose rate.

Distinct increases in bonds sensitive to the helical conformer of the PP backbone are also visible (for example at 997 and 841 cm^{-1}). These changes have previously been observed in photooxidized samples and correlate with the backbone-scission-induced restructuring of the polymer. γ-Irradiation of PP film also caused a sharp drop in the elongation at break of the films (Table 1). Tensile strength was little effected. For the same total dose, both the oxidation product quantities and relative proportions as well as the elongation at break were markedly dependent on dose rate (Table 1).

In the presence of the tertiary hindered amine (StNCH$_3$), the hindered N-oxyl (StNO·) and the phenol (AOX), the extent of oxidation and embrittlement were sharply reduced, with AOX being most effective (Table 1). The thioester (DSTDP) had little effect on the 3400 cm^{-1}, -OH absorption but appeared to accelerate embrittlement. Further study of this additive was abandoned. The γ-irradiation of an unstabilized PP sample wrapped in a thin, carefully extracted film of natural rubber showed a lower overall oxidation level by IR than a PP film irradiated alone (Table 1). However, when working at the higher dose rate (and so much shorter irradiation times), this effect was barely detectable.

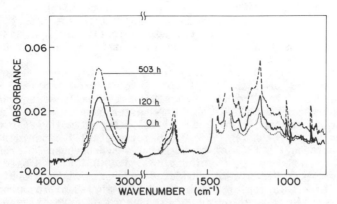

Figure 1   Difference IR Spectra of PP Films
Films all γ-irradiated to 2.8 Mrad. at 0.14 Mrad·h⁻¹; times refer
to the period of storage at 23°C following the irradiation.

TABLE 1.   Immediate Effects of γ-and UV radiation on PP Film

Additive	Exposure[a]	IR Absorbance		$\frac{E_o^b}{E_{irr}}$
		3400 cm⁻¹	1715 cm⁻¹	
---	2 Mrad. fast	0.015	0.007	1.5
---	2 Mrad. slow	0.025	0.010	5.
---	as above + NR[c]	0.018	0.012	N.D.
---	44 h Xe UV	0.015	0.006	1.5
StNCH₃[d]	2 Mrad. fast	0.005	0.006	1.1
StNO₃[e]	2 Mrad. fast	0.004	0.010	1.1
AOX[e]	2 Mrad. fast	0.003	0.006	1.1
DSTDP[e]	2 Mrad. fast	0.015	N.D.	6.

a) "Slow" refers to 0.14 Mrad·h⁻¹ and "fast" to 1.35 Mrad·h⁻¹.
b) E_o/E_{irr} = initial elongation at break/elongation after irradiation.
   E_o = 1000%.
c) PP film in intimate contact with a natural rubber film.
d) 9 x 10⁻³ mol·kg⁻¹.
e) 2 x 10⁻² mol·kg⁻¹.

The e.s.r. spectrum of the γ-irradiated film showed the presence of peroxy radicals, as indicated by the well known asymmetric singlet in Figure 2 (3,5). Macro-alkyl radicals were not detected in thin films; the complex multiplet resulting from PP macro-alkyl radicals formed by γ-irradiation under vacuum is overlayed in Figure 2 for comparison purposes. For thicker film samples (≳ 300 μm) irradiated at the higher dose rate (1.35 Mrad h⁻¹) in air, a weak macro-alkyl signal was visible under the dominant peroxyl signal immediately after irradiation, but converted quite quickly (undetectable after ~ 10 minutes at 60°C or ~ 2 hours at 23°C) to peroxyl radicals.

Photo-oxidation of the additive-free PP film gave similar oxidation product yields and embrittlement as exposure to the 1.35 Mrad.h⁻¹ γ-cell (Table 1).

Post-γ-Degradation. During storage of a pre-irradiated film, oxidation continues steadily as can be seen from the IR changes in Figure 1. The peroxyl radical population decays monotonously to zero (Figure 2). The decay of peroxyl radicals under a variety of storage conditions is shown in Figures 3 and 4. Radicals decayed somewhat faster in atactic PP as compared to the isotatic film. Film immersion in hexane caused the peroxyl signal to decay quite quickly but immersion in a hexane solution of the radical-scavenging nitroso compound caused the fastest decay at 23°C. The rate of peroxyl radical decay showed a pronounced temperature dependence, being extremely rapid at 60°C yet essentially zero at -5°C (not shown). Some peroxyl radical decay data are combined with data for the accumulation of -OH species in Figure 4 for 23°C storage. Changes in the carbonyl region absorbance and in elongation at break for the same samples stored at 60°C are collected in Figures 5 and 6 for additive free and additive containing samples. That the protective effect of AOX, StNO· and StNCH₃ continues in the post-γ period is clearly visible, with AOX the most effective but also seriously discoloured (yellow). These studies were made at 60°C to reduce the protracted lifetime at room temperature for the stabilized samples. As compared to 23°C, the 60°C aging caused about a x 10 acceleration in the rate of oxidative deterioration of the unstabilized film as measured both by changes in the IR and elongation at break. IR data for the -OH absorptions at 3400 cm⁻¹ fell into two distinct patterns. In the additive-free samples and with DSTDP, the 3400 cm⁻¹ absorption increased rapidly and was always numerically about twice the 1715 cm⁻¹ absorption. For the StNCH₃, StNO· and AOX containing samples, the low initial 3400 cm⁻¹ absorption (Table 1) either stayed roughly constant, or even decreased slightly up to 1000 hours at 60°C. It is also worth mentioning that the carbonyl yield (Figure 5) appears to bear little relationship to the elongation at break for the stabilized films (Figure 6). At 60°C the pre-photo-oxidized film was also extremely unstable as shown by the growth of carbonyl species (at ~ 1715 cm⁻¹) and the drop in elongation at break (Figures 5 and 6).

The oxidation of a series of additive free samples is shown in Figure 7. The post-UV-irradiated and post-γ-irradiated, but nitroso treated, films show very similar oxidation rates. However, γ-

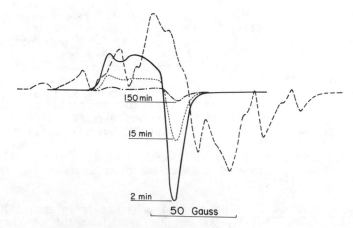

Figure 2   E.S.R. Spectra of PP Films
Films all γ-irradiated to 2.0 Mrad. at 1.35 Mrad·h⁻¹.
---- Film irradiated under vacuum.
All other curves refer to film irradiated in air at ~ 35°C then
stored in air for the shown times at 60°C.

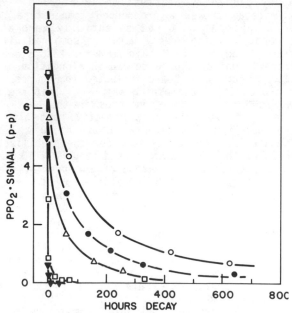

Figure 3   Peroxyl Radical Decay in PP Film
All films γ-irradiated in air at 0.14 Mrad·h⁻¹ to 2.8 Mrad. dose.
Spectra measured at peak-to-peak maximum for 0.030 g samples.
Isotactic PP film, spectrometer gain 2 x 10²: ● Film stored at
23°C in air. ▼ Film stored at 60°C in air. △ Film immersed
in hexane immediately after irradiation. □ Film immersed in
2-methyl-2-nitrosopropane/hexane solution immediately after
irradiation. Atactic PP film, spectrometer gain 5 x 10³:
○ Film stored in air at 60°C.

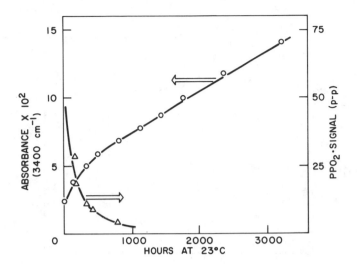

Figure 4   Comparison of Peroxyl Decay and Oxidation Growth
After γ-Irradiation
Film irradiated in air at 0.14 Mrad·h⁻¹ to 2.8 Mrad. total dose.

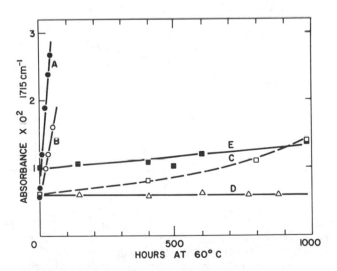

Figure 5   Film Oxidation After γ-Irradiation
Films γ-irradiated in air to 2.0 Mrad. dose at 1.35 Mrad·h⁻¹,
then aged in air at 60°C. ● No additive. △ AOX (2 x 10⁻² mol·kg⁻¹).
□ StNCH₃ (9 x 10⁻³ mol·kg⁻¹), ■ StNO· (2 x 10⁻² mol·kg⁻¹).
Film pre-photo-oxidized in air (Xe arc, 44 h), then aged at
60°C:- ○

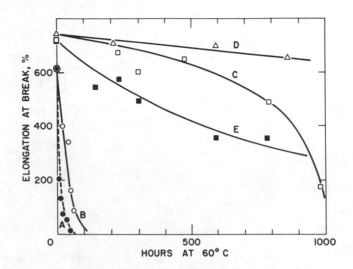

Figure 6   Film Embrittlement After γ-Irradiation
Samples and details as in Figure 5.

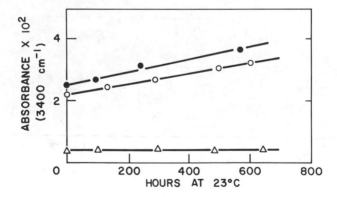

Figure 7   Film Oxidation After Irradiation Exposures
γ-Irradiated Films (2.8 Mrad., at 0.14 Mrad·h⁻¹): ◯ Radicals
destroyed by a brief immersion of the film in 2-methyl-
2-nitrosopropane solution. △ Hydroperoxides destroyed by a brief
(15 h) exposure of the film to gaseous SO_2. Photo-oxidized
film (95 h, Xe arc exposure in air): ●

irradiated film which had been exposed to either gaseous SO_2 or SF_4 (not shown) immediately after irradiation showed negligible further oxidation.

Discussion

Polymer (RH) oxidation, irrespective of the precise initiation process, can be expressed largely in terms of the classical, thermal oxidation scheme (reactions 1-3) (9). In the case of PP, tert.-

$$R\cdot + O_2 \longrightarrow RO_2\cdot \tag{1}$$
$$RO_2\cdot + RH \longrightarrow ROOH + R\cdot \tag{2}$$
$$2\ RO_2\cdot \longrightarrow \text{some non-radical products} \tag{3}$$

peroxyl radicals are dominant and their termination is quite complete (reacton 4) (7). Only peroxide formation is truely a termination step, whereas β-scission to form a ketone is expected to

$$\tag{4}$$

be an important backbone scission process. In a γ-initiated, thermal oxidation, free radicals will be generated throughout the sample cross-section after a complex, rapid cascade of reactions involving ions and excited states (reaction 5). Even at the higher degrees of oxidation ($\sim$ 0.2 mol·kg^{-1}) γ-interaction with oxidation products is much less likely than C-H bond scission based on the sheer concen-

$$RH \longrightarrow R\cdot + H\cdot \tag{5a}$$
$$\longrightarrow R'\cdot + R\cdot " \tag{5b}$$

tration of C-H sites. We have followed the increase of -OH and >C=O species by IR for films irradiated progressively up to 5 Mrad total dose. The buildup of both products was accurately linear with dose up to this point, consistent with the complete insensitivity of the oxidation products to the γ-irradiation process.

Thus in contrast to photo-initiated or thermally initiated oxidation in which hydroperoxide scission must occur to generate the first radicals, and, through the intermediacy of the macro-alkoxyl radicals, some chain scission, chain scission during γ-irradiation only occurs as the result of the irradiation itself (reaction 5b) and the termination process (reaction 4).

The complexity of the carbonyl envelope (Figure 1) indicates that reactions other than β-scission of tert.-alkoxyl radicals contribute to this absorption. Other sources include termination of

primary or secondary peroxyl radicals and further oxidation of unstable products such as aldehydes. The smaller effect on elongation
at break of the 2.5 Mrad. dose at the faster dose rate (Table 1) is
not expected from reaction 1 to 5. The higher dose rate should
generate more radicals, lead to a higher rate of peroxyl self-
termination and so generate more chain scission and scission products (carbonyl species) but less oxidation product (-OOH). The
discrepancies with this scheme may result from other termination
reactions (such as R· + R or R· + RO$_2$· etc.) becoming important at
the higher dose rate as a result of some O$_2$ depletion in more highly
oxidized domains.

The stabilization effect during irradiation of the piperidyl
compounds and the phenol most likely result from the scavenging of
peroxyl radicals but the phenol is converted to yellow quinone products (10). Piperidyl compounds are much less efficient scavengers
of peroxyl radicals, but can also compete with oxygen for the macro-
alkyls (reaction 1) (11). None of the stabilizer systems completely
suppressed the γ-initiated oxidations, which is not surprising in
view of the high number of randomly dispersed initiation processes
from γ-irradiation. The acceleration of embrittlement by the thioester (DSTDP), which is normally believed to stabilize via a hydroperoxide decomposition mechanism rather than radical scavenging, is
unexpected but may result from the rapid oxidation of the thio
additive to unstable intermediates. Horng and Klemchuk have also
reported that DSTDP is ineffective against γ-deterioration (6).

The γ-irradiation of samples in air occurs in the presence of
some ozone formed by the radialysis of the air itself. (The smell
of ozone is always detectable when the Gamma cell is opened.) Ozone
can attack saturated polyolefins to generate radicals and initiate
oxidative chains (12). The small but reproducible retardation of
the oxidation of the PP films intimately wrapped in a natural rubber
film is then consistent with an O$_3$ component to the oxidation, the
unsaturated rubber shielding the PP film to some extent from O$_3$
attack by means of the rapid O$_3$-unsaturation reaction. As well as
free radical generation, the O$_3$-PP reaction may also lead to unstable ozonide and peroxide intermediates.

The post-γ oxidation is markedly retarded by the three additives (Figures 5 and 6). The phenolic additive is outstanding but
its performance is marred by the yellowing of the film as the phenol
is converted to quinone products. Horng and Klemchuk have found
that phenolic, phosphite and a secondary hindered amine were essentially unchanged by a 2.5 Mrad. γ-dose, and so available to suppress
the post-γ deterioration (6).

The protracted thermal oxidation which follows the end of
γ-irradiation could stem from two sources:
1) Macro-alkyl radicals trapped in the crystalline domains.
2) Decomposition of unstable oxidation products.

The former mechanism has frequently been observed, but usually
only for samples exposed to electron beam or γ-irradiation at 77°K
under vacuum followed by annealing at ambient, then O$_2$ admission
(2-4). Under our experimenal conditions our data appear to largely
refute the occurrence of process 1. The detection solely of peroxyl
radicals in thin films and the rapid post-γ conversion of the macro-

alkyl radicals implies that all of the detected radicals are in O_2 accessible (amorphous or defective crystalline) regions. A macro-alkyl radical/hydrogen atom pair or macro-alkyl pair generated in the crystalline domains (reaction 5) probably recombines quickly in the rigid domain. Alternatively, H· may hydrogen abstract close to the geminate macro-alkyl radical and rapid combination of the two macro-alkyls then takes place. Furthermore, the peroxyl radicals in the γ-irradiated atactic sample (completely non-crystalline) (Figure 3) decayed in a similar fashion to those in the semi-crystalline isotatic film. In both the atactic and isotactic samples, PPO_2· decay appears to fit a good second order relationship as expected from reaction 4. However, the relative second order rate constant for peroxyl decay is about a factor of 6 faster in the atactic polymer as compared to the isotactic film. Furthermore, from an ongoing study of the effect of film morphology on PPO_2· decay we do find a slower decay as morphological perfection increases. This may result from a progressive increase in restriction of peroxyl mobility within the amorphous domains. Finally, the fact that a low molecular weight alkane and, even more dramatically, a solution of a radical trap can accelerate the PPO_2· decay implies that the radical sites are in the amorphous domains. The effect of hexane appears to be analogous to the effect of a "mobilizer" (an alkane oil) reported previously (3). From the e.s.r. of films treated with the nitroso solutions, there was an indication that a new radical species was formed, but was always weak in comparison with the residual peroxyl signal. Nitroso compounds are known to scavenge peroxyl radicals, but the peroxy N-oxyl product is not stable at ambient temperature and decays through a series of unstable intermediates (reaction 6) (14).

$$RO_2\text{·} + \text{---NO} \longrightarrow \text{---N-O-O-R} \xrightarrow{\Lambda} \text{non radical products} \tag{6}$$
$$\overset{|}{O}\text{·}$$

Of the products from γ-initiated thermal oxidation, peroxides, hydroperoxides and ozonides are the prime candidates to decompose quickly enough at room temperature to initiate the post-γ oxidation. From Figure 4, the rate of post-γ oxidation slows and becomes approximately constant after 5-600 h at 23°C, which corresponds to the point at which the peroxyl signal has extensively decayed. It is tempting to associate the faster, post-γ oxidation stage to a combination of the residual peroxyl radical effect and oxidation product effect but the subsequent, steady oxidation region solely to oxidation product decomposition. However, from Figure 7, post-γ oxidation is stopped completely by SO_2 (or SF_4) treatment of the film. Both treatments destroy -OOH groups (8) yet were found to cause negligible change in the peroxyl signal level. This result clearly points to the residual peroxyl level contributing little to the post-γ oxidation.

When the peroxyl population was quickly destroyed by 2-methyl-2-nitrosopropane treatment, post-γ oxidation still occurred (Figure 7). In addition, film which had been pre-oxidized by xenon-arc photo-initiated oxidation showed an identical, slow, thermal oxidation during storage, as well as a rapid embrittlement in the accelerated aging at 60°C (Figure 6). The photo-oxidation of PP is well documented to produce hydroperoxide and lesser amounts of peroxides, but only low concentrations of peroxyl radicals. The thermal instability of peroxidic groups in several polymers has, in fact, been invoked previously as a source of oxidation initiation at moderate temperatures. Clough and Gillen suggested that thermal decomposition of -OOH sites promoted the degradation of polyethylene and poly(vinyl chloride) during exposure to low dose rates of γ-irradiation in a nuclear power station (15). Citovicky et al and Kishore have also invoked peroxidation of PP by ozone or γ-exposure as a source of initiation of thermal degradation (16-17).

Conclusions

The post-γ oxidation appears to result predominantly from the slow decomposition of hydroperoxides and/or some other unstable peroxidic product. The more rapid, early rate of oxidation (Figure 4) might include a component from an extremely unstable oxidation product (ozonide, peroxide, etc.) not present after photo-oxidation and destroyed by the nitroso treatement. Work is continuing on the characterization of the products from γ-initiated oxidation, but is made difficult by the problems of quantifying the peroxide species. Degradation both during γ-irradiation and during post-γ storage can be prevented by phenolic and piperidyl stabilizers. The latter were less effective than the phenol, but did not cause discolouration. Other hindered amines with a higher radical scavenging efficiency would be extremely valuable.

Acknowledgment

Issued as NRCC 23466.

Literature Cited

1. Geymer, D.O., "The Radiation Chemistry of Macromolecules", Dole, M. Ed., Academic Press, New York, 1973, vol. 2, p.4.
2. Dole, M., Adv. Rad. Chem. 1974, 4, 307.
3. Dunn, T.S. and Williams, J.L., J. Ind. Irrad. Technol. 1983, 1, 33.
4. Bohm, G.G.A., J. Polym. Sci. 1967 A2, 5, 639.
5. Tsuji, K., Adv. Polym. Sci. 1973, 12, 131.
6. Horng, P. and Klemchuk, P, Plast. Eng. 1984, (April), 35.
7. Decker, C. and Mayo, F.R., J. Polym. Sci., Polym. Chem. Ed. 1973 11, 2847.
8. Carlsson, D.J. and Wiles, D.M., Macromolecules 1969, 2, 597.
9. Garton, A., Carlsson, D.J. and Wiles, D.M., Dev. Polym. Photochem., 1980, 1, 93.
10. Carlsson, D.J. and Wiles, D.M., J. Macromol. Sci., Rev. Macromol. Chem., 1976. C14, 155.
11. Wiles, D.M., Tovborg Jensen, J.P. and Carlsson, D.J., Pure Appl. Chem., 1983, 55, 1651.
12. Krisyuk, B.E., Popov, A.A. Griva, A.P. and Denisov, E.T., Dokladi Akad. Nank S.S.S.R. 1983, 269, 400.
13. Vieth, W. and Wuerth, W.F., J. Appl. Polym. Sci., 1969, 13, 685.
14. Chatgilialoglu, C., Howard, J.A. and Ingold, K.U., J. Amer. Chem. Soc., 1982, 47, 4361.
15. Clough, R.L. and Gillen, K.T., J. Polym. Sci., Polym. Chem. Ed., 1981, 19, 2041.

16. Citovicky, P., Mikulasova, D. and Chrastova, V., Euro. Polym. J. 1976, 12, 627.
17. Kishore, K, J. Macromol. Sci., Chem., 1983, A19, 937.

RECEIVED October 26, 1984

Comparison of Chemiluminescence with Impact Strength for Monitoring Degradation of Irradiated Polypropylene

G. D. MENDENHALL[1], H. K. AGARWAL[1], J. M. COOKE[2], and T. S. DZIEMIANOWICZ[2]

[1]Department of Chemistry and Chemical Engineering, Michigan Technological University, Houghton, MI 49931
[2]Plastics Technical Center, Himont U.S.A., Inc., Wilmington, DE 19808

The loss of impact strength of polypropylene was followed from sheets stored in air at 25°C and 60°C after irradiation with electron beams. A marked difference in efficacy of phenolic and thioether-based stabilizers at the two temperatures was found, with the thioether active alone at 60°C but only synergistically at 25°C. This difference was also reflected qualitatively in differences in chemiluminescence emission from the samples.

Irradiation of polymers with γ- or electron beams is an attractive alternative to chemical sterilization because of its speed, ease of control, and the absence of residue. Radiation treatment of polypropylene, however, also initiates chemical changes which lead ultimately to embrittlement. These changes in physical properties may not become apparent until some time after the treatment. The ability of antioxidants to prevent radiation damage does not always follow the trends observed in thermal oxidation, which has stimulated efforts to develop new stabilizers or optimized combinations of existing ones.

The testing of polymers stabilized against radiation damage raises the familiar questions (a) whether the use of elevated temperatures for accelerated aging is valid, and (b) whether new analytical techniques under actual use conditions can give the same information in comparable time.

In this study we measured chemiluminescence of polypropylene stabilized with different combinations of antioxidants and irradiated to different extents, and made correlations with conventional impact strength measurements of the same materials.

Chemiluminescence has been used by a number of workers to characterize the thermal oxidation of polypropylene (1,2). This study allowed an opportunity to use fiber-optics for transmitting chemiluminescence from the heated sample to the detector, which promises to simplify greatly the apparatus required for the technique.

0097–6156/85/0280–0373$06.00/0

The light emission from autoxidizing organic materials can arise from self-reaction of primary or secondary alkylperoxyl or alkoxyl radicals. In the present case, the predominent site of peroxyl radical attack with hydroperoxide formation in PP involves the tertiary centers, so that the chemiluminescence probably arises (1) from primary or methylperoxyl radicals formed from β-scission, eg.

Since the light emitted by the oxidizing polymer is extremely faint, it is however possible that a minor reaction pathway with a relatively high quantum yield could be the predominent source of excited states.

The mechanism in Scheme 1 makes polypropylene an attractive substrate for study, because it implies a proportionality between rates of light emission and scission of polymer chains.

Experimental

Himont polypropylene (controlled rheology type with a melt flow rate of 12 dg m^{-1}) containing 0.0100% of a phenolic processing stabilizer, was combined with calcium stearate, inhibitors as required, and extruded to give 9" x 0.040" sheet with the following compositions:

	0.1% Ca stearate	0.030% Goodrite 3114[a]	1.0% DLTDP[b]
Neat	x		
Phenol	x	x	
Thio	x		x
Comb	x	x	x

a. 1,3,5-Tris-(4-hydroxy-3,5-di-tert-butylbenzyl)cyanuric acid.
b. Di-n-dodecyl 2,2'-thiodipropionate.

Irradiation procedure. The sheets were irradiated with a Van de Graaf electron accelerator (High Voltage Engineering, Model AK) with an electron energy of 2Mev and a dose rate of about 0.3MR s^{-1}. A nylon matrix radiochromatic film (Far West Technology) was used for dosimetry. Calculations indicated that the dosage at the lower surface of the sheet was about 20% higher than at the upper surface.

Impact measurement. The impact tests were conducted in the usual manner on single sheets with a Gardner Laboratory Impact Tester (Model IG-1120) with a 0.625" diameter punch hammer. At least ten drops were performed in the center 50% of the sheets and with points of impact at least 1" apart. The failure criterion was a brittle (not ductile or tear) break. The values reported in Table I are 50% probabilities (energy at which 50% of failures are brittle). The standard deviation of the values is about 10%.

Chemiluminescence. The PP sheets were stored at ambient temperature in the dark. Chemiluminescence emission from the initial batch of samples was measured 4, 19, 38, 52, 65, 80, and 94

days after irradiation from circular pieces (3/8" diam., wt. 60.0 –
71.5 mg) of PP freshly cut from a sheet with a cork borer. The
pieces were weighed to ±0.1mg and placed in a sample holder shown
in Figure 1. The steel-jacketed fiber-optic cable (Dolan-Jenner
Inc., high-temperature variant Model BXT424, 1/4" fiber bundle) was
screwed into the top of the sample holder, which was then placed in
a 1 x 6" copper tube which had been immersed in an ethylene glycol
bath at 150.0 ± 0.1°C (Poly Temp Model 80). A minor problem with
the cable was the coating of the ends of the glass fibers with dark
material which had to be sanded off after several months (this
deposit probably arose from different materials concurrently under
study). The other end of the fiber-optic cable was connected to an
end-on photomultiplier system described elsewhere (3). The data
were normalized to the average sample weight of 70.0 mg. For
plotting, the text files for individual experiments were loaded
into a Sperry mainframe computer and plotted with modified
commercial software (TELAGRAF; Issco Graphics, Inc.). Control
experiments showed that removal of the surface layer with a blade
just before examination at 150°C did not change the shape of the
chemiluminescence curve. Several samples were weighed before and
after the high temperature examination, and no significant weight
differences were found.

Chemiluminescence at ambient temperature in this study was
obtained from circular samples 1.0" in diameter (average wt. 0.5g),
irradiated to 5MR only, which were placed in the sample well of an
apparatus with automated counting function (Turner Designs, Inc.
Model 20 Luminometer). The light emission was measured for several
time periods of 120 seconds each. Average values and standard
deviations were then obtained with pocket calculators.

Care was taken in the chemiluminescence work to keep the
samples clean, unexposed to room lights for more than a few
minutes, and they were handled gently with gloves or tweezers.

Oven aging. Plaques were physically separated from each other
during accelerated aging on a rack in an oven maintained at 60±1 C
with forced air circulation.

Results

Chemiluminescence (150°C, 50 min) from irradiated samples
after 4 and 65 days are presented in Figures 2-3. The irradiation
dose, stabilizers, and ambient storage times before examination are
indicated on each figure. Smoothed, 3D plots derived from complete
sets of aging data for individual compositions and dosages appear
in Figures 4-6. Note the contour lines on the xy plane of each
figure, and that the z axis has been expanded in some figures to
reveal differences between the more highly stabilized samples.

In spite of losses of light through the cable and the absence
of any focussing lenses, the chemiluminescence intensity was fully
sufficient for precise monitoring. The curves show a rise in
intensity from zero time whose magnitude reflects the amount of
peroxidic initiators present, and the ability of added stabilizers
to prevent the chemiluminescent reactions. In general, the light
emission at 150°C increased with increasing number of days from

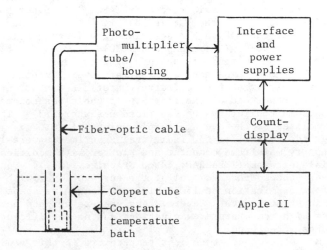

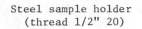

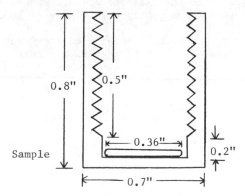

Figure 1. Schematic of apparatus for measurement of chemilum-inescence at 150°C.

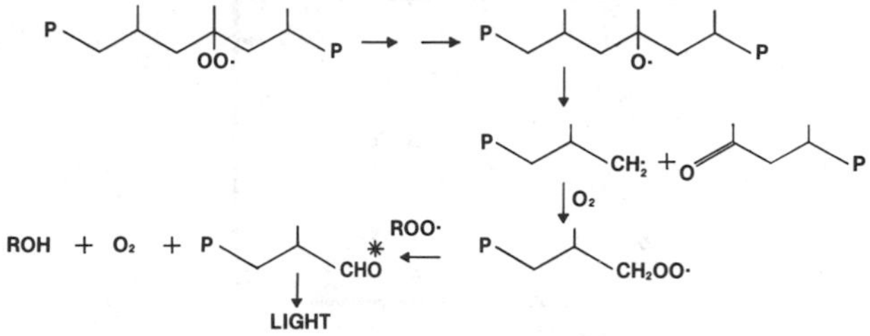

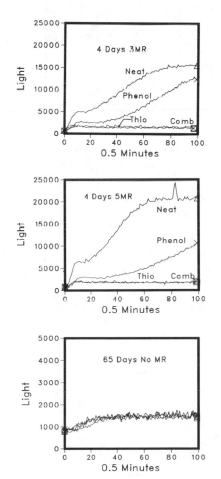

Scheme 1. Mechanism.

Figure 2. Representative plots of chemiluminescence, in counts per 30-second intervals, vs. time at 150°C.

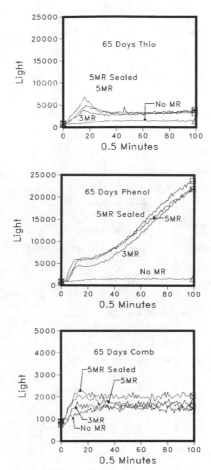

Figure 3.   Representative plots of chemiluminescence, in counts
per 30-second intervals, vs. time at 150°C.

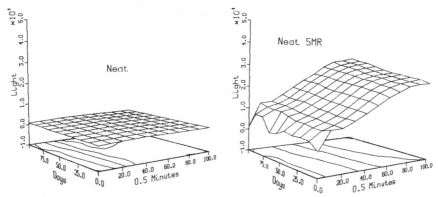

Figure 4.   3D-Plots of chemiluminescence vs. time at 150°C, vs.
aging time at 25°C.

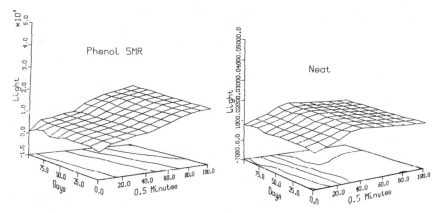

Figure 5.   3D-Plots of chemiluminescence vs. time at 150°C, vs. aging time at 25°C.

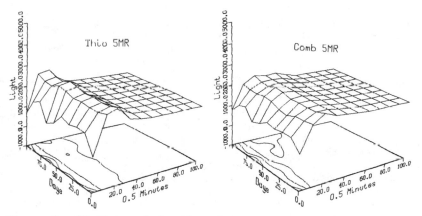

Figure 6.   3D-Plots of chemiluminescence vs. time at 150°C, vs. aging time at 25°C.

the time of irradiation, and otherwise equivalent samples showed
more light with higher radiation dose. The unirradiated samples,
as expected, showed little change on standing. There was no
pronounced difference whether or not a sample was wrapped in
aluminum foil during irradiation, except in one case, where the
foil-wrapped, sample dosed with 5MR showed a chemiluminescence
curve essentially identical to the unwrapped sample which received
a dose of 3MR.

The stability order deduced from the relative intensities in
the chemiluminescence curves at 150°C (Figures 2-6) is comb > thio
>> phenol > neat, the former two showing little difference between
irradiated and unexposed samples. The ordering is in good
agreement with the ranking derived from the impact strength
measurements (Table I and Figure 7) of the samples stored at 60°C,

Table I. Gardner impact strengths (in-1b) of polypropylene

Aging T,°C	Time, days	Neat	Phenol	Thio	Comb
25	1	22.5	22.0	22.6	22.2
	7	11.4	23.0	25.4	25.0
	18	4.6	14.0	7.8	21.8
	32	2.0	3.5	2.0	19.8
60	3	3.0	10.3	23.8	21.5
	7	2.0	5.0	20.2	24.2
	18	2.0	2.0	21.0	19.4
	32	2.0	3.7	20.6	19.4
Before irradiation		26.2	27.8	31.0	26.6

but not at 25°C. When the latter data became available, it was
hypothesized that better agreement between chemiluminescence and
impact data might be obtained if the light emission also could be
measured at ambient temperature. Examination of the neat and
irradiated (5MR) neat samples in hand 110 days after irradiation,
with the commercial apparatus gave values of 0.4 ± 0.2 and 2.7 ±
0.3 counts/minute, respectively, at room temperature. Although the
emission rates were at the threshhold of detectability, the result
and high precision were sufficiently encouraging that a second
batch of polypropylene samples was irradiated, and the
chemiluminescence was measured repeatedly from the same samples
during ambient storage for several weeks. The results are given in
Table II and Figure 7, in which the values reflect the average of
ten 2-minute counting intervals. As we had hoped, this approach
restored a qualitative agreement between the two methods. The
ranking with light emission after six days corresponds to the order
from impact strength after about a month. The plots of chemiluminescence in Figure 7 incorporate single points obtained over weeks
of intermittant observation and do not resemble the ones obtained
during much shorter examination times at 150°C.

Table II. Chemiluminescence (counts/2 min.) at ambient temperature from polypropylene irradiated at day 0 with a dose of 5MR.

Day	Neat	Phenol	Thio	Comb
6	11.7 ± 2.0	8.7 ± 1.4	11.8 ± 1.2	1.2 ± 0.8
7	10.8 ± 3.8	6.2 ± 3.8	11.2 ± 1.6	1.7 ± 1.3
11	5.8 ± 3.0	1.5 ± 0.5	5.3 ± 0.5	1.7 ± 0.6
12	4.0 ± 0.7	1.2 ± 1.3	4.4 ± 0.6	1.0 ± 1.0
14	2.8 ± 1.9	0.6 ± 0.5	4.6 ± 0.5	0.5 ± 0.5
16	2.0 ± 1.2	0.8 ± 0.8	4.0 ± 0.4	0.5 ± 0.5
21	2.2 ± 0.8	0.6 ± 0.5	4.0 ± 1.4	0.5 ± 0.5

Discussion

At first glance, the data seem a little inconsistent because the chemiluminescence was measured (at 25°C or 150°C) from samples which had stood at ambient temperature after irradiation in each case. Apparently the irradiation and ambient-temperature storage left the thioether ester essentially unchanged, so that upon heating to 150°C after various storage periods, nearly the full complement of antioxidant became available to retard autoxidation. The resulting chemiluminescence data correlate therefore with aging results at 60°C, at which temperature the thioether ester is an active antioxidant.

The reversal in the order of effectiveness of the stabilizers toward impact strength on aging at 60°C vs 25°C is unusual because the temperature difference is a relatively modest one. The stabilizing effects and synergism of thio compounds with phenols at high temperatures has been investigated previously in some detail (4-6).

The initial, progressive decline of chemiluminescence at ambient temperature (Table II and Figure 7) is consistent with a decrease in initiating species, such as trapped radicals or labile peroxides, which are produced in the irradiation process in amounts above the steady-state concentrations. The light emission from the "thio" sample exceeded that from the "neat" sample after two weeks, suggesting a sulfide prooxidant effect (7). The short burst of light in the thioether-stabilized runs at 150°C (Figure 6) may similarly arise from induced decomposition of peroxidic species by the thioether to produce radicals, even though reactions of peroxidic compounds with sulfides are generally thought to be predominently non-radical in nature (8). Any detailed mechanistic interpretation of the data is rather limited by the presence of processing stabilizer in all samples, and by the unknown details of the light-emitting reactions.

Inhibitors reduce oxidative chemiluminescence by reducing the rate of light-producing peroxyl self-terminations (Scheme 1), by quenching electronically excited states (9) or simply by absorption of emitted light. The second factor may be much less important in polymers because of lowered rates diffusion (10). The rate of diffusion of 2,4-dihydroxybenzophenone in polypropylene has been measured (11) as 5.5 x 10^{-11} cm^2 s^{-1} at 44°C. If we assume that

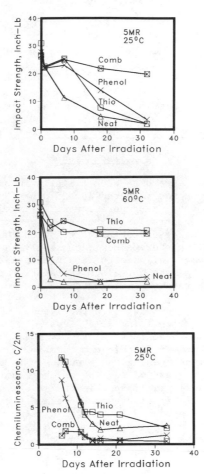

Figure 7. Top and Middle: Loss of impact strength of polypropylene, irradiated to 5MR, vs. storage time at 25 and 60°C.
Bottom: Ambient chemiluminescence from irradiated (5MR)
polypropylene samples vs. storage time at 25°C.

the constant for the stabilizers in our systems is of similar order
of magnitude, we can calculate from the Smoluchowski equation
(12,13) the corresponding diffusion-limited rate constant for
self-encounter, k_{diff} = 4 x 10^4 M^{-1} s^{-1}. For a second-order
quenching process k_q = k_{diff} [P*][Q], where [Q] is the quencher, we
can then very roughly estimate the pseudo first-order term k_{diff}
[Q] from the initial concentrations of our additives. For phenol
and thioether these are 2 x 10^4 * 3 x 10^{-4} = 6 s^{-1} and 2 x 10^4 *
0.02 M =400 s^{-1}, if we correct the above diffusion-limited rate
constant for difference in approximate molecular diameters. For
molecular oxygen in polypropylene we similarly calculate k_{diff}
(25°C) = 8 x 10^7 M^{-1} s^{-1} starting with a recently published
equation (13), and for air-saturated polypropylene, we estimate
k_{diff} [O_2] = 8000 s^{-1}. These estimated, maximum rates do not
appear to be competitive with the excited state lifetimes of most
simple aliphatic carbonyls, and the estimates also suggest that
quenching by oxygen would predominate even if they were. Singlet
state lifetimes are rarely longer than a few nanoseconds, while
aliphatic triplet carbonyl decay constants are usually about 10^5
s^{-1} (15). The triplet lifetime of excited carbonyls in solid
ethylene-carbon monoxide copolymer has been estimated as only 10^{-8}
s (16). Recently a lowering of quenching rates toward excited
states in dissolved macromolecules has also been demonstrated (17).
     The role of physical quenching of excited states in polymer
photooxidation is a closely related question. Although we
acknowledge that many details such as sample homogeneity are not
known in our system, Wiles and Carlsson concluded in an earlier
study that physical quenching by photoprotective agents in
commercial (solid) polymers was of lesser importance than other
modes of protection (18). To the extent that this generalization
is true, the chemiluminescence intensities in solids will not in
general be reduced by physical quenching.
     The above considerations bear on the ambient chemiluminescence
from polypropylene, although at 150°C one would still expect that
the fluorescence emission from singlet carbonyls would be relatively
unaffected by quenching from stabilizers in our samples. One can
also infer that physical quenching alone is not sufficient to
explain the decrease in light from the polypropylene sample with
both phenol and thioether (Table II, comb), since the initial
reduction in intensity is more than can be accounted for by the
combined effects of the additives separately.
     The differences in chemiluminescence intensities of the
samples at 25°C are discernible at a much earlier point than the
differences in impact strengths. This observation is reasonable,
since the former technique measures a dynamic process at the
molecular level, while the latter responds to the cumulative damage
to the sample up to the time of testing. The data are not
completely in accord with this generalization, since the light

emission from the neat and thio samples (Figure 7) is identical up
to about two weeks after irradiation, whereas the impact strengths
(Figure 6) of these two samples are quite different after only one
week. Although the comparison involved separate batches of irradiated polymers, the impact strengths showed good reproducibility in
routine studies.

Conclusions

1. Chemiluminescence at ambient temperature and at 150°C of
several irradiated (electron-beam) polypropylene formulations can
be qualitatively correlated with loss of impact strength. The
correlation with chemiluminescence is preserved at high vs. low
temperatures, even though the ranking changes.
2. The thioether ester (DLTDP) shows a very strong
synergistic effect with the phenolic stabilizer at room temperature, and is active alone at 60°C for protection from loss of
impact strength of the irradiated polypropylene.
3. The thioether alone at room temperature does not protect
polypropylene from loss of physical properties after electron-beam
treatment, although irradiated samples that are subsequently heated
to 150°C are apparently protected from thermal oxidation.
4. Fiber-optics is a convenient technique to isolate a heated
sample from cooled photon-detecting equipment.
5. The differences of the relative efficacy of the stabilizers
at 25°C vs 60°C emphasizes the importance of monitoring changes in
physical properties after irradiation under use conditions.
6. Irradiation of samples of polypropylene sealed in aluminum
foil makes little difference in their subsequent properties compared
with unsealed samples under the same conditions.
Although the number of samples was rather limited, and the
results tend to raise rather than settle mechanistic questions, the
study indicates the possible utility of chemiluminescence for
nondestructive evaluation of radiation-treated polyolefins.

Acknowledgements

The authors are grateful to Himont, Inc. for a fellowship to
H.K.A., support from 3M Co. for hardware and software development,
and to Mr. George Turner of Turner Designs, Inc., for loan of a
Luminometer. The assistance of Mr. Burt Blanch with the Gardner
impact tests is gratefully acknowledged.

Literature Cited

1. George, G.A., Develop. Poly. Degrad., 1981, 3, 173 and
    references therein.
2. Flaherty, K.F., M.S. Thesis, Michigan Technological
    University, Houghton, Michigan, 1982.
3. Ogle, C.A.; Martin, S.W.; Dziobak, M.P.; Urban, M.P.;
    Mendenhall, G.D., J. Org. Chem., 1983, 48, 3728.

4. De Paolo P.A.; Smith, N.P.; In "Stabilization of Polymers and Stabilizers Processes"; Gould, R.F., Ed.; ADVANCES IN CHEMINSTRY SERIES No. 85, American Chemical Society: Washington, D.C., 1968, p. 202.

5. Ray, W.C.; Isenhart, K., Poly. Sci. and Eng., 1975, 15, 703

6. de Jonge, C.R.H.I.; Giezen, E.A.; van der Maeden, F.P.B.; Huysmans, W.G.B.; de Klein, W.J., Mijs, W.J. In "Stabilization and Degradation of Polymers"; Allara, D.L.; Hawkins, W.L., Eds.; ADVANCES IN CHEMISTRY SERIES No. 169, American Chemical Society: Washington, D.C., 1978; p. 399.

7. Scott, G., Communication at the time this paper was presented.

8. Swern, D., "Organic Peroxides"; Wiley-Interscience: New York, 1971; Vol. II, p.73 and references therein.

9. Vasil'ev, R.F., Prog. React. Kin., 1967, 4, 305.

10. Pratte, J.F.; Webber, S.E., Macromolecules, 1983, 16, 1188.

11. Westlake J.F.; Johnson, M., J. Appl. Poly. Sci., 1975, 19, 319.

12. Smoluchowski, M.V., Z. Physik, 1916, 17, 557.

13. Frost, A.A.; Pearson, R.G., "Kinetics and Mechanism"; John Wiley and Sons: New York, 1961; 2nd Ed., p. 271.

14. Kiryushkin, S.G.; Gromov, B.A., Vysokomol. Soedin., Ser. A., 1972, 14, 1746.

15. Wilson, T.; Halpern, A.M., J. Amer. Chem. Soc., 1980, 102, 7279.

16. Heskins, M.; Guillet, J.E., Macromolecules, 1970, 31, 3224.

17. Scaiano, J.C.; Lissi, E.A.; Stewart, L.C.; J. Amer. Chem. Soc., 1984, 106, 1539.

18. Wiles, D.M.; Carlsson, D.J., Poly. Degr. and Stab. 1980-81, 3, 61.

RECEIVED December 7, 1984

Chemiluminescence in Thermal Oxidation of Polymers: Apparatus and Method

L. ZLATKEVICH

Pola Company, Skokie, IL 60077

A chemiluminescence multi-sample apparatus
and method are described for determining
polymer stability by measuring the intensity of the light emitted during thermal
oxidation. The chemiluminescence technique
is shown to provide essential advantages
over the other methods for studying thermal
oxidative stability of polymers (DSC, oxygen
uptake, oven aging). Depending on the
nature of a material analyzed the chemiluminescence experiments are performed either
under O_2 atmosphere at a constant temperature or under N_2 atmosphere at a constant
heating rate. In the former case applicable
to polypropylene (PP) and acrylonitrile-
butadiene-styrene copolymers (ABS) parameters such as induction time and oxidation
rate can be evaluated. In the latter case
applicable to nylon the extent of oxidation
in a certain temperature region can be
evaluated by measuring the area under the
intensity of light - temperature curve.
Along with providing a great deal of knowledge on thermal oxidative stability, the
chemiluminescence approach gives the
additional information concerning polymer
quality. The appearance of the low temperature pulses on the chemiluminescence curve
observed before the onset of autocatalytic
oxidation is associated with the history
and processing of the sample and with the
natural aging of the polymer.

It is often desirable for polymer producers, end-use manu-
facturers, additive suppliers, academicians, and others

0097-6156/85/0280-0387$07.00/0

to establish quality control tests concerning antioxidant concentration or oxidative stability. Numerous techniques have been developed over the years to study the oxidative stability of polymers. Among various methods, chemiluminescence accompanying the thermal oxidation has been referred to by a number of authors (1-9). It was pointed out that the intensity of emitted light could be a convenient criterion for the estimation of thermal oxidative stability of polymers. The relationship

$$I_t = C \quad [ROOH]_t \tag{1}$$

has been proposed where I_t is time dependent light intensity, C is a constant and $[ROOH]_t$ is hydroperoxide concentration (3). In spite of the fact that the first publications concerning the possibility of using the chemiluminescence technique as the method for evaluating polymer thermal oxidative stability appeared about 20 years ago and, after that many separate findings in this field were published, neither a standard method nor a commercial instrument of this kind has so far been offered.

There are several reasons explaining this discrepancy:
1. Chemiluminescence technique is still largely a matter of discovering conditions under which the light emission relates to the properties of interest.
2. Although several methods for the evaluation of oxidation initiation, propagation and termination rate constants and activation energies of these processes have been proposed (8,9), they did not provide the ground for samples comparison on routine basis and thus were of limiting practical value.
3. The chemiluminescence test may require many hours, especially when performed at relatively low temperatures and applied to the analysis of highly stabilized polymer systems. Thus, the productivity of the instruments used was low and could not satisfy the demands.

It is therefore desirable to have a method for the evaluation of chemiluminescence results which will provide information on induction time, oxidation rate, and extent of oxidation, relevant to the thermal oxidative stability of materials. It is also advantageous to have an instrument which would be able to analyze numerous polymer samples simultaneously.

The object of this paper is to present a new chemiluminescence instrument and method having the above mentioned features.

The apparatus and method described are patented and patent pending in several countries.

Apparatus

The apparatus developed (Fig. 1), comprises a dark chamber 1 with a sliding stage 2 which holds numerous

individual test cells 3. The test cells are maintained
on a metal support plate 4 which provides even temperature distribution to the cells. A heater 5 is placed
under the metal plate. The lower part of the dark
chamber is separated from its upper part by a metal plate
6 with a number of holes equal to the number of test
cells. Each hole in the separating plate is covered by a
glass window. When the sliding stage is in "in" position,
each of the holes in the separating plate is strictly
above one of the test cells. The light emitted by the
samples placed in the test cells is sequentially
measured by a rotating photomultiplier 7 placed in the
upper part of the dark chamber. The rotation of the
photomultiplier is provided by an electric motor 8. The
electronic part of the apparatus 9 consists of a photo-
meter, a temperature programmer/controller, a digital
data processing board, control knobs and pilot lights.

The temperature is registered by an iron-constantan
thermocouple located in the metal support plate 4. The
instrument gives the opportunity of measuring the intensity of emitted light vs. temperature from room
temperature up to 300°C. Izothermal as well as various
heating rate experiments can be carried out with the
precision of the temperature control of $\pm1°C$.

In order to avoid the photomultiplier overheating
during the experiments constant outside air circulation
is provided by a fan placed in the upper part of the
dark chamber.

Light emitted by the samples is registered by a
general purpose side-on photomultiplier tube
(Hamamatsu 1P28, the spectral response from 185 to 700nm,
the peak sensitivity at 450 nm),and recorded indepen-
dently for each of eight cells by a multichannel recorder
(Hewlett Packard 7418A)

The test cells (Fig. 2), have a construction
contributing to the warm up of the gas before reaching
the test samples. Each cell contains metal shavings
which have a large surface area for heat exchange. The
gas flow to each sample is evenly distributed by a mani-
fold with individual flow adjustment. Each test cell is
covered by a glass cover to prevent cross contamination.
Glass covers also restrict reaction volume of each cell
and promote fast replacement of one gas by another.

Samples in a powder form as well as plaques can be
analyzed. In the latter case a spring ring is put on
the top of the sample to assure good contact between the
sample and the cuvette.

Method
─────

It was established by Bolland and Gee that organic hydroperoxides appear as one of the first products of oxidation (10, 11). Subsequent oxidation of the polymer is
autocatalysed by the decomposition of hydroperoxides
which produce free radical chain carriers for the chain
reaction.

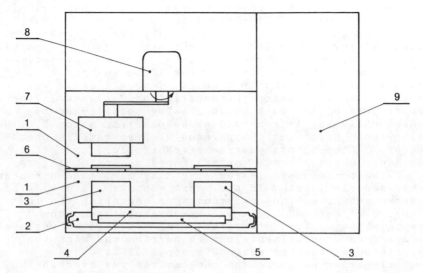

Figure 1. The diagram of the multi-sample chemiluminescence apparatus. The numbers are identified in the
text.

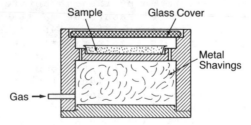

Figure 2. The diagram of the cell used in the
chemiluminescence apparatus.

The following reaction scheme has been offered:

Initiation $2ROOH \xrightarrow{K_1} RO_2^{\bullet} \quad + \quad R^{\bullet} + H_2O$ (2)

Propagation $R^{\bullet} + O_2 \xrightarrow{K_2} RO_2^{\bullet}$ (3)
$RO_2^{\bullet} + RH \xrightarrow{K_3} ROOH + R^{\bullet}$ (4)

Termination $R^{\bullet} + R^{\bullet} \xrightarrow{K_4} R-R$ (5)
$R^{\bullet} + RO_2^{\bullet} \xrightarrow{K_5} ROOR$ (6)
$RO_2^{\bullet} + RO_2^{\bullet} \xrightarrow{K_6} ROOR + O_2$ (7)

where ROOH and RH are hydroperoxide and polymer, respectively; $R^{\bullet}$ and $RO_2^{\bullet}$ are free radicals. (The $R^{\bullet}$ shown on the right side of equation (2) is assumed to result from either a chain transfer step of $RO^{\bullet}$ with RH or by self-dismutation of $RO^{\bullet}$ to $R^{\bullet}$ (12).

Initial Stages of Oxidation

Under mild conditions of oxidation the chain lengths are long and the amount of oxygen participating in the reaction is approximately equal to the amount of hydro-peroxides formed. Under this condition the amount of hydroperoxides which decompose to initiate further oxidation is very small and neglect of the reaction (2) in writing the expression for oxidation rate is valid. At high oxygen pressures, steps (5) and (6) can be neglected and the solution of the equations (2) - (7) using the steady state approximation is

$$- \frac{d[O_2]}{dt} = \frac{d[ROOH]}{dt} = K_3 \ (K_1/K_6)^{\frac{1}{2}} [ROOH] \ [RH] \tag{8}$$

At low oxygen pressures, steps (6) and (7) are neglected to give

$$- \frac{d[O_2]}{dt} = \frac{d[ROOH]}{dt} = K_2 \ (K_1/K_4)^{\frac{1}{2}} \ [ROOH] \ [O_2] \tag{9}$$

For long chain lengths, many molecules of hydroperoxide are formed per free radical initiating the reaction before termination occurs and hence variations of over-all rate constant K essentially reflects changes in the rate of propagation, i.e. in eq. (8)

$K = K_3 (K_1/K_6)^{\frac{1}{2}} \approx K_3$ and in eq. (9) $K = K_2 (K_1/K_4)^{\frac{1}{2}} \approx K_2$

Let us consider the two cases presented by eqs. (8) and (9) separately and try to estimate what kind of chemi-luminescence response one should expect for each of them.

High oxygen pressure. Isothermal conditions. Eq. (8) can be rewritten:

$$\frac{d[B]}{dt} = K_3 [A] [B] \tag{10}$$

where [A] is polymer and [B] is hydroperoxide concentration, respectively. K_3 is the oxidation rate constant.

(10) presents the autocatalytic reaction with regard to the substrate and hydroperoxide and as it is typical for autocatalytic reactions, induction and acceleration periods should be expected. Designating the increase in [B] during the oxidation as $X = [B] - [B]_0$ and noting that the increase in [B] is equal to the decrease in [A], $\big([B] - [B]_0 = [A]_0 - [A]\big)$,

$$\frac{dx}{dt} = K_3 \big([A]_0 - X\big) \big([B]_0 + X\big) \tag{11}$$

where $[A]_0$ and $[B]_0$ are the initial polymer and hydroperoxide concentrations, respectively.

Integration of eq. (11) gives

$$K_3 \big([A]_0 + [B]_0\big) t = \ell n \left(\frac{[B]_0 + X}{[A]_0 - X} \cdot \frac{[A]_0}{[B]_0}\right) \tag{12}$$

Since the chemiluminescence emission intensity is proportional to hydroperoxide concentration (see eq. (1)

$$I_t = C \big([B] - [B]_0\big) = CX \tag{13}$$

When the chemiluminescence intensity reaches the maximum

$$Imax = C [A]_0 \tag{14}$$

Two cases should be considered:
(1) $[B]_0 \ll [A]_0$ (long induction time): Eq. (12) can be rewritten as

$$K_3 [A]_0 \ t = \ell n \left[\frac{[A]_0 X}{([A]_0 - X)[B]_0}\right] \tag{15}$$

Substituting $[A]_0$ and X from eqs. (13) and (14) into eq. (15)

$$\ell n \left(\frac{I_t}{Imax - I_t}\right) = \ell n \frac{[B]_0}{[A]_0} + K_3 [A]_0 \ t \tag{16}$$

Equation (16) provides a good method of estimating $\ell n \big([A]_0 / [B]_0\big)$ and $K_3 [A]_0$ values. The first expression is proportional to induction time, whereas the second is proportional to oxidation rate. A plot of $\ell n [I_t / (Imax - I_t)]$ against t has slope $K_3 [A]_0$ and intercept $\ell n \big([B]_0 / [A]_0\big)$

The advantage of the evaluation according to eg. (16) is that the results do not depend on the amount of the material analyzed, and the induction time and oxidation rate values obtained for various samples of different weight can be directly compared. However, if the weight of the samples or more correctly, the surface of the samples emitting light is kept constant, equation (16) can be simplified and only the initial exponential portion of the curve need be utilized. When $I_t \ll$ Imax

$$\ln I_t = \ln(C[B]_o) + K_3 [A]_o \, t \tag{17}$$

In eq. (17) $-\ln(C[B]_o)$ is proportional to induction time, whereas oxidation rate is expressed similarly to eq. (16) by $K_3 [A]_o$

(2) $[B]_o \ll [A]_o$ (short induction time): When $[B]_o$ is not significantly smaller than $[A]_o$, it can be found if one knew I_1 and I_2 values corresponding to two moments of time t_1 and t_2 (13). Choosing $t_2 = 2t_1$, it can be shown that

$$\ln\left(\frac{[B]_o}{[A]_o}\right) = \ln\left[\frac{I_1 / Imax}{Z-1-(I_1 / Imax) \, Z}\right]$$

$$K_3 [A]_o = \frac{\ln Z}{(1+ [B]_o / [A]_o) \, t_1} \tag{18}$$

where Z is the larger radical of the quadratic equation

$$\frac{I_1}{Imax}\left(1 - \frac{I_2}{Imax}\right)Z^2 - \frac{I_2}{Imax}\left(1 - \frac{I_1}{Imax}\right)Z + \left(\frac{I_2}{Imax} - \frac{I_1}{Imax}\right) = 0$$

Low oxygen pressure.   Constant heating rate.   Eq. (9) can be rewritten

$$\frac{d[B]}{dt} = K_2 \, [B] \, [O_2] \tag{19}$$

where $[O_2]$ is oxygen concentration and K_2 is the oxidation rate constant.

Designating the increase in [B] during the oxidation as $X=[B] - [B]_o$ and noting that the increase in [B] is equal to the decrease in $[O_2]$, ($[B] - [B]_o = [O_2]_o - [O_2]$),

$$\frac{dX}{dt} = K_2 \, ([O_2] - X) \, ([B]_o + X) \tag{20}$$

Taking into account eq. (13) and introducing the constant heating rate $T=T_o + \alpha t$ and the Arrhenius type equation for the change of K_2 with temperature $K_2 = K_o \exp(-E/RT)$ eq. (20) can be rewritten for the initial stages of the reaction ($[O_2]_o \gg X$, $[B]_o \gg X$)

$$\frac{dI}{dT} = \frac{K_o C}{\alpha} [O_2]_o [B]_o \exp (-E/RT) \tag{21}$$

Thus, one should expect the exponential increase of the chemiluminescence intensity with the temperature. Since the extent of oxidation in a certain temperature region ($T_2 - T_1$) is proportional to the amount of hydroperoxides formed, it can be expressed as $\int IdT$ and evaluated by measuring the area under the intensity of light-temperature curve.

Advanced Stages of Oxidation

Under relatively severe conditions of oxidation (high temperatures, long time intervals, presence of metallic activators and light) the decomposition of the hydroperoxides becomes appreciable, and the rate of oxidation can no longer be equated to the rate of hydroperoxide formation as represented by the first two terms in equation (8). In this case the disappearance of hydroperoxides by eq. (2) should be included in writing the equation for the rate of change of hydroperoxide concentration with time at high oxygen pressure:

$$\frac{d[ROOH]}{dt} = K_3 (K_1/K_6)^{\frac{1}{2}} [ROOH][RH] - K_1[ROOH]^2 \quad (22)$$

The advanced stages of oxidation must be marked by appreciable disappearance of substrate and this factor may be introduced into the above equation if one assumes at a first approximation that the unoxidized substrate present at any given time is equal to that present initially less the concentration of hydroperoxides formed, i.e. [RH] = [RH]$_0$ - [ROOH] (14). This assumption leads to the following equation:

$$\frac{d[ROOH]}{dt} = K_3(K_1/K_6)^{\frac{1}{2}}[RH]_0 [ROOH] - \left[K_3(K_1/K_6)^{\frac{1}{2}}+K_1\right][ROOH]^2 (23)$$

Eq. (23) can be integrated to give

$$[ROOH] = \frac{[ROOH]_\infty}{1 - \left(1 - \frac{[ROOH]_\infty}{[ROOH]_0}\right) e^{-at}} \quad (24)$$

where [ROOH]$_\infty$ is the steady state value of hydroperoxide (the concentration approached at long times as $t \to \infty$) and [ROOH]$_0$ is the initial concentration of hydroperoxides;

$a = K_3 (K_1/K_6)^{\frac{1}{2}} [RH]_0 = K [RH]_0$, [ROOH]$_\infty$ = $a/\left[K_3(K_1/K_6)^{\frac{1}{2}} + K_1\right]$ = $(K/K+K_1)$ [RH]$_0$

As it follows from eq. (24), the concentration of hydroperoxides approaches a steady state value which is a maximum value if [ROOH]$_0$ < [ROOH]$_\infty$ and a minimum value if [ROOH]$_0$ > [ROOH]$_\infty$

Three situations can exist:

(1) [ROOH]$_0$ ⩽ [ROOH]$_\infty$
Replacing [RH]$_0$, [ROOH]$_0$, [ROOH] and [ROOH]$_\infty$ by [A]$_0$, [B]$_0$, [B] and $(K/K+K_1)$ [A]$_0$

$$[B] = \frac{(K/K+K_1)\quad [A]_o}{1 - \left[1 - \frac{(K/K+K_1)\quad [A]_o}{[B]_o}\right] e^{-K\ [A]_o\ t}} \qquad (25)$$

Since $I_t = C[B]$ and $Imax = C[A]_o\ (K/K+K_1)$

$$\ln\left(\frac{I_t}{Imax - I_t}\right) = \ln\left[\frac{[B]_o}{(K/K+K_1)\quad [A]_o - [B]_o}\right] + K[A]_o\ t \qquad (26)$$

When $[A]_o \gg [B]_o$ and $K \gg K_1$ eq. (26) transforms into eq. (16) for initial stages of oxidation

(2) $[ROOH]_o < [ROOH]_\infty$

$I_t = C[B]$, $I_o = C[B]_o$, $Imax = C(K/K+K_1)[A]_o$

$$\ln\left[\frac{I_t}{I_o}\frac{(Imax - I_o)}{(Imax - I_t)}\right] = K[A]_o\ t \qquad (27)$$

(3) $[ROOH]_o > [ROOH]_\infty$

$I_t = C[B]$, $I_o = C[B]_o$, $Imin = C(K/K+K_1)[A]_o$

$$\ln\left[\frac{I_t}{I_o}\frac{(I_o - Imin)}{(I_t - Imin)}\right] = K[A]_o\ t \qquad (28)$$

Experimental Results and Discussion

Experimental Conditions

All materials analyzed were ground to 40 mesh particle size, the standard amount of powder (0.1 g) was poured into the metal cuvette (1" diameter) and carefully spread to a uniform thickness. The cuvettes were placed in the separate test cells of the chemiluminescence apparatus, and covered by glass covers at room temperature under nitrogen atmosphere. Two different procedures have been utilized:

(1) Isothermal in oxygen atmosphere. Heating under nitrogen from room temperature up to a chosen temperature followed by replacement of nitrogen by oxygen and start of the isothermal experiment. Polypropylenes (PP) and acrylonitrile-butadiene-styrene (ABS) copolymers have been evaluated under these conditions. Isothermal experiments have been carried out also with nylon samples.

(2) Constant Heating Rate in Nitrogen Atmosphere. Heating under nitrogen from room temperature up to 300°C with constant heating rate (10 degrees/min.). Constant heating rate experiments have been performed with nylon

samples. In all cases two samples of the same material
were studied and the average results of the two measurements taken. The reproducibility of the experimental
data was found to be good (15%)

Initial Stages of Oxidation

It was found that light is emitted by PP and ABS only
under an oxygen atmosphere. In the case of PP, switching
from nitrogen to oxygen atmosphere was not accompanied by
a burst of light and it required some time before the
intensity of chemiluminescence started to increase
steadily.
        A typical light intensity-versus-time curve for the
chemiluminescence produced by autoxidation of PP consists
of four regions (Fig. 3). There is an induction period
during which there is practically no light emitted by the
sample, oxidation is slight and buildup of peroxides and
hydroperoxides is slow. The unstabilized sample exhibited a very short induction period, whereas for
stabilized material this period was longer. Following
the induction period, there is an autocatalytic stage in
which the hydroperoxides catalyze further oxidation and
the intensity of emitted light increases quite rapidly.
The induction and acceleration periods are not separate
phenomena, but parts of a typical autocatalytic reaction.
The light intensity next reaches the highest level (peak
hydroperoxide concentration). Finally, there is a period
of light decay (a deceleration of the rate of oxidation).
The decrease in oxidation rate after passing the maximum
has been observed previously (14). Possible explanations
for the rate drop could involve a decrease in the per-
meability of the outer surface of oxidized sample to
oxygen or the formation of reaction products which tend
to inhibit the oxidation reaction either by interaction
with chain carriers or by nonradical induced decomposition of hydroperoxides. For our purposes the first three
regions are of most importance since they represent the
autocatalytic process.
        Fig. 4 presents the plot according to eq. (16) of
the chemiluminescence results obtained for PP samples A
and B. The experimental results are well approximated by
a straight line for each of the samples studied from
which $\ln([A]_o/[B]_o)$ and $K[A]_o$ values can be evaluated.
Results for PP samples A and B together with the results
obtained for two other PP samples are shown in Table I.
Chemiluminescence data are presented together with the
conventional oven aging test results. One can conclude
that both techniques give correlative results: long oven
lives correspond to long induction times and low oxida-

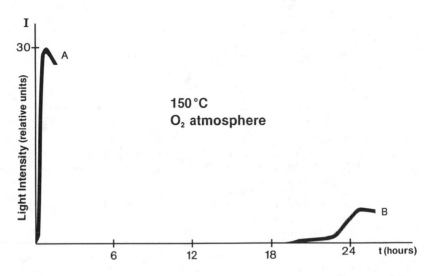

Figure 3.   The chemiluminescence curves of unstabilized (A) and stabilized (B) polypropylene samples.

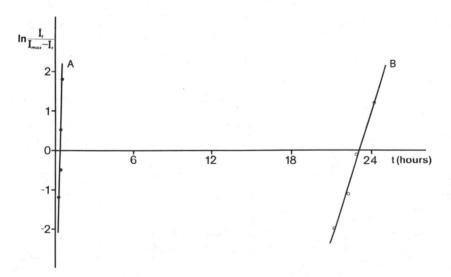

Figure 4.   Plot of ln [I_t/($I_{max}-I_t$)] vs. t for unstabilized (A) and stabilized (B) polypropylene
samples.

Table I.   The evaluation of polypropylene thermal oxida-
            tive stability by the chemiluminescence and
            oven aging methods at 150°C

| | Chemiluminescence Analysis | | |
Sample	Induction time (relative units)	Oxidation rate (relative units)	Oven life (days)
A	1.9	0.05	2
B	12.2	0.006	106
C	3.3	0.02	12
D	5.2	0.02	74

tion rates. It has to be emphasized that the time
required for chemiluminescence analysis was 2 hours for
sample A and 25 hours for sample B, compared to 48 hours
and 106 days, respectively, in the case of the oven aging
test. The other important advantage of the chemiluminescence technique is the possibility of obtaining quantitative information (induction time and oxidation rate),
whereas the failure point in the oven aging test is
defined as the first observation of powdery disintegra-
tion or brittleness and thus is essentially qualitative.
    In order to estimate the activation energy (E) of PP
autoxidation in the solid state, the analysis of sample A
was performed at five different temperatures and the
values of $K[A]_o$, $\ln([A]_o/[B]_o)$ and I_{max} obtained (Table II)

Table II.   Parameters of autoxidation for polypropylene
             Sample A

Temp. (°C)	$\ln([A]_o/[B]_o)$ (relative units)	$K[A]_o \cdot 10^2$ (relative units)	I_{max} (relative units)
150	1.97	5	30
140	3.17	2.9	18.9
130	4.13	1.7	8.2
120	4.71	0.85	3.7
110	5.12	0.43	1.9

Two different approaches in evaluating E have been used:
the conventional plot of $\ln(K[A]_o)$ vs. $1/T$ and the
method originally developed for isothermal solid-state
decomposition reactions studied by DTA (15) where it was
suggested that the slope of $\ln(hmax)$ vs. $1/T$ plot yields
the activation energy (hmax is the maximum peak height of
the isothermal DTA trace). The activation energy in the
110-150°C temperature interval was 19.4 and 23.3
kcal/mole, respectively. Although the second value precisely coincides with the activation energy of PP
oxyluminescence reported by M.P. Schard and C.A. Russell
(4), whereas the first value is slightly lower, we be-

lieve that the direct method of activation energy evaluation provided by the logarithm of reaction rate vs.
reciprocal temperature plot is more reliable.

It has been reported that the induction period for
linseed oil (16) and polybutadiene (14) oxidation
decreases logarithmically as the temperature is raised.
The attempt to use the same approach for PP oxidation
failed. Both $\ln([B]_0/[A]_0)$ value and the slope of the
curve plotted in $\ln([B]_0/[A]_0)$ – T coordinates monotoni-
cally increase with the temperature showing that at least
at high temperatures the induction time decreases with
temperature faster than is predicted by $\ln([B]_0/[A]_0)$ vs.
T linearity.

Similarly to PP, ABS does not emit light under
nitrogen atmosphere. The only difference in the character of chemiluminescence vs. time curves between PP and
ABS is the initial burst of emission observed for all
ABS samples when the atmosphere is changed from nitrogen
to oxygen. The initial increase in the light intensity
immediately after the introduction of oxygen has been
observed previously and attributed to the presence of
easily oxidizable centers in a polymer at the beginning
of the chemiluminescence experiment, when accumulated
hydroperoxides are not as yet present (17).

The experimental chemiluminescence results obtained
for two ABS samples at 150°C and 190°C are presented in
Fig. 5. The chemiluminescence experiments at 190°C have
been performed in order to be able to compare the data
with the DSC and oxygen uptake results since the sensitivity of the DSC and oxygen uptake techniques is not
sufficient enough for their application at 150°C. Besides
exhibiting the initial burst of emission, the character
of the chemiluminescence vs. time curves for the ABS
samples was similar to that for PP and has been treated
by applying eg. (16) for the 150°C experiment. The
chemiluminescence data together with the DSC and oxygen
uptake results are shown in Table III.

Table III.   The evaluation of ABS thermal oxidative stab-
             ility by the chemiluminescence, DSC and
             oxygen uptake methods

Sample	DSC 190°C	Oxygen uptake 190°C	Chemiluminescence 150°C		190°C	
	Induction Time (min)		Induction Time (relative units)	Oxidation Rate	t_{max} (min)	I_{in}/I_{max}
A	V.Short	V.Short	5.3	0.014	7	0.46
B	34	43	10.7	0.012	35	0.17
C	13	11	3.8	0.017	9	0.24
D	11	5	3.8	0.020	6	0.26
E	13	10	4.2	0.020	8	0.21

Since the chemiluminescence experiment at 190°C was completed within several minutes, the kinetic approach
according to eq. (16) was not used. Instead the time to
reach the maximum intensity (t max) and the ratio of the
initial burst of emission (I_{in}) to the maximum intensity
(I_{max}) were measured (Table III). As it follows from
the Table III, there is basically a good correlation between the induction time values obtained by the DSC and
oxygen uptake methods and the time to reach the maximum
intensity (the chemiluminescence method). The only
exception was sample A. The induction time for this
sample was evaluated as "very short" by the DSC and oxygen uptake techniques and was smaller than for the other
samples. On the other hand, t max was similar for the
samples A, C, D and E when estimated by the chemiluminescence method. There is a better correlation between
the DSC, oxygen uptake and chemiluminescence results
when instead of t max the value I_{in}/I_{max} is used. At a
first approximation the I_{in}/I_{max} ratio is indicative of
the amount of easily oxidizable sites on polymer surface.
Thus, it seems that "very short" induction time obtained
by the DSC and oxygen uptake methods for sample A in this
particular case is really not the induction time of an
autocatalytic process but rather is associated with the
content of unstable products which, however, do not
autocatalyze oxidation. Further indication of this was
obtained by the chemiluminescence experiments performed
at 150°C (Table III). At this temperature sample A
exhibited longer induction time and smaller oxidation
rate than the samples C, D, and E, although the sample B
remained the most stable. It should be emphasized that,
as it was indicated above, the induction and acceleration
periods are not separate phenomena but parts of a typical
autocatalytic process. Thus both these parameters should
be considered together and the utilization of the
induction time only by the DSC and oxygen uptake methods
may not be appropriate.

It is known that ABS exhibits dramatic loss of
impact resistance when aged in an air oven even at 130°C.
The DSC and oxygen uptake methods are probably useful in
estimating the high temperature performance but not
necessarily directly applicable to predict life times at
service temperatures (18). Thus the ability of chemiluminescence to be applied for evaluation of ABS thermal
oxidative stability at 150°C and even lower temperatures
seem to be important.

In contrast to PP and ABS, nylon emits weak light
even when heated in nitrogen, although the level of light
emitted by this polymer under nitrogen is 1-2 orders of
magnitude smaller than the chemiluminescence intensity of
PP and ABS under oxygen. Since nylon glows under
nitrogen, the experiments were performed under this
atmosphere at constant heating rate.

Both unstabilized and stabilized nylon samples
exhibit very weak, practically constant light emission in

the temperature interval 40-150°C (Fig. 6). At higher
temperatures, the light intensity increases exponentially.
When the melting temperature is reached, there is a sharp
decrease in the light emission. The area under the light
intensity vs. temperature curve was larger for the
unstabilized sample indicating that the value $1/\int IdT$ can
be taken as a measure of the degree of oxidative stability. At the same time a low and steady level of light
emission in the 40-150°C temperature region shows that up
to 150°C nylon auto-oxidation is not significant.

Since a certain (probably very low) concentration of
oxygen is necessary for the reaction responsible for the
emission of light, the source of oxygen in the system
must be established. In this regard the results obtained
by L. Matisova-Rychla et. al. ($\underline{7}$) are of interest. It
was shown that preoxidized polypropylene exhibits
chemiluminescence when heated under nitrogen, whereas
pure polypropylene glows only under oxygen. The increase
in polarity should promote oxygen adsorption and it was
assumed that during storage of a polymer an equilibrium

$$O_2 \rightleftharpoons O_2 \begin{array}{l} \text{physically} \\ \text{adsorbed} \end{array} \rightleftharpoons O_2 \begin{array}{l} \text{chemically} \\ \text{adsorbed} \end{array}$$

is set up between gaseous oxygen and oxygen physically or
chemically adsorbed on the surface of the polymer.
Oxygen in its adsorbed form can interact with hydroperoxides directly in a bimolecular reaction according to
equation (19). The ability of the surface to chemisorb
oxygen depends essentially on the nature of a polymer,
i.e. it should display a polar effect when the electron
affinity of oxygen results in the formation of an O_2^-
radical-ion in the presence of suitable electron donors

$$e + O_2 \rightleftharpoons O_2^-$$

This equilibrium is shifted to the left side with
increasing temperature and thus can also be a source of
oxygen when thermal treatment is performed under inert
atmosphere. That raises the question concerning the
possibility to deal with purely thermal degradation for
polymers with electron supplying groups.

Further experiments with both unstabilized and
stabilized nylon samples showed that any kind of
additional heat treatment is accompanied by a rise in the
emitted light intensity especially at low temperatures.
The example of this kind is shown on Fig. 7 for
stabilized nylon. Both annealed and quenched samples
were heated under nitrogen from room temperature up to
250°C and held at this temperature for one hour. Then
the annealed sample was slowly cooled to room temperature
when still under nitrogen whereas the quenched sample was
quickly immersed in ice water. As is shown in Fig. 7,
both samples exhibited an intense maximum around 80°C on

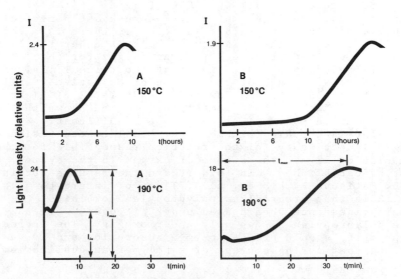

Figure 5. The chemiluminescence curves of two acry-lonitrile-butadiene-styrene copolymer samples A and B obtained at 150 and 190°C.

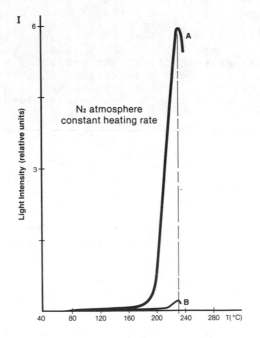

Figure 6. The chemiluminescence curves of unstabil-ized (A) and stabilized (B) nylon samples.

the chemiluminescence curve although this effect is more pronounced for the annealed sample. At the same time the quenched and annealed samples showed similar increase in the light intensity of the exponential high temperature portion of the curve (150-220°C). These results are in agreement with the data obtained by G. A. George (19) where it was shown that the intensity of the initial increase in light emission at 100°C after admission of oxygen depends on the previous time of heating the nylon sample in nitrogen. Thus it can be concluded that nylon undergoes oxidation which is most probably provided by physically and/or chemically adsorbed oxygen even when exposed to high temperatures under nitrogen. The activation energies evaluated using the method widely applied in luminescence experiments, the so called method of initial rises (20) was 37 kcal/mole at temperatures of 50-80°C and 18 kcal/mole at temperatures of 160-210°C. The latter value correlates well with the activation energy of 15.4 kcal/mole reported for nylon oxylumin-escence (4), whereas the value of 37 kcal/mole is close to 42 kcal/mole obtained for the low temperature chemiluminescence observed for preoxidized PP heated under nitrogen (6). A large activation energy of the reaction responsible for the appearance of the chemiluminescence maximum around 80°C together with the fact that it takes place at relatively low temperatures certainly indicates that this is a neighboring group-associated reaction where the hydroperoxide clusters provide a very high value of the pre-exponential factor. Neighboring hydro-peroxides have been shown to decompose with greater ease than the isolated hydroperoxides responsible for the auto-catalytic oxidation (21). Thus, the low temperature chemiluminescence and its intensity seem to be associated with the history and processing of the sample and with the natural aging of the polymer.

It should be underlined that the chemiluminescence response depends not only on the sample thermal treatment but also on the time it was kept at room temperature after aging prior to the analysis (Fig. 8). The intensity of the emitted light in the low temperature region increases with the time of exposure to ambient conditions for both samples aged at 120 and 160°C. At the same time the high temperature part of the chemiluminescence curve remains practically unchanged (Fig. 9). Such a behavior is understandable if one bears in mind that the solubility and chemisorption of oxygen in a polymer decreases with increasing temperature and that chemisorption is a relatively slow process. Thus it takes some time to reach an equilibrium level of chemisorbed oxygen at room temperature.

The stable level of the light emission at high temperatures independent of the time of sample exposure to ambient conditions indicates that physically adsorbed oxygen participates in the reaction with isolated hydroperoxides. The low concentration of oxygen

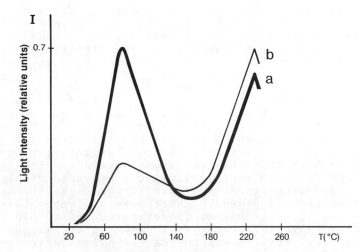

Figure 7.   The chemiluminescence curves of annealed
(a) and quenched (b) stabilized nylon.

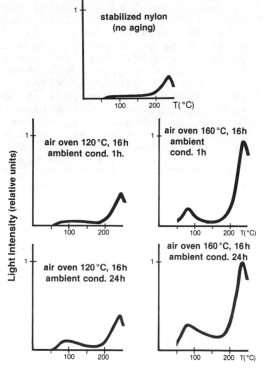

Figure 8.   The chemiluminescence curves of stabilized
nylon sample aged in the air oven at 120 and 160°C for
16 hours and then exposed to ambient conditions for
various time.

sufficient to promote the reaction with isolated hydroperoxides may also explain the chemiluminescence response
at high temperatures.

The sensitivity of the chemiluminescence technique
to the changes in the concentration of chemisorped
oxygen can be probably utilized in kinetic studies of
oxygen adsorption in polymers.

Advanced Stages of Oxidation

At low temperatures the isothermal chemiluminescence
curve for nylon produced under oxygen atmosphere has a
sigmoidal shape (8) and can be evaluated by utilizing
the approach developed for the initial stages of
oxidation. However, at relatively high temperature nylon
oxidation does not exhibit a noticeable induction period.
When the initial concentration of hydroperoxides in a
polymer does not essentially vary from its limiting value
either an increase ($[ROOH]_0 < [ROOH]_\infty$) or a decrease
($[ROOH]_0 > [ROOH]_\infty$) in the light emission with time
can be expected. Fig. 10 presents the results obtained
for unstabilized and stabilized nylons at three different
temperatures. All samples exhibit a burst of emission
when oxygen is introduced into the system (zero time).
Then the light emission from unstabilized nylon increases
at 150 and 170°C and decreases at 190°C. The equilibrium
level of chemiluminescence for stabilized nylon is
reached very fast at 150 and 170°C, whereas at 190°C
similarly to unstabilized material there is a decay in
light emission.

Several problems in evaluating advanced stages of
oxidation by chemiluminescence should be noted:
1. Comparative evaluation of different samples can be
made only if they exhibit relatively prolonged growth or
decay of light emission before reaching the equilibrium
(for example, unstabilized and stabilized nylons can not
be compared at 150 and 170°C because the equilibrium for
the stabilized sample is reached at these temperatures
practically instantly).
2. In some cases it is difficult to establish a reliable
equilibrium level of light emission since the light growth
or decay may continue over a long period of time. This
might be a serious obstacle because the evaluation
according to eqs. (27) and (28) is sensitive to Imax and
Imin values.
3. At the best only $K_3 (K_1 /K_6)^{\frac{1}{2}} [A]_0$ and
$K_3 (K_1 /K_6)^{\frac{1}{2}} [A]_0 / [K_3 (K_1 /K_6)^{\frac{1}{2}} + K_1]$ values can be
obtained and thus complete elucidation of the oxidation
process at its advanced stages (independent evaluation of
K_1 , K_3 and K_6) can not be accomplished.

When the growth and decay to the steady state is
plotted according to egs. (27) and (28), a poor fit is
observed for unstabilized nylon at 150 and 190°C. A
better fit is obtained for stabilized nylon at 190°C.
This is shown in Fig. 11. The failure to express the

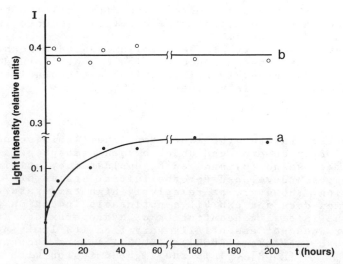

Figure 9. The dependence of the intensity of emitted
light at 80°C (a) and 230°C (b) vs. the time of
exposure to ambient conditions for stabilized nylon
aged in the air oven at 120°C for 16 hours.

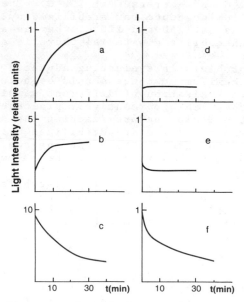

Figure 10. The growth and decay of light emission for
unstabilized (a, b, c) and stabilized (d, e, f) nylon
samples at 150 (a, d), 170 (b, e) and 190°C (c, f)
after heating in nitrogen and then admitting oxygen
at zero time.

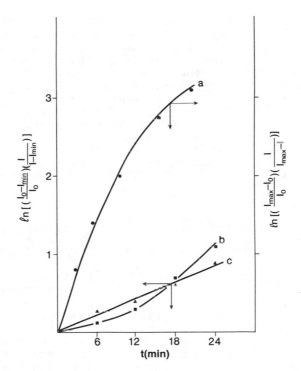

Figure 11. Analysis of the chemiluminescence emission growth and decay according to egs. (27) and (28). a- unstabilized nylon (150°C), b - unstabilized nylon (190°C), c - stabilized nylon (190°C)

experimental results by equations (27) and (28) indicates
that advanced stages of oxidation are complicated by
processes not taken into account. Some of the complications are: (1) the products of oxidation may play an
important role as inhibitors or activators of the
oxidation; (2) impurities may be very important as
inhibitors or activators; and (3) a side reaction or the
chain carrying radicals themselves may cause destruction
of hydroperoxides.

Thus it can be concluded that the application of
chemiluminescence for the evaluation of advanced stages
of oxidation is restricted by both the ability of the
technique and the complex nature of the process. It is
doubtful, however, whether the rate constants measured
in the region where chain lengths are close to unity and
the hydroperoxide concentration is reaching its limiting
value are of significance (19). The essential changes
reflected in a material's mechanical properties occur
during the induction and auto-acceleration periods and
the evaluation of the useful lifetime of a polymer should
be based on the determination of the parameters of
oxidation in these two regions.

Conclusions

The chemiluminescence apparatus and method described provide a sensitive technique for thermal oxidative
stability evaluation of materials with many important
advantages:
1)   The chemiluminescence analysis requires a very small
amount of material.
2)   The data are instantly and permanently recorded.
3)   For initial stages of oxidation the chemiluminescence
experiment permits the evaluation of induction time and
oxidation rate values when performed at a constant
temperature and under oxygen atmosphere, and the extent
of oxidation in a certain temperature region when performed at a constant heating rate and under nitrogen
atmosphere.
4)   The chemiluminescence method is much faster and less
tedious than the air oven aging test.
5)   In contrast to the DSC and oxygen uptake methods, the
high sensitivity of chemiluminescence enables testing to
be done at relatively low temperatures closely associated
with actual temperatures the material is exposed to in
the field.
6) The chemiluminescence apparatus furnishes the opportunity to analyze numerous samples simultaneously and does
not require the attention of the operator after the
experiment is set up.
7)   The multi-sample chemiluminescence apparatus is low
in cost and up-keep expenses, simple in operation, light-
weight and flexible in use.
Along with providing a great deal of knowledge concerning the thermal oxidative stability, the chemilumin-

escence approach gives additional information related to
polymer quality. The appearance of the low temperature
pulses on the chemiluminescence curve observed before the
onset of the autocatalytic process is associated with the
history and processing of the sample and with the natural
aging of the polymer.

Literature Cited

1. G. E. Ashby, J. Polym. Sci., 1961, L, 99.
2. N. S. Allen and J. F. McKeller, Polymer, 1977, 18.968
3. L. Reich and S. S. Stivala, J. Polym. Sci., 1965, A3,
    4299
4. M.P. Schard and C. A. Russell, J. Appl. Polym. Sci.,
    1964, 8, 997
5. L. Reich and S. S. Stivala, J. Polym. Sci., 1965, A3,
    55
6. L. Matisova-Rychla, J. Rychly and M. Vavrekova,
    Eur. Polym. J., 1978, 14, 1033
7. L. Matisova-Rychla, Zs. Fodor and J. Rychly, Polym.
    Degrad. Stab., 1981, 3, 371
8. G. A. George, G. T. Egglestone and S. Z. Riddell,
    Polym. Eng. Sci., 1983, 23, 412
9. G. D. Mendenhall, Angew, Chem. Int. Ed. Engl., 1977,
    16, 225
10. J. L. Bolland, Proc. Roy. Soc. (London), 1946, 186A,
    218
11. J. L. Bolland and G. Gee, Trans. Faraday Soc., 1946,
    42, 236
12. A. V. Tobolsky, D. J. Metz and R. B. Mesrobian,
    J. Am. Chem. Soc., 1950, 72, 1942
13. L. Zlatkevich, J. Polym. Sci., Polym. Letters Ed.,
    1983, 21, 571
14. R. G. Bauman and S. H. Maron, J. Polym. Sci., 1956,
    22, 1
15. S. R. Dharwadkar, A. B. Phadnis, M. S. Chandrasek-
    haraiah and M. D. Karkhanavala, J. Thermal Anal.,
    1980, 18, 185
16. P. S. Hess and G. A. O'Hare, Ind. Eng. Chem., 1950,
    42, 1424
17. L. Matisova-Rychla, P. Ambrovic, N. Kulickova, and
    J. Rychly, J. Polym. Sci., 1976, Symposium No. 57,181
18. D.M. Chang, Polym. Eng. Sci., 1982, 22, 376
19. G. A. George, Polym. Deg. Stab., 1979, 1, 217
20. V. G. Nikolskii and G. I. Burkov, Khim. Vys. Energ.,
    1971, 5, 416
21. J. C. W. Chien, J. Polym. Sci., A-1, 1968, 6, 375

RECEIVED December 7, 1984

Physical Techniques for Profiling Heterogeneous Polymer Degradation

R. L. CLOUGH and K. T. GILLEN

Sandia National Laboratories, Albuquerque, NM 87185

Three general techniques for studying heterogeneous degradation in polymers are described, and an example of heterogeneous effects on degradation behavior is illustrated. Heterogeneous degradation frequently occurs in materials subjected to high dose rate irradiation in air and results in oxidation only near surfaces. Optical examination of cross-sectioned, metallographically-polished samples provides qualitative information on oxidation depth. Quantitative profiles of heterogeneous material property changes are provided by relative hardness measurements across cross-sectioned surfaces using either of two experimental apparatuses. Quantitative profiles of oxidation are obtained using density gradient columns. Viton is found to become embrittled when irradiated at high dose rate but becomes soft when irradiated at low dose rate; this is shown to result from differences in oxidative penetration depth at different dose rates.

Degradation in polymeric materials frequently takes place in a heterogeneous manner, such that the nature of the degradation in the interior of a sample may be very different from that near the edges (1 4). Some examples where degradation can be strongly heterogeneous are: 1) photochemical aging, where light is absorbed near the surface, 2) thermal aging, where plasticizer might be volatilized from near the surface or 3) chemical aging, where some corrosive gas or liquid is diffusing into a material. However, for polymers aged in air, probably the most common cause of heterogeneous effects is oxygen diffusion-limited degradation. This mechanism is relevant in many different environments where oxidation may be the predominant degradation mechanism. Examples include high-energy radiation, UV light, elevated temperature and mechanical stress.

A long-standing goal in polymer science has been the development of accelerated aging tests for predicting polymer degradation rates in long-term applications. Design of meaningful accelerated aging tests must begin with replication of degradation mechanisms and modes which occur in the environment to be simulated (5–8). Hetero-

0097–6156/85/0280–0411$06.00/0

geneous oxidative degradation is frequently an impediment to this goal, due to the fact that at the high stress levels characteristic of accelerated tests, the oxidation rate may be sufficiently high that oxygen diffusion becomes a dominant rate-determining step. Thus in studying polymer degradation and particularly in attempting accelerated aging simulations, the ability to identify and characterize heterogeneous effects is of fundamental importance. In this report we describe three techniques which we have developed and used for investigating heterogeneous oxidative degradation in irradiated polymers. We believe these techniques should be widely applicable to heterogeneous degradation studies involving a variety of environmental conditions. We also present an illustration of the dramatic effects that differing oxidation depths can exert over mechanical property changes.

Oxygen Diffusion Effects

If an air-saturated polymer sample is placed in a radiation environment, homogeneous oxidation will take place initially. At sufficiently high oxidation rates, the initially dissolved oxygen may be used up faster than it is replenished from the atmosphere. This gives rise to heterogeneous degradation with the oxidation rate in interior regions of the material decreasing to zero (or to some fixed rate lower than that near the surfaces). For materials exposed to high-energy radiation conditions which result in heterogeneous degradation, it is possible to estimate an absorbed dose by which strongly heterogeneous degradation is already taking place. This can be accomplished by calculating the amount of oxygen which will be dissolved in a polymer at equilibrium with the surrounding atmosphere, and then dividing by the amount of oxygen consumed (by chemical reaction) per rad of radiation absorbed by the polymer. We have obtained the following general expression for the equivalent dose, R, in rads, required to use up all the oxygen initially dissolved in a given polymer:

$$R = \frac{S \cdot P}{G(-O_2)} \cdot A \cdot (1.6 \times 10^{-12}) \qquad (1)$$

where S is oxygen solubility in the polymer in mol/g·bar, P is oxygen pressure (in bars) in the atmosphere surrounding the polymer sample, $G(-O_2)$ is the oxygen consumption yield (in molecules per 100 eV absorbed energy), and A is Avogadro's number. The numerical constant serves to convert eV to rads, and has the units of rad·g/100 eV. For the majority of common polymeric materials, we have calculated that the transition from homogeneous to strongly heterogeneous degradation will occur at quite low doses—generally less than a few tenths of a megarad. For virtually all polymers, measurable degradation occurs only after substantially higher doses. Thus, where oxidation inhomogeneities occur in high-energy radiation environments, the degradation can typically be treated as coming entirely from a heterogeneous mechanism.

If the oxygen permeation constant and the rate of oxygen consumption for the material remain relatively constant as a function of total absorbed dose, a steady state in heterogeneous oxidation will be approached. If the rate of oxygen consumption by reaction with free radicals generated by the radiation exceeds the rate of supply of oxygen from the edges of the polymer sample, distinct regions of oxidized and nonoxidized polymer can result. In comparing polymer samples irradiated at different dose rates, those exposed at the highest dose rate may degrade heterogeneously, becoming oxidized only near the surfaces. At successively lower dose rates, oxidation depth will become progressively larger. Eventually, at sufficiently low dose rates, the oxygen can fully penetrate the sample, giving rise to oxidation which is homogeneous throughout the material.

Rapid Identification of Heterogeneous Degradation: Metallographic Polishing

The first technique is a qualitative or semiquantitative method for rapid identification of oxidative inhomogeneities. By this method, samples are mounted in epoxy and a cross-sectional surface polished using standard metallographic techniques (10).
     In the degraded material, the physical and mechanical properties have undergone changes due to chain scission, crosslinking, plasticizer loss, etc. Material regions having different degradation have different physical properties, and so take on different lusters upon polishing. Oxidized and nonoxidized regions have different reflectivities, resulting in visual bands when examined under an optical microscope. We refer to the widths of such bands as the depth of oxidation, although strictly speaking, a step function change from oxidized to nonoxidized regions rarely occurs, as will be clear from some of the examples given below.
     The photographs shown in Fig. 1 were obtained on a series of clay-filled ethylene-propylene rubber (EPR) materials which were irradiated and then subjected to cross-sectional polishing. The occurrence of heterogeneous oxidation is clearly illustrated in photo B for a sample irradiated at high dose rate (6.7×10^5 rad/h). Photo C represents a sample irradiated to similar total dose, but at lower dose rate (1.1×10^5 rad/h); here, oxygen permeates throughout the sample giving rise to essentially homogeneous oxidation. Photo A represents an unirradiated sample. Photo D represents a sample irradiated at high dose rate (1.1×10^6 rad/h) but in the absence of oxygen.
     The two samples (B and C) which were irradiated at different dose rates but to similar total doses showed significant differences in ultimate tensile strength; this stems from differences in the extent of oxidation at the different dose rates. Sample B had a tensile strength that was 85% ($\pm$7%) of the tensile strength of unaged material, while C exhibited a tensile strength of 53% ($\pm$7%) compared with unaged.
     As might be expected, the optical rings became progressively fainter on going to successively lower doses; rings were not visible at doses below about 50 Mrad. However, the size of the rings in this material did not change significantly as a function of dose at constant dose rate. This observation indicates that oxygen permeation and consumption rates in this material are not significantly dependent on dose.
     The visual rings seen in cross-sectioned, polished samples correspond to areas having large differences in the extent of oxidation, and are useful for identification of heterogeneous degradation and for a qualitative or semiquantitative determination of the depth of oxidation. However, the exact shape of the degradation profile

will depend on the underlying oxidation kinetics. Equations for
calculating gradients in systems having simultaneous diffusion and
chemical reaction have been described (11-14). Using free radical
reaction kinetics, oxidation profiles ranging from gentle "parabolic"
shapes to step-function transitions can be obtained (14). The shapes
depend on the relative rates of termination and propagation steps.
Thus oxidative rings visible from cross-sectional polishing may
represent situations ranging from sharp transitions between oxidized
and nonoxidized regions, to more gradual transitions between heavily
oxidized regions and regions having either no oxidation or light
oxidation. The term oxidation depth has quantitative meaning in the
former limit, but has a somewhat more qualitative meaning in the
latter.

Degradation Profiling in Terms of Relative Hardness

More quantitative profiles of heterogeneous degradation can be obtained by measuring changes in relative hardness across polished
cross-sectional surfaces of degraded samples. Relative hardness
profiles are obtained using one of two types of instrumentation.
For relatively hard plastics, a commercial Knoop Hardness Tester
gives good results. This apparatus employs a thin, convex diamond
blade which is pressed into the sample under constant weight for a
set period of time. The softer the material, the deeper the blade
penetrates into the sample. Data is obtained by measuring the length
of the impression left by the blade. For this experiment, the sample
is placed on a calibrated translational microscope stage, and measurements made at regular intervals. We can readily obtain 20 to 30
data points over a distance of 1 mm. By obtaining measurements
over a cross-section of a heterogeneously degraded material, a useful
profile of the heterogeneity in material property changes is obtained
(10). Material hardness is related to modulus: increased penetration
corresponds to decreased modulus. (In fact, modulus can be calculated
from such penetration experiments, dependent upon tip geometry and
experimental procedure.) (15).

Figure 2 gives an example of a profile of changes in relative
hardness for a clear polypropylene material exposed to UV light in a
Rayonet chamber. The data indicate that the polypropylene becomes
progressively harder near the edges with UV exposure, yielding a broad
profile. The material in the interior of the sample has undergone
essentially no change in relative hardness.

For softer, rubbery polymers, a different experimental procedure
gives better results. For these experiments we have measured directly
the penetration distance of a tiny weighted probe into the cross-
sectioned surface of polymer samples. For this purpose we have made
use of a Perkin-Elmer Thermomechanical Analyzer equipped with a tip
modified to be small enough to provide measurements of the desired
resolution. (We are currently using a conical diamond phonograph
needle having a tip angle of 60°.)

Figure 3 shows profiles indicating changes in relative hardness
on cross-sectioned samples of the EPR material of Fig. 1. For unirradiated material, the profile is essentially flat (x's). For the
samples irradiated at 6.7×10^5 rad/h, a distinct flat-bottomed,
U-shaped profile is seen (circles). The boundary position between
optical bands (photo B, Figure 1) corresponds to the steep part of

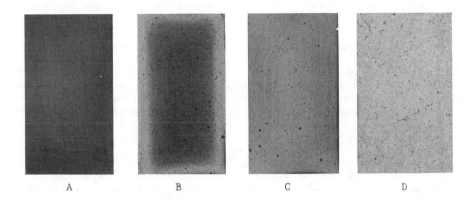

Figure 1. Cross-sectioned, polished samples of gamma irradiated EPR. A: Unirradiated material. B: 6.7×10^5 rad/h (in air) to 165 Mrad. C: 1.1×10^5 rad/h (in air) to 175 Mrad. D: 1.1×10^6 rad/h (in vacuum) to 253 Mrad. All irradiations carried out at 70°C. Sample thickness = 3.15 mm.

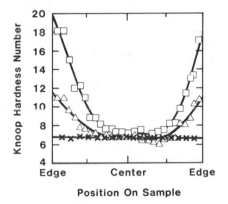

Figure 2. Hardness profile for polypropylene exposed to UV light in a Rayonet. □ = 6 day exposure △ = 3 day exposure, X = unexposed material. A Knoop Hardness Tester was employed.

the profile (slightly less than 20 percent of the way in on both
sides). The irradiated material has become significantly harder
throughout (i.e., increased modulus) with the largest increase occurring at the interior portion where oxygen is absent. For a sample irradiated at a lower dose rate (1.1 x 10^5 rad/h) the profile approaches
a homogeneous condition, showing only a slight, shallow curvature
(squares).

Degradation Profiling in Terms of Density

In another technique, the densities of pieces of degraded samples are
profiled using a salt gradient' column (7). This technique provides
a profile on oxygen uptake in the sample--information which is complementary to that provided by the techniques based on changes in
mechanical properties. Oxidation of samples normally leads to increases in sample density.

Figure 4 shows density data on two EPR samples (B and C of Fig.
1) that were irradiated to similar dose, but at different dose rates.
Strongly heterogeneous oxidation is again indicated for the high-dose-
rate sample, in contrast to nearly homogeneous oxidation for the low-
dose-rate sample. The results correlate well with results obtained
by optical examination of polished samples and by relative hardness
measurements.

Figure 5 shows another example of density gradient results
obtained with an EPR insulation material which had been extruded onto
a copper conductor. For the degradation studies, the insulation had
been stripped from the conductor and irradiated in air as a hollow
tube. At high dose rates, a U-shaped gradient characteristic of oxygen
diffusion was obtained (data not shown). However, at low dose rates,
a profile showing dramatically enhanced oxidation only near the interior surface (i.e., the surface which had been adjacent to the
copper) was found. The profiling results led to the conclusion that
oxidation catalyzed by copper ions was a predominant degradation
mechanism in this material (7). Supporting evidence was obtained by
demonstrating a large copper concentration in the region near the
interior surface by means of emission spectroscopy. This example
illustrates a case of heterogeneous degradation which resulted from
a mechanism different from oxygen diffusion.

Effect of Heterogeneous Degradation on Macroscopic Property Changes: Viton

The degradation behavior of a Viton material provides a good example
of the effects of heterogeneous oxidation. Figure 6 provides a plot
of mechanical property changes in Viton as a function of absorbed
dose in air at three different dose rates. The data show that at high
dose rate (5.5 x 10^5 rad/h, open squares) the elongation drops markedly while the tensile strength undergoes a more modest decrease. At
low dose rate (1.3 x 10^4 rad/h, circles) the elongation changes very
little, whereas the tensile strength decreases sharply. Data obtained
at an intermediate dose rate (9.2 x 10^4 rad/h, diamonds) show intermediate behavior. In terms of visual examination, samples irradiated
at the high dose rate are found to become progressively harder and
eventually so embrittled that they break readily when flexed lightly
by hand. Samples irradiated at the low dose rate degrade in just the

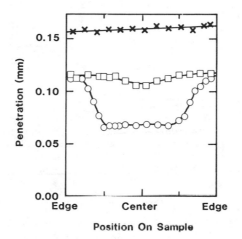

Figure 3. Profiles of relative hardness in terms of penetration distance of a weighted probe into cross-sectioned samples of irradiated EPR. **X** = unirradiated material, ◯ = 6.7×10^5 rad/h to 165 Mrad. ☐ = 1.1×10^5 rad/h to 175 Mrad. Probe load was 5 g.

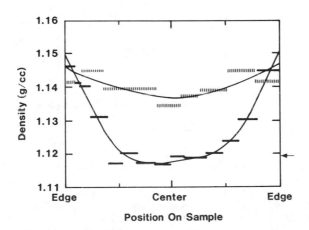

Figure 4. Density profiles for irradiated EPR samples. Solid line symbols are for 165 Mrad at 6.7×10^5 rad/h. Dotted line symbols are for 175 Mrad at 1.1×10^5 rad/h. The arrow indicates the density of unirradiated material, which gives a flat profile.

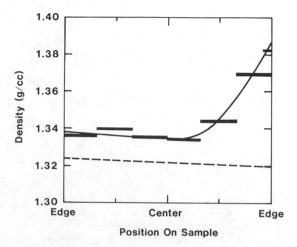

Figure 5. Density profile for an EPR cable insulation material
which had been stripped from the copper conductor and irradiated
as a hollow tube. At low dose rate (1.6×10^4 rad/h) the material shows enhanced oxidation near the interior surface (solid
line symbols). The dotted line symbols show the flat profile
of unirradiated material.

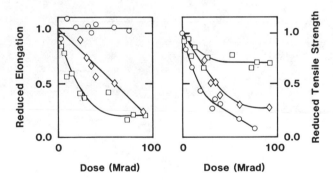

Figure 6. Change in ultimate tensile properties for a Viton
material irradiated at 70°C using three different dose rates.
☐ = 5.5×10^5 rad/h (in air), ◇ = 9.2×10^4 rad/h (in air),
○ = 1.3×10^4 rad/h (in air). Left: reduced elongation (e/e_o),
Right: reduced tensile strength (T/T_o).

opposite sense, becoming progressively softer and weaker and very easily stretchable.

The differences in degradation behavior as a function of dose rate can be readily explained in terms of differences in oxidation depth. Table I gives data obtained by optical examination of cross-sectioned, polished samples. At the high dose rate, strongly heterogeneous degradation results. Oxidation occurs only near the edges, and degradation in the interior takes place under anaerobic conditions. Solvent-swelling experiments indicate that samples irradiated at high dose rate have undergone net crosslinking. At progressively lower dose rates, oxidation depth increases; essentially homogeneous oxidation throughout is found at the lowest dose rate studied. Solvent swelling experiments indicate that samples irradiated at this lowest dose rate have suffered extensive net chain scission.

Table I.   Oxidation depth as determined by cross-sectional polishing of Viton material irradiated at 70°C, using different dose rates

Dose Rate (rad/h)	Oxidation as % of Total Thickness
5.5×10^5	35%
1.8×10^5	44%
9.2×10^4	82%
1.3×10^4	100%

Relative hardness profiling of the Viton material provides further confirmation of oxygen-diffusion control of the degradation behavior. The x's in Fig. 7 show the uniform hardness of an unirradiated sample. The data represented by open circles were obtained on a material irradiated at an intermediate dose rate (1.8×10^5 rad/h) in air. Strongly heterogeneous degradation is clearly seen. The sample regions near the edge have become softer, while at the interior the material has become harder. The flat profile defined by the solid black squares was obtained on a sample irradiated to similar total dose but under inert atmosphere. This sample has become uniformly harder throughout; the relative hardness value corresponds closely to that of the interior of the sample irradiated at high dose rate in air.

Experimental

Test specimens were cut from samples of commercial polymeric materials. Irradiation was carried out in Sandia's Co-60 facility which has been described elsewhere (16). A slow, steady flow of air was supplied to the sample chambers during the irradiation. Tensile tests were performed at ambient temperature using a Model 1130 Instron at a strain rate of 12.7 cm/min with an initial jaw gap of 5.1 cm.

For optical determinations of oxidation depth, irradiated samples were cut in cross section, potted in Shell Epon 828 epoxy, and cured overnight at 90°C with a diethanolamine catalyst. (Potting with a

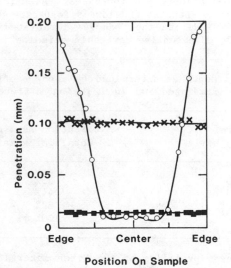

Position On Sample

Figure 7. Hardness profiles for irradiated Viton samples. X = unirradiated material, ○ = 1.8×10^5 rad/h to 186 Mrad, ■ = 9.0×10^5 rad/h to 191 Mrad (under vacuum). Probe load was 3 g. The samples had a thickness of 1.9 mm.

room-temperature-cure epoxy system was also carried out in some cases, and led to similar results.) The samples were polished using standard metallographic techniques (9). Measurements of the oxidized areas in cross-sectioned samples were obtained using a microscope fitted with a filar eyepiece.

For relative hardness determinations, polymer samples were placed in a tiny specially-designed vise. Three samples of cross-sectioned polymer were held together and the entire vise assembly, with samples in place, was subjected to metallographic polishing. Measurements were performed across the (cross-sectioned) surface of the central of the three samples by determining indentation distances of a weighted probe into the material. The samples were placed on a calibrated X-Y translational microscope stage, and typical measurements were made at regular intervals of between 0.05 and 0.15 mm. For experiments on hard materials, the Knoop hardness tester (17) was employed using a load of 25 g applied for a period of 15 sec. For measurements on rubbery materials, which employed a TMA, a probe tip consisting of a conical diamond phonograph needle having a tip angle of 60° was employed. This diamond tip was affixed to the shank of a micro drill bit, which was itself attached to the shaft of a commercial probe used for the Perkin Elmer TMS-1 Thermomechanical Analyzer. The LVDT, loading stage, and electronics of this commercial instrument were utilized to determine penetration distances. The diamond-tipped probe was preloaded with approximately 0.5 g for 30 sec, then an additional load of either 3 g or 5 g was added, depending on the sample being profiled. The change in penetration, 30 sec following the addition of the second load, was taken as the experimental penetration distance.

For density profiling, 100-cm long density gradient columns were prepared using calcium nitrate-water solutions. The columns were calibrated with standard glass calibration balls. Polymer specimens were sectioned into pieces whose thicknesses were measured with a micrometer, and the density of each piece determined from the equilibrium floating position in the column (7).

Conclusions

Heterogeneous degradation can be an overriding factor in material property changes. An example occurs with Viton, which becomes hard and brittle when degraded under conditions leading to low oxidation depth, yet becomes soft and stretchable when degraded under conditions where oxidation penetrates throughout. The occurrence of heterogeneous oxidation in samples exposed to severe environments (such as high radiation dose rates) is an important concern in the design or interpretation of accelerated aging experiments carried out for the purpose of estimating degradation rates and trends under low level, longer term environmental exposures (18). Circumventing the problem of heterogeneous effects' may require identifying sufficiently low level conditions for accelerated tests such that approximately homogeneous oxidation is obtained.

We have developed three general techniques which are useful for studying heterogeneous degradation in polymers. Optical examination of metallographically polished, cross-sectioned samples allows rapid identification of strong heterogeneous effects. A more quantitative profile of material property changes may be obtained in terms of differences in material hardness over a cross-sectioned sample. For hard plastics, a Knoop Hardness Instrument can be used; for rubbery materials a penetration device consisting of a modified TMA instrument with very small probe tip is appropriate. The other quantitative technique, density profiling, provides information that is complementary to relative hardness. This technique makes use of density gradient columns and reflects differences in extent of oxidation.

Acknowledgment

This work was supported by the U. S. Nuclear Regulatory Commission as a part of the Qualiication Testing Evaluation Program conducted by Sandia National Laboratories under contract DOE 40-550-70, NRC FIN No. A-1051. The authors are grateful to C. A. Quintana for technical support.

Literature Cited

1. Chapiro, A. "Radiation Chemistry of Polymeric Systems"; Interscience: New York, 1963, p 361.
2. Sisman, A.; Parkinson, W.; Bopp, C. In "Radiation Effects on Organic Materials"; Bold, R.; Carroll, J.; Eds.; Academic: New York, 1963; p 171.
3. Schanbel, W. In "Aspects of Degradation and Stabilization of Polymers"; Jellinek, H. H. G., Ed.; Elsevier: Amsterdam, 1978; p. 169.
4. Dole, M. "The Radiation Chemistry of Macromolecules"; Academic: New York, 1973; Chap. 13, p. 263.
5. Clough, R. L.; Gillen, K. T. J. Polym. Sci., Polym. Chem. Ed., 1981, 19, 2041.
6. Clough, R. L.; Gillen, K. T. Radiat. Phys. and Chem. 1983, 22, 527.
7. Gillen, K. T.; Clough, R. L. Radiat. Phys. and Chem. 1983, 22, 537.
8. Clough, R. L.; Gillen, K. T. Nucl. Technol. 1983, 59, 344.
9. Samuels, L. E. "Metallographic Polishing by Mechanical Methods" Second Edition, American Elsevier: New York, 1971.
10. Clough, R. L.; Gillen, K. T. Polymer Preprints April 1984, 83.
11. Crank, J. "The Mathematics of Diffusion"; Oxford: London, 1956.
12. Giberson, R. C. J. Phys. Chem. 1962, 66, 463.
13. Billingham, N.; Walker, T. J. Polym. Sci., Polym. Chem. Ed. 1975, 13, 1209.
14. Cunliffe, A. V.; Davis, A. Polym. Deg. and Stab. 1982, 4, 17.
15. Gillen, K. T. J. Appl. Polym. Sci. 1978, 22, 1291.
16. Gillen, K. T.; Clough, R. L.; Jones, L. H. SAND81-2613, Sandia National Laboratories, 1981.
17. Bikales, N., Ed.; "Mechanical Properties of Polymers"; Wiley: New York, 1971.
18. Clough, R. L.; Gillen, K. T.; Campan, J. L.; Gaussens, G.; Schonbacher, H.; Seguchi, T.; Wilski, H.; Machi, S. Nuclear Safety 1984, 25, 2, 238.

RECEIVED November 27, 1984

Synthesis of Biodegradable Polyethylene

WILLIAM J. BAILEY and BENJAMIN GAPUD

Department of Chemistry, University of Maryland, College Park, MD 20742

Although polyethylene is essentially nonbiodegradable, it was found that copolymers of ethylene and 2-methylene-1,3-dioxepane prepared by a high pressure free radical solution polymerization were indeed biodegradable. Most synthetic polymers are not biodegradable because they have not been on the earth long enough to have microorganisms evolve to utilize them as food. One exception to this rule is the low melting polyesters since poly(β-hydroxybutyrate) is a naturally occurring material which many bacteria and fungi use to store energy in the same manner that animals use fat. Since the free radical ring-opening polymerization of cyclic ketene acetals, such as 2-methylene-1,3-dioxepane, made possible the introduction of an ester group into the backbone of an addition polymer, this procedure appeared to be an attractive method for the preparation of biodegradable polymers. Copolymers of ethylene containing 10% ester-containing units were shown to be highly biodegradable while copolymers containing 2% ester-containing units were only slowly biodegradable.

For many applications in the medical, agricultural, and ecological fields, it is desirable to have a biodegradable polymer that can be easily fabricated by injection molding, melt spinning, and melt extrusion into films. Of course, all naturally occurring polymers, such as starch, cellulose, proteins, nucleic acids, and lignin, are biodegradable, since they have been existing on earth for a length of time sufficient for microorganisms to evolve that can utilize these materials as food. Unfortunately, almost all of these materials contain very polar groups and, therefore, decompose on heating before they melt. This means that these materials can be fabricated only by solution processes, but not by the relatively inexpensive commercial processes mentioned above. Furthermore, if

0097–6156/85/0280–0423$06.00/0

one attempts to modify these polymers in order to lower the melting point, such as the conversion of cellulose into cellulose butyrate acetate, the materials become nonbiodegradable.

Biodegradable Polyesters

On the other hand, synthetic polymers have not, in general, been in existence long enough to have microorganisms evolve that can utilize them as food. Thus, there are very few biodegradable synthetic polymers currently known. Potts, et al., (1) found that only low melting and low molecular weight polyesters showed any reasonable degree of biodegradation. It is really not surprising that some synthetic polyesters are biodegradable since poly(β-hydroxybutyrate) is a naturally occurring material that many bacteria and fungi use for energy storage in the same way that animals use fat (2). For example, a polytetramethylene succinate with a reduced viscosity of 0.08 was biodegradable, but a sample of the same polymer with a reduced viscosity of 0.59 was essentially not degraded.

A typical use of a biodegradable or bioabsorbable polyester in the medical field is the new polyglycolate absorbable suture. Historically the surgeon used "catgut," which is a biodegradable protein. However, many people are allergic to this proteinaceous material. For this reason the surgeon often shifted to a nylon suture, which, although nonbiodegradable, is essentially nonallergenic. American Cyanamid's Dexon polyglycolate has the best of both worlds; it is bioabsorbable in the body and therefore does not have to be physically removed, and at the same time it is nonallergenic (3).

An example of an agricultural application of a biodegradable polyester is Union Carbide's use of polycaprolactone (mp 60°C) for reforestation by injection molding of a bullet-shaped container in which a seed is planted. The plastic container, which takes about one year to biodegrade in the soil, protects the seedling from attack during its first year of life. This procedure makes possible the mechanization of seedling planting and gives a higher yield of mature trees (1).

Biodegradable Polyamides

In an effort to find another class of polymers which would be biodegradable, it appeared that the polyamides had many attractive features. In our laboratories, a new synthesis of polypeptides through the polymerization of the amino acid azide hydrobromides had been developed. Thus, it was feasible to extend this procedure to the synthesis of a dimer of glycine and ϵ-aminocaproic acid followed by the polymerization of this dimer through the amino acid azide hydrobromide to the desired regularly alternating copolyamide of nylon 2/nylon 6 (4).

$$\text{HBr} \cdot \underset{\displaystyle \text{NH}_2}{}\text{CH}_2\text{—}\overset{\displaystyle \overset{O}{\|}}{C}\text{—NH—}(\text{CH}_2)_5\text{—}\overset{\displaystyle \overset{O}{\|}}{C}\text{—N}_3 \qquad \xrightarrow[\text{78\%, -15°C}]{\text{Et}_3\text{N, DMF}}$$

$$\left[\text{—NH—CH}_2\text{—}\overset{\displaystyle \overset{O}{\|}}{C}\text{—NH—}(\text{CH}_2)_5\text{—}\overset{\displaystyle \overset{O}{\|}}{C}\text{—}\right]_x$$

The regularly alternating copolyamide of nylon 2/nylon 6, mp 268-270° C and [η] (m-cresol) at 25° C of 0.46 dl/g, was highly crystalline as evidenced by an x-ray determination. The melting point is intermediate between that of nylon 6, 220° C, and polyglycine, which decomposes at about 400° C. It is hydrophilic, absorbing about twice as much water as nylon 6, since it has the same average concentration of amide groups as nylon 4, but is much more thermally stable, less flammable, and lower melting.

Since the copolyamide contained two different amide bonds, both of which were similar, but not identical to a peptide bond, it was hoped that the extracellular enzymes of the bacteria or fungi would cleave the polymer to small fragments that could be utilized by the microorganisms as a food. In a fast screening technique, a culture of bacteria and/or fungi was fed a standard amount of hydrolyzed casein and the amount and rate of carbon dioxide liberated was monitored. In order to determine whether a material was biodegradable on the time scale used, the material was added to a similar flask and any increase in the rate or amount of carbon dioxide over the control was noted (5,6).

It was not surprising that glycine and ε-aminocaproic acid were readly consumed by the microorganisms as food, while nylon 6 was inert as was polyglycine. When the regularly alternating copolyamide (nylon 2/nylon 6) was tested, it was degraded by both fungi and bacteria. The fungus Aspergillus niger completely degraded the copolymer in about 3 weeks. Examination of the photomicrographs of the polymer during degradation indicated that certain portions, presumably the amorphous regions that were swollen in water, were attacked first and the other regions, presumably the more crystalline parts, were attacked more slowly (7).

Biodegradable Addition Polymers

Addition polymers have been generally quite resistant to biodegradation since the carbon-to-carbon bonds in the backbone are not very susceptible to biological cleavage. As a consequence, many hydrophilic addition polymers have been banned from medicinal use because the high molecular weight polymers are not degraded or eliminated from the body in a reasonable length of time. We reasoned that if an easily hydrolyzable group could be introduced into an addition copolymer by a free radical process, a wide variety of biodegradable polymers could be prepared.

Although there are microorganisms which will degrade linear hydrocarbons, the degradation of polyethylene is very slow (8). Recently, Corbin (9) monitored the biodegradation of ^{14}C-labelled polyethylene in soil by means of the carbon dioxide evolved. Since he observed a minimum conversion rate of 2% per year, he concluded that polyethylene film did not degrade to any significant extent the bioactive system studied. This slow degradation results because the mechanism of degradation of linear hydrocarbons involves the oxidation of the terminal methyl group to a carboxylic acid group and then, degradation of the resulting fatty acid by step-wise β-oxidation, two carbon units at a time. In linear polyethylene of high molecular weight, there are very few methyl groups and even they are located in the bulk of a hydrophobic medium not readily accessible to the microorganism. It was reasoned, therefore, if ester groups could be introduced into the backbone of polyethylene, it would become biodegradable.

In a search to find monomers which would undergo free radical ring-opening polymerization, several unsaturated heterocyclic monomers were investigated (10). One of the monomers that we studied was 2-methylene-1,3-dioxepane (I). It was reasoned that since the resulting ester carbonyl group was thermodynamically more stable by at least 40 kcal than the carbon-to-carbon double bond in the starting material, there would be a driving force for the ring-opening polymerization. Treatment of this monomer with benzoyl peroxide gave a high molecular weight polymer by a free radical ring-opening polymerization with 100% ring opening even at room temperature (10). This procedure represented the first synthesis of a polyester by a free radical mechanism as well as the first general method to introduce an ester group into the backbone of an additional polymer.

Since the monomer I would copolymerize with a wide variety of comonomers with the introduction of an ester group into the main chain, this appeared to make possible the preparation of biodegradable addition polymers. Copolymerization of ethylene and the ketene acetal I at 120°C produced a series of copolymers containing ester groups in the backbone of the copolymer, again with quantitative ring opening.

$$CH_2=CH_2 \quad + \quad CH_2=C\begin{array}{c} O-CH_2-CH_2 \\ | \\ O-CH_2-CH_2 \end{array} \qquad \xrightarrow[120°C]{\text{di-}\underline{\text{tert}}\text{-butyl peroxide}}$$

I

$$-CH_2-\overset{O}{\overset{\|}{C}}-O-(CH_2)_4-(CH_2-CH_2)_x-CH_2-\overset{O}{\overset{\|}{C}}-O-(CH_2)_4-(CH_2-CH_2)_y-$$

In order to ensure that the copolymers were reasonably homogeneous in view of the fact that the 2-methylene-1,3-dioxepane (I) is only slightly more reactive than ethylene, all conversions were held to below 2% as indicated in Table I. A series of copolymers containing from 2.1 to 10.4 mol-% of the ester-containing units was prepared. All of these copolymers were about 5,000 molecular weight and had melting points varying from 100–105°C for the copolymer containing 2.1 mole-% of I to 84–88°C for the copolymer containing 10.4 mole-%.

Determination of Biodegradability

The rate of biodegradability was determined by a method based on the determination of the carbon dioxide produced by the microbial metabolism of the polymer samples. Our screening test, which was the method developed by Kramer and Ennis (5,6), consisted of feeding a standard amount of hydrolyzed casein and the polymer to a culture containing a mixed micro-flora from soil or known microorganisms in a modified Warburg apparatus. The amount of CO_2 liberated was monitored by a Fisher-Hamilton gas partitioner. The increase in the amount of carbon dioxide in polymer-containing cultures over that of the control was used as a measure of the rate of biodegradation on the time scale used. Statistical analysis involved analysis of variance and the statistical difference was determined with a Duncan's test at a 5% level. The results are listed in Table II and are plotted in Figure 1. It is obvious that the copolymers containing at least 6.7 mol-% of the ester-containing units are highly degradable producing CO_2 at a rate 108 to 118% of the hydrolyzed casein control. The copolymers containing the lower amount of the ester groups biodegrade at quite low rates but much greater than polyethylene.
Inoculation of a sandy loam to nutrient broth and incubating for 4 days provided the mixed soil micro-flora source for 1 ml inoculation of the artificial medium. Thus 20 mg of test material was added to 20 ml of the medium in 50 ml flasks, which contained 5 g of a pancreatic digest of casein plus 1 g of dextrose per liter. After sterilization at 121° C for 15 min., the samples were inoculated with 1 ml of the soil micro-flora source and the headspace was flushed with oxygen. Controls consisted of samples tested indentically but containing no test material.

Table I. Copolymerization of Ethylene with 2-Methylene-1,3-dioxepane (I)

Amount of I (mol-%)		Conversion (weight-%)	Copolymer (mp, °C)	Intrinsic* Viscosity (dl/g)	Analysis			
In Feed	In Copolymer				Calculated		Found	
					C	H	C	H
5.00	2.1	0.80	100–105	0.13	83.90	13.84	83.78	14.00
10.00	4.8	1.6	95–99	0.14	81.90	13.35	81.85	13.62
15.00	6.7	1.2	90–95	0.122	80.61	13.04	80.55	13.38
19.00	9.3	1.5	89–96	0.16	79.25	12.71	78.98	13.03
22.00	10.4	1.7	84–88	0.144	78.46	12.51	78.18	12.80

*In xylene at 75° C.

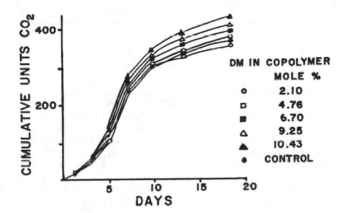

Figure 1. Cumulative headspace of CO_2 from copolymer of ethylene and 2-methylene-1,3-dioxepane.

Table II. Biodegradation of Copolymers of Ethylene and
2-Methylene-1,3-dioxepane (I)

Amount of I in Copolymer (mol-%)	Cumulative CO_2[a] after	
	7 days	20 days
2.1	103	98
4.8	107	103
6.7	116	108[b]
9.3	116	113[b]
10.4	121	118[b]

[a]Expressed as percent of control.
[b]Significantly different from the control.

Thus it has been demonstrated that the introduction of ester groups into the backbone of polyethylene will render the copolymer highly biodegradable. The copolymers are sufficiently high melting to make them useful for a wide variety of applications in the form of injection molded plastics, extruded films, and melt spun fibers. For example, the copolymer containing 9.3% ester-containing units and 90.7% ethylene units has a melting point of 89-96° C and has many of the characteristics of high pressure polyethylene. At the same time, this copolymer biodegrades at a rather fast rate.

Experimental

Copolymerization of Ethylene with 2-Methylene-1,3-dioxepane (I)-- Typical copolymerizations were carried out as follows: To a 10.125 X 1.5-inch steel pressure vessel (300-ml capacity) was added 50 ml of a solution of 1.1 g (1.0 mol-%) of di-tert-butylperoxide and 5.0 g (0.043 mol) of 2-methylene-1,3-dioxepane, bp 49-50°C (20 mm), in purified cyclohexane to give 50 ml of the reaction mixture. The sealed vessel was flushed twice with ethylene gas (99.9% pure) and finally filled with ethylene to an equilibrium pressure of 1000 psi. If one neglects the amount of ethylene that dissolves in the organic layer, the amount of ethylene gas with a volume of 250 ml at 1000 psi was calculated to be 21 g (0.75 mole). The copolymeriza-tion was allowed to proceed to low conversions (less than 2%) at 120°C for 30 minutes. After the reaction vessel was quickly cooled in Dry Ice, it was opened and methanol was added to it to facilitate the removal of the product as a white precipitate. After the solid was collected by filtration with suction and washed with methanol, the polymer was purified further by dissolution in hot chloroform and addition of the resulting solution into the nonsolvent methanol. The polymer was collected by filtration and dried in vacuo at 40°C for 24 hours to give a white powder.

Acknowledgments

The authors are grateful to the Frasch Foundation, the Polymer Progam of the National Science Foundation, and the Goodyear Tire and Rubber Company for partial support of this work.

Literature Cited

1. Potts, J. E.; Clendinning, R. A.; Ackart, W. B.; Niegisch, W. D. In "Polymers and Ecological Problems"; Guillet, J., Ed.; Plenum Press: New York, 1973, p. 61.
2. Shelton, J. R.; Agostini, D. E.; Lando, J. B. Am. Chem. Soc., Div. Polymer Chem., Preprints 1971, 12(2), 483.
3. Frazza, E. J.; Schmitt, E. E. J. Biomed. Mater. Res. Symp. 1970, 1, 43.
4. Bailey, W. J.; Okamoto, Y.;Kuo, W. C.; Narita T. In "Proc. Third International Biodegradation Symposium"; Sharpley, J. M.; Kaplan, A. M., Eds.; Applied Science Publishers: Essex, England, 1976, p. 765.
5. Ennis, D.; Kramer, A. Lebensm. -Wiss. u. Technol. 1974, 7(4), 214.
6. Ennis, D.; Kramer, A. J. Food Science 1975, 40, 181.
7. Bailey, W. J. Preprints, IUPAC Post-congress Symposium, Biomedical Materials, Kyoto, Japan, September 13, 1977, p. 10.
8. Albertson, A. -C.; Ranby, B. In "Proc. Third International Biodegradation Symposium"; Sharpley, J. M.; Kaplan, A. M., Eds. Applied Science Publishers: Essex, England, 1976, p. 743.
9. Corbin, D. G., cited by Henman, T. J. Proc. Third International Conference on Advances in the Stabilization of Polymers, Lucerne, Switzerland, June 1981, p. 116.
10. Bailey, W. J.; Chen, P. Y.; Chiao, W. -B.; Endo, T.; Sidney, L.; Yamamoto, N.; Yamazaki, N.; Yonezawa, K. In "Contemporary Topics in Polymer Science"; Shen, M., Ed.; Plenum Publishing Company: New York, 1979, Vol. 3, p. 29.
11. Bailey, W. J.; Ni, Z.; Wu, S. -R. J. Polym. Sci., Polym. Chem. Ed. 1982, 20, 3021.
12. Bailey, W. J.; Gapud, B. Am. Chem. Soc., Div. Polym. Chem., Preprints 1984, 26(1), 58.

RECEIVED February 21, 1985

Author Index

Subject Index

Indexing and production by Deborah H. Steiner
Jacket design by Pamela Lewis

Elements typeset by Hot Type Ltd., Washington, D.C.
Printed and bound by Maple Press Co., York, Pa.